Membranes and Cell Signaling

PRINCIPLES OF MEDICAL BIOLOGY
A Multi-Volume Work, Volume 7A

Editors: **E. EDWARD BITTAR,** *Department of Physiology, University of Wisconsin, Madison*
NEVILLE BITTAR, *Department of Medicine, University of Wisconsin, Madison*

Principles of Medical Biology
A Multi-Volume Work

Edited by **E. Edward Bittar,** *Department of Physiology, University of Wisconsin, Madison and*
Neville Bittar, *Department of Medicine*
University of Wisconsin, Madison

This work provides:

* A holistic treatment of the main medical disciplines. The basic sciences including most of the achievements in cell and molecular biology have been blended with pathology and clinical medicine. Thus, a special feature is that departmental barriers have been overcome.

* The subject matter covered in preclinical and clinical courses has been reduced by almost one-third without sacrificing any of the essentials of a sound medical education. This information base thus represents an integrated core curriculum.

* The movement towards reform in medical teaching calls for the adoption of an integrated core curriculum involving small-group teaching and the recognition of the student as an active learner.

* There are increasing indications that the traditional education system in which the teacher plays the role of expert and the student that of a passive learner is undergoing reform in many medical schools. The trend can only grow.

* Medical biology as the new profession has the power to simplify the problem of reductionism.

* Over 700 internationally acclaimed medical scientists, pathologists, clinical investigators, clinicians and bioethicists are participants in this undertaking.

Membranes and Cell Signaling

Edited by **E. EDWARD BITTAR**
Department of Physiology
University of Wisconsin
Madison, Wisconsin

NEVILLE BITTAR
Department of Medicine
University of Wisconsin
Madison, Wisconsin

JAI PRESS INC.

Greenwich, Connecticut *London, England*

Library of Congress Cataloging-in-Publication Data

Membranes and cell signaling / edited by E. Edward Bittar, Neville Bittar.
p. cm. — (Principles of medical biology ; v. 7A-7B)
Includes index.
ISBN 1-55938-812-9
1. Cell membranes. 2. Cellular signal transduction. I. Bittar, E. Edward. II. Bittar, Neville. III. Series.
[DNLM: 1. Cell Membrane—physiology. 2. Signal Transduction—physiology. 3. Biological Transport—physiology, QH 601 M532935 1997]
QH601.M4812 1997
611'.0181—dc21
DNLM/DLC 97-5100
for Library of Congress CIP

55 Old Post Road No. 2
Greenwich, Connecticut 06836

JAI PRESS LTD.
38 Tavistock Street
Covent Garden
London WC2E 7PB
England

ISBN: 1-55938-812-9
Library of Congress Catalog No.: 97-5100

Manufactured in the United States of America

CONTENTS (Volume 7A)

CONTENTS (Volume 7B)

LIST OF CONTRIBUTORS

Ata A. Abdel-Latif	Department of Biochemistry and Molecular Biology Medical College of Georgia Augusta, Georgia
Paul Burn	Hoffmann-La Roche Inc. Nutley, New Jersey
Bruno H. Dalle Carbonare	Department of Biology Pharmaceutical Research-New Technologies F. Hoffmann-La Roche Ltd. Basel, Switzerland
Michael Thomas Clandinin	Department of Foods Science & Nutrition Department of Medicine The University of Alberta Edmonton, Alberta, Canada
Carl E. Creutz	Department of Pharmacology University of Virginia Charlottesville, Virginia
LLoyd Culp	Department of Molecular Biology and Microbiology Case Western Reserve University School of Medicine Cleveland, Ohio
N.W. Davies	Department of Physiology University of Leicester Leicester, England

John R. Dedman — Department of Molecular and Cellular Physiology
University of Cincinnati Medical School
Cincinnati, Ohio

Nomi Eshhar — Chemistry Department
Life Science Research,
Israel, Inc.
Ness-Ziona, Israel

W. Howard Evans — Department of Medical Biochemistry
University of Wales College of Medicine
Cardiff, Wales, United Kingdom

Dalia Gordon — Institute of Life Sciences
Department of Cell and Animal Biology
The Hebrew University
Jerusalem, Israel

Alan Hall — Institute of Cancer Research
Royal Cancer Hospital
London, England

Yusuf A. Hannun — Department of Medicine
Duke University Medical Center
Durham, North Carolina

Morley D. Hollenberg — Department of Pharmacology and Therapeutics
Faculty of Medicine
The University of Calgary
Calgary, Alberta, Canada

John Hjort Ipsen — Department of Physical Chemistry
The Technical University of Denmark
Lyngby, Denmark

Simon M. Jarvis — Biological Laboratory
The University
Canterbury, Kent, England

Timothy Jegla
Department of Anatomy and Neurobiology
Washington University School of Medicine
St. Louis, Missouri

Kent Jørgensen
Department of Physical Chemistry
The Technical University of Denmark
Lyngby, Denmark

Marcia Kaetzel
Department of Molecular and Cellular Physiology
University of Cincinnati Medical School
Cincinnati, Ohio

Doris Koesling
Institut für Pharmakologie
Freie Universität Berlin
Berlin, Germany

G.J. Law
Laboratory of Neural and Secretory Signalling
Department of Neurobiology
BBSRC Babraham Institute
Cambridge, England

Roseanna Lechner
Department of Molecular Biology and Microbiology
Case Western Reserve University
School of Medicine
Cleveland, Ohio

P.M. Lledo
Laboratory of Neural and Secretory Signalling
Department of Neurobiology
BBSRC Babraham Institute
Cambridge, England

Mark O. Lively
Biochemistry Department
Wake Forest University Medical Center
Winston-Salem, North Carolina

Thomas D. Madden — Department of Pharmacology and Therapeutics
University of British Columbia
Vancouver, British Columbia, Canada

Anthony I. Magee — Laboratory of Eukaryotic Molecular Genetics
National Institute for Medical Research
London, England

W.T. Mason — Department of Neurobiology
Babraham Institute
Cambridge, England

Catherine E. Morris — Loeb Institute
Ottawa Civic Hospital
Ottawa, Ontario, Canada

Ole G. Mouritsen — Department of Physical Chemistry
The Technical University of Denmark
Lyngby, Denmark

Matthias Müller — Institute for Physiological Chemistry, Physical Biochemistry, and Cell Biology
University of Munich
Munich, Germany

Kathleen L. O'Connor — Department of Molecular Biology and Microbiology
Case Western Reserve University
School of Medicine
Cleveland, Ohio

Ole H. Petersen — The Physiological Laboratory
The University of Liverpool
Liverpool, England

Stephen Rothman — Department of Physiology
University of California, San Francisco
San Francisco, California

M. Rupnik
Laboratory of Neuroendocrinology
Institute of Pathophysiology
School of Medicine
Ljubljana, Slovenia

Lawrence Salkoff
Department of Anatomy and Neurobiology
Washington University
School of Medicine
St. Louis, Missouri

David L. Severson
Department of Pharmacology and Therapeutics
Faculty of Medicine
The University of Calgary
Calgary, Alberta, Canada

N.B. Standen
Department of Physiology
University of Leicester
Leicester, England

P.R. Stanfield
Department of Physiology
University of Leicester
Leicester, England

Thomas Wileman
Institute for Animal Health
Pirbright Laboratory
Pirbright, England

Philippe Zlatkine
Laboratory of Eukaryotic Molecular Genetics
National Institute for Medical Research
London, England

R. Zorec
Laboratory of Neuroendocrinology
Institute of Pathophysiology
School of Medicine
Ljubljana, Slovenia

PREFACE

It should not come as too much of a surprise that biological membranes are considerably more complex than lipid bilayers. This has been made quite clear by the fluid-mosaic model which considers the cell membrane as a two-dimensional solution of a mosaic of integral membrane proteins and glycoproteins firmly embedded in a fluid lipid bilayer matrix. Such a model has several virtues, chief among which is that it allows membrane components to diffuse in the plane of the membrane and orient asymmetrically across the membrane. The model is also remarkable since it provokes the right sort of questions. Two such examples are: Does membrane fluidity influence enzyme activity? Does cholesterol regulate fluidity? However, it does not go far enough. As it turns out, there is now another version of this model, the so-called post-fluid mosaic model which incorporates two concepts, namely the existence in the membrane of discrete domains in which specific lipid-lipid, lipid-protein and protein-protein interactions occur and ordered regions that are in motion but remain separate from less ordered regions. We must admit that both are intriguing problems and of importance in guiding our thinking as to what the next model might be.

We have chosen not to include the subject of membrane transport in the present volume. This obviously represents a break with convention. However, the intention is to have the topic covered in subsequent volumes relating to organ systems. It would be right to regard this as an attempt to strengthen the integrated approach to the teaching of medicine.

We are most grateful to the authors for their scholarly contributions and patience. We are also grateful to Ms. Lauren Manjoney and the staff members of JAI Press for their skill and courtesy.

E. EDWARD BITTAR
NEVILLE BITTAR

Chapter 1

Model Membrane Systems

THOMAS D. MADDEN

Principles of Medical Biology, Volume 7A
Membranes and Cell Signaling, pages 1–17.

ISBN: 1-55938-812-9

INTRODUCTION

The ubiquitous nature of membranes is illustrated by the fact that our very definition of a cell encompasses the idea that it is surrounded by an outer membrane. This membrane allows the cell to optimize its internal medium and protect sensitive components from the vagaries of a potentially harsh external environment. Functional compartmentalization may also occur within the cell in membrane bound organelles such as the mitochondria, lysosomes, Golgi, and so forth. Biological membranes, however, are not simply inert impermeable barriers but, as is described in subsequent chapters, play a crucial role in almost all cellular events. This requires that they selectively allow molecules to pass into or out of the cell or subcellular organelle and permit both intracellular and extracellular communication. In addition, they need to be sufficiently versatile to accommodate dynamic changes in structure and function during membrane trafficking, membrane fusion and cell division. Given such a diversity of roles it is perhaps surprising that in general all biological membranes have a similar structure. As described by Singer and Nicholson in 1972, this consists of a fluid lipid bilayer in which intrinsic proteins are either partly inserted or which they completely traverse. Both lipids and proteins are free to diffuse within the plane of the membrane unless constrained, for example, by extrinsic proteins associated with the cytoskeleton.

While the basic architecture of biological membranes may be fairly straightforward, they generally contain an almost bewildering variety of components. In addition to dozens (possibly hundreds) of different proteins, even the simplest membrane will contain several phospholipid classes e.g., phosphatidylcholine, phosphatidylethanolamine, phosphatidylserine, etc. each of which can be divided into different subclasses on the basis of fatty acid composition. In addition, a variety of neutral lipids may also be present. This complexity makes it exceedingly difficult to define the properties and roles of individual components in the native structure. In a membrane containing dozens of different proteins, for example, it may be very difficult to identify which one is responsible for a particular biological activity. Similarly, defining the role of individual lipid species, or their influence on membrane properties, is virtually impossible in the presence of a host of other lipids. For this reason techniques have been developed to allow isolation and purification of both proteins and lipids from native membranes. These individual components can then be more readily studied in simple model systems. This chapter will describe some of the membrane models used in research and indicate the type of information they can provide. Further, given that the ideal model should mimic the biological membrane in all but its complexity and that information obtained is of value only if it accurately portrays the native situation, I will try to critically evaluate each model system highlighting its advantages and limitations. For convenience, model systems have been divided into two categories, those which examine lipid properties and those in which membrane proteins or lipid-protein interactions are studied.

LIPID MODEL SYSTEMS

In order to study a particular lipid species it is necessary to either extract and purify it from a natural membrane or to synthesize it chemically. Most extraction techniques rely on the solubility of lipids in organic solvents. Different lipid classes can then be separated and purified using techniques such as thin layer chromatography (TLC) or high performance liquid chromatography (Kates, 1975). Chemical synthesis is complicated by the fact that many natural lipids are single stereoisomers. Stereospecificity can be maintained, however, by performing partial chemical syntheses using natural precursors and/or employing enzymes to convert one lipid class e.g., phosphatidylcholine to another e.g., phosphatidylserine (Comfurius and Zwaal, 1977). Single lipid species or defined lipid mixtures can then be incorporated into a number of different model membranes. These include vesicle systems, which as described below are subdivided into different classes on the basis of their size and structure, black lipid membranes (BLMs) and monolayers.

Large Multilamellar Vesicles

It was first shown by Bangham and co-workers in 1965 that dry phospholipids upon hydration in excess aqueous medium spontaneously form closed spheres which are relatively impermeable to most ions. These liposomes, as they were termed, are large, 1–20 microns in diameter, and consist of concentric spheres of lipid bilayers, one inside the other, separated by narrow aqueous channels. To differentiate them from other lipid model systems, they are now termed large multilamellar vesicles (MLVs). A schematic representation of MLV structure is presented in Figure 1 together with a freeze-fracture electron micrograph in which the characteristic "onion-like" structure is apparent. Generally MLVs are prepared by drying lipid dissolved in an organic solvent such as chloroform onto the sides of a glass flask to form a thin film. An aqueous solution is then added and the flask gently agitated to produce a suspension of MLVs.

Applications

While MLVs differ from most biological membranes (with the exception perhaps of the myelin sheath surrounding nerve axons) as a result of their multiple internal bilayers, they are nevertheless very useful model systems, particularly in studies which employ biophysical techniques to examine the physical and structural properties of phospholipids.

Phospholipid bilayers can adopt various phases differing in the motional freedom and packing properties of the fatty acyl chains. At low temperatures, for example, lipids may exist in the gel phase where molecules are closely packed together and only relatively slow rates of lateral and rotation diffusion are possible. As the temperature is increased a transition may occur to the liquid-crystalline state. This phase is characterized by a reduced packing density and greatly increased rates of

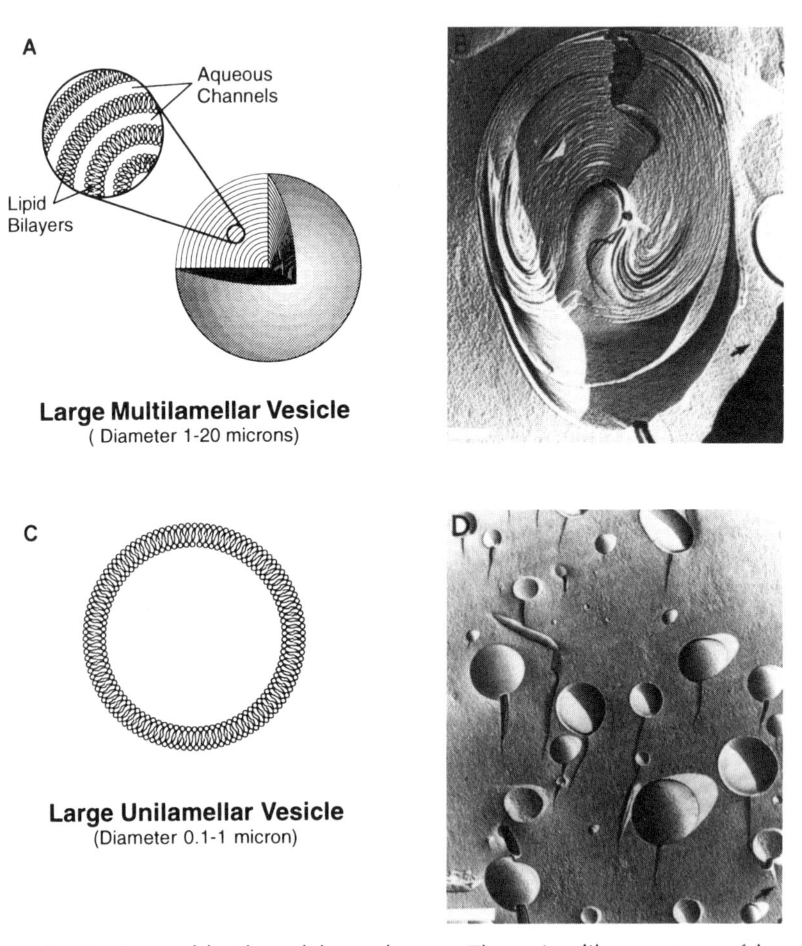

Figure 1. Structure of lipid model membranes. The onion-like structure of large multilamellar vesicles is illustrated schematically in (A) and is also evident in a freeze-fracture electron micrograph (B) where the fracture plane has passed through the MLV exposing the interior. The structure of large unilamellar vesicles is illustrated in (C) and a corresponding freeze-fracture micrograph shown in (D). For both electron micrographs the bar shown represents 200 nm and the direction of shadowing is indicated by the arrow.

lipid diffusion. The temperature at which transitions occur between such phases is dependent on a number of factors including the nature of the polar head group and the length and degree of saturation of the fatty acyl chains. Lipid thermotrophic properties can be conveniently studied using calorimetric techniques (Blume, 1988) and due to their simplicity and ease of preparation MLVs are the favored membrane model.

Another important area of membrane research concerns the structural properties of lipids. While biological membranes are normally arranged in a bilayer organization, during membrane fusion, for example, transitory non-bilayer structures are generated. To determine the role of specific lipids in generating either bilayer or non-bilayer phases, the structural properties of individual lipids in isolation can be determined using such physical techniques as X-ray crystallography and nuclear magnetic resonance (NMR) spectroscopy. For different reasons MLVs provide the most suitable model for such studies. In the case of X-ray crystallography the closely spaced bilayers in MLVs provide an ideal "repeating" structure yielding good X-ray diffraction. On the other hand, it is the size of MLVs which is important for NMR measurements. Their rate of "tumbling" and the change in orientation of lipids as they diffuse around the vesicle are sufficiently slow to permit the use of ^{31}P-NMR and ^{2}H-NMR to determine macromolecular structures.

In addition to the studies mentioned above, MLVs are frequently used, due to their ease of preparation, where more sophisticated membrane models are not required.

Small Unilamellar Vesicles

As their name suggests, small unilamellar vesicles (SUVs) are vesicles with only a single bilayer having diameters of about 20–30 nm. Mechanical disruption of MLVs by ultrasonic irradiation (Huang, 1969) or using a French Press (Barenholtz et al., 1979) will produce SUVs. Generally such preparations are then centrifuged to remove any remaining MLVs. A consequence of their size is that the bilayer is highly curved and phospholipids have difficulty in packing uniformly so that defects are not created in the membrane.

Applications

One of the potential advantages of unilamellar vesicles over MLVs is that they should be more suitable for permeability studies. Unfortunately the lipid packing problems experienced by SUVs make them relatively leaky and therefore not an appropriate model for most biological membranes which do not exhibit such pronounced curvatures. SUV systems are commonly used in reconstitution studies with integral membrane proteins; such studies will be described in a later section.

Large Unilamellar Vesicles

While similar in structure to SUVs, large unilamellar vesicles, as the name implies, are bigger having diameters of about 0.1 to 1.0 microns. As a consequence of their greater diameter, they do not experience the lipid packing problems exhibited by SUVs and, as described later, this makes them an ideal model for membrane permeability studies.

Large unilamellar vesicles (LUVs) can be formed directly from a suspension of MLVs by forcing the mixture, under high pressure, through filters with a small, well-defined pore size. This technique produces LUVs of approximately the same size as the filter pore. A significant advantage of this procedure is that it avoids the use of organic solvents or detergents (see below) which are difficult to completely remove and which may influence the properties of the vesicles. In addition, LUVs can be produced by either dispensing phospholipids in an appropriate detergent and then removing the detergent by, for example, dialysis or column chromatography or by dissolving the lipids in ether or ethanol and injecting the solution into water. These preparative techniques are reviewed in detail elsewhere (Szoka and Papahadjopolous, 1980; Hope et al., 1986).

Applications

The major advantage of LUV systems over MLVs is that like most biological membranes they consist of a single bilayer separating two aqueous spaces (see Figure 1). This makes them a much superior model for membrane permeability studies, especially because the solute has to cross only one bilayer to enter or leave the vesicles. In addition, the aqueous volume encapsulated by LUVs at a given phospholipid concentration, termed the trapped or captured volume, is generally greater than for MLVs or SUVs. As a result, much more of a compound can be entrapped during LUV formation and its efflux then followed in subsequent permeability studies.

Many biological membranes exhibit transmembrane gradients where the concentration of ions such as protons, sodium, potassium, or calcium is much greater on one side than the other. These chemical gradients are often used to drive energy-dependent processes such as ATP synthesis or the transport of a second molecule, e.g., glucose, across the membrane against a concentration gradient. Similar ion gradients can be created in LUV systems and their influence on the transmembrane distribution of other molecules determined (Redelmeier et al., 1989).

In addition, many biological membranes are asymmetric with internal and external leaflets of the bilayer composed of different lipid species. The mechanism whereby this asymmetry is generated and its influence on membrane properties is largely unknown. Using LUV systems, it has been shown that an asymmetric distribution of the acidic phospholipid, phosphatidylglycerol, can be achieved when a transmembrane proton gradient is created (Hope et al., 1989). These studies will hopefully provide an insight into asymmetry in biological membranes. Recently LUVs have also been employed as drug delivery vehicles, particularly for anticancer agents. This selection was based on the stability of such systems, their permeability properties and high trapped volume. For a review of the therapeutic advantages of such liposomal delivery systems see Ostro and Cullis (1989).

Giant Unilamellar Vesicles

These systems, which vary in diameter from about 20–40 microns, are distinguished from LUVs because their size allows them to be readily observed using ordinary light microscopy (see Figure 2). As a result, they have been widely used to study vesicle osmotic properties and also to characterize the factors influencing vesicle shape. In addition, individual giant unilamellar vesicles (GUVs) can be examined using micromechanical techniques to obtain direct measurements of such physical properties as surface cohesion, elasticity, rigidity and surface attraction properties (Evans and Needham, 1987).

Planar Bilayers

These model systems, which are also known as black lipid membranes (BLMs), are most often used to study the electrical properties of membranes. They are formed by applying a solution of lipid in an organic solvent such as hexane to a small aperture in a Teflon sheet. As the solvent evaporates the film thins and ultimately forms a lipid bilayer that is non-reflective to light, hence its name. The Teflon sheet normally separates two aqueous compartments, allowing measurements of membrane conductance, capacitance, and so forth. One disadvantage of BLMs is that residual solvent may affect the properties of the bilayer.

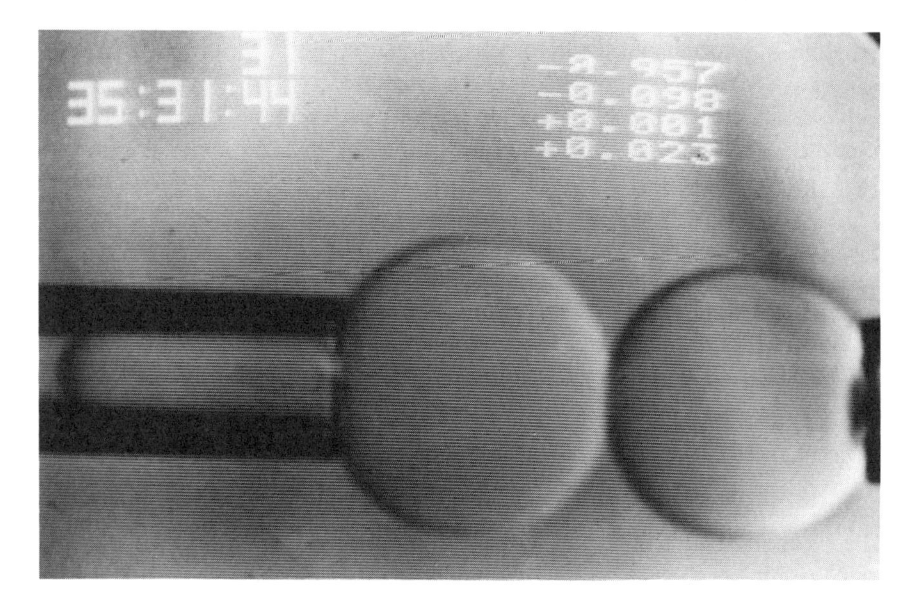

Figure 2. Videomicrograph of giant unilamellar vesicles aspirated in opposing micropipettes. Such systems can be used to quantitate membrane interactive forces. (Courtesy of Dr. Evan Evans, University of British Columbia).

Several techniques are available to incorporate proteins into planar membranes and these reconstituted systems have allowed direct measurements of the electrophysiological properties of various channel-forming proteins (Hartshorne et al., 1985).

Lipid Monolayers

In the same manner that oil forms a thin film on the surface of water, phospholipids dissolved in an organic solvent such as petroleum ether and applied to a water surface form a stable monolayer. Due to their amphipathic nature, the polar headgroup is in contact with water while the hydrocarbon chains are exposed to air.

Lipid monolayers are generally studied in an apparatus containing a movable barrier permitting the area occupied by the film to be varied. The effect of compression on film pressure is then measured using a device termed the Langmuir surface balance. The area occupied by a single lipid molecule can be determined from such measurements. Studies on phospholipid and cholesterol monolayers showed that the area occupied by the mixture was less than the sum of the areas for both components. This condensing effect of cholesterol is indicative of a strong, specific interaction with phospholipids (Demel and de Kruijff, 1976) and results in a considerable increase in the mechanical strength of cholesterol-containing bilayers. These studies on model systems, therefore, provided insight into the biological role of cholesterol. Monolayers can also be used to study the influence of surface pressure on protein interaction or insertion (Boguslavsky et al., 1994).

LIPID-PROTEIN MODEL SYSTEMS

Protein Reconstitution

As mentioned in the Introduction, most biological membranes contain numerous different proteins. This complexity can make it very difficult to examine a particular enzyme function and in 1971, therefore, Kagawa and Racker developed a procedure allowing individual membrane proteins to be studied in simple model systems. In this technique, termed reconstitution, the biological membrane is first disrupted using a detergent. This results in the intrinsic proteins being solubilized in detergent micelles. Like phospholipids, detergents are amphiphilic molecules and bind to the hydrophobic membrane-spanning domains of intrinsic proteins shielding these regions from contact with water and thereby preventing protein denaturation. From this solubilized mixture the desired protein can be purified using largely conventional techniques. Once purified the protein can be incorporated into a simple artificial membrane by adding it to a solution containing phospholipids and detergent. When the detergent is then removed, for example by dialysis or gel exclusion chromatography, the phospholipids spontaneously form vesicles with the protein

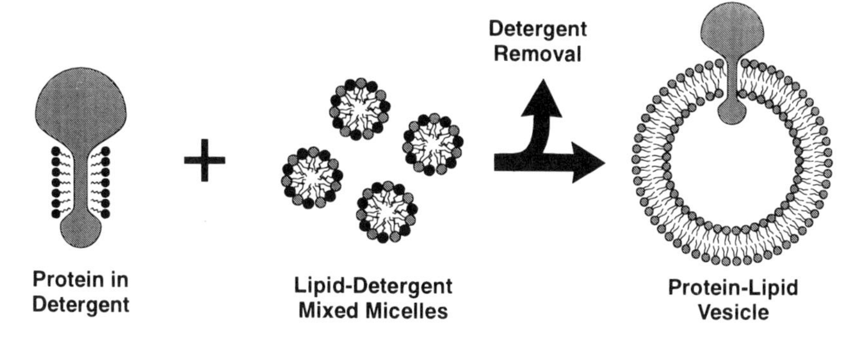

Figure 3. Schematic representation of protein reconstitution into lipid vesicles from a detergent suspension.

either incorporating during formation or inserting into pre-formed vesicles. This procedure is illustrated diagrammatically in Figure 3.

While the general principle behind reconstitution is straightforward, the physical characteristics of lipid-protein systems produced can be influenced by the detergent and lipid employed as well as the technique used for detergent removal. The next section outlines some of these characteristics and provides information on how reconstituted systems can be optimized for specific research studies. In addition, a brief review of the type of information generated by reconstituted vesicles is included.

Characterization of Reconstituted Systems

Protein Insertion

Clearly the first question we need to address in characterizing a reconstituted system is whether the protein is truly inserted into the vesicles. If during detergent removal the protein denatures or aggregates, then functional reconstitution will not occur. Of several techniques available to verify incorporation, density gradient centrifugation is one of the most frequently used. In this procedure solutions of varying densities are prepared in a centrifuge tube and the reconstituted vesicles then layered on top. If the protein is incorporated into the vesicles, then upon high speed centrifugation the lipid and protein will migrate together to the same position on the gradient.

Another technique which can be used to directly visualize the insertion of certain proteins is freeze-fracture electron microscopy. This involves splitting a frozen sample of the reconstituted system. Where the fracture plane meets a vesicle it passes between the two leaflets of the bilayer exposing the interior. Incorporated proteins can often be seen as small projections on the otherwise smooth vesicle surface. Freeze-fracture electron micrographs of reconstituted vesicles containing

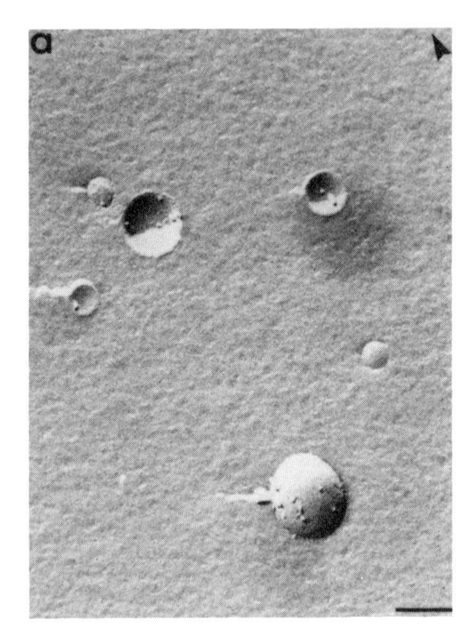

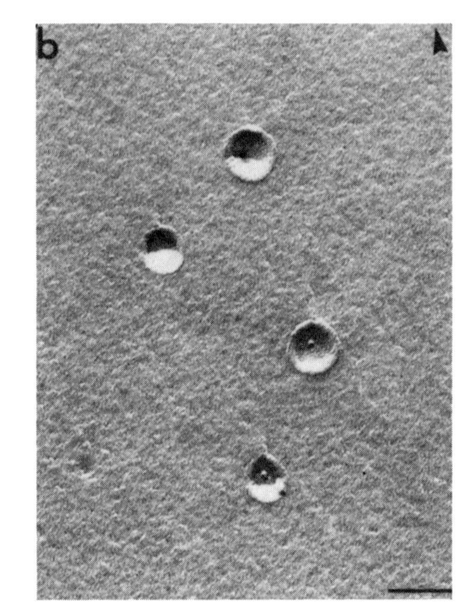

Figure 4. Freeze-fracture electron micrographs of reconstituted cytochrome oxidase vesicles. Reconstituted systems were prepared using a lipid to protein molar ratio of either 5000:1 (a) or 50,000:1 (b). The arrowheads indicate the direction of shadowing while the bars represent 100 nm.

the mitochondrial enzyme cytochrome oxidase are shown in Figure 4. By adjusting the initial ratio of protein and lipid, reconstituted systems containing either several cytochrome oxidase molecules per vesicle (Figure 4a) or just one per vesicle (Figure 4b) can be prepared.

Size of Reconstituted Vesicles

As already discussed for lipid model membranes, the size of reconstituted vesicles may influence their suitability for particular studies. Reconstitution from the detergents cholate and deoxycholate typically yields small unilamellar vesicles. While they are convenient for many studies they are often unsuitable for experiments involving transport proteins as a result of both their "leakiness" and small trapped volume. At the other extreme, reconstitution procedures which generate multilamellar vesicles (MLVs) in which the protein may be trapped within inner lamellae and therefore enzymically "silent" would also be unsuitable.

For most applications large unilamellar vesicles of diameter 100 nm or greater provide ideal reconstituted systems. Their permeability properties and large trapped volume allow studies of ion and solute transport, countertransport and measurements of intravesicular pH and membrane potential. Three techniques are available to reconstitute membrane proteins into LUVs. First, they may be generated spontaneously using an appropriate detergent such as the non-ionic surfactant octyl glucopyranoside (Mimms et al., 1981). Second, LUVs can be prepared using the techniques described for lipid model membranes and then mixed with protein in the presence of detergent concentrations which do not solubilize the lipid vesicles. The vesicles therefore retain their original size during detergent removal and protein insertion. A third technique, which has been widely used in the reconstitution of transport enzymes, involves reconstitution of the protein into SUVs which are then fused together by freezing and thawing the mixture to obtain LUVs which can be greater than 1 micron in diameter (Karlish and Pick, 1981).

Protein Orientation

The intrinsic proteins present in biological membranes exhibit a defined orientation. In the inner mitochondrial membrane, for example, the catalytic domain of ATP synthase faces the matrix while in contrast the cytochrome *c* binding site on cytochrome oxidase faces the intermembrane space. A similarly asymmetric orientation is desirable in reconstituted vesicles. In the case of an enzyme, for example, this would ensure that all of the catalytic sites, assuming that it faced outwards, were available to substrate added to the external medium. In addition, if the protein catalyzed a vectorial transport reaction all substrate movement would be either into or out of the vesicles, ensuring that different transport proteins in the same vesicle were not working against one another. Further, studies on the structure of membrane proteins and the arrangement of different subunits are greatly simplified in asymmetric reconstituted vesicles.

Transmembrane protein orientation in reconstituted systems can be assessed by a number of techniques. The hydrophilic domains of membrane proteins can often be visualized in negatively stained specimens by electron microscopy. An assessment of protein orientation can therefore be made based on the distribution of domains at the inner or outer membrane surface. In addition, the availability of reconstituted proteins to externally added proteases or specific antibodies (Herzlinger et al., 1984) can be used to quantify the percentage of outwardly and inwardly directed protein.

Reconstitution can yield symmetrical vesicles, in which the protein is inserted in both possible orientations, or asymmetric systems depending upon both the specific protein and the particular reconstitution conditions. When the spike glycoprotein of Semliki Forest virus, for example, is reconstituted from octylglucoside a population of vesicles is obtained in which greater than 95% of the protein faces outwards (Helenius et al., 1981). Another protein that spontaneously inserts asymmetrically is bacteriorhodopsin although in this case vesicles are obtained with the opposite orientation to that of the native membrane (Bayley et al., 1982).

The majority of reconstituted systems, however, contain proteins with both possible orientations, and some researchers, therefore, have attempted to separate

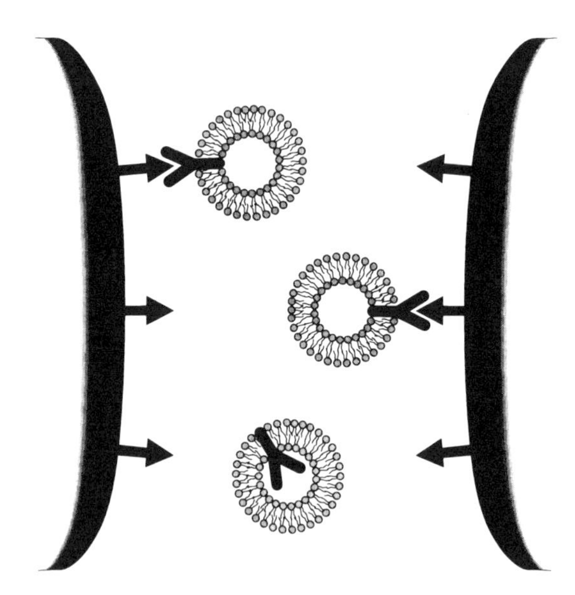

Figure 5. Schematic representation of the use of affinity chromatography to separate reconstituted vesicles in which the protein active site either faces inward or outward.

this mixture into vesicle populations with defined protein orientation. One approach involves reconstitution at high lipid-to-protein ratios so that the majority of vesicles produced contain only a single protein. In the case of cytochrome oxidase, vesicles containing the outwardly facing enzyme can then be separated from protein-free vesicles and those containing inwardly oriented enzyme by ion-exchange chromatography, or affinity chromatography employing cytochrome c covalently attached to Sepharose beads (Madden and Cullis, 1985). This latter procedure is illustrated diagrammatically in Figure 5.

Applications of Reconstituted Systems

Enzyme Identification

While a few highly specialized membranes may contain predominantly a single protein (examples are the purple membrane of *Halobacterium* and muscle sarcoplasmic reticulum) most biological membranes contain numerous different polypeptides. As a result the assignment of an observed enzyme activity to a particular polypeptide usually requires that the protein be isolated from the membrane and purified to verify that it is responsible for that activity. In many cases, however, intrinsic enzymes are inactive in detergent micelles or catalyze a transport reaction which can only be assayed in a sealed vesicular system. This requires that following purification the protein be reconstituted into lipid vesicles. The erythrocyte membrane proteins, for example, were initially resolved on the basis of their size and charge by polyacrylamide gel electrophoresis (PAGE), each protein being named from its position on the gel. Using the technique of protein reconstitution it could be shown that the protein responsible for anion transport across the erythrocyte membrane was Band 3 (Ross and McConnell, 1978) while the hexose transport activity could be assigned to Band 4.5 (Kasahara and Hinkle, 1977).

Another important application of reconstitution is in determining the minimum protein requirement for a particular biological activity. A single polypeptide, for example, was shown to be sufficient for lactose transport (Newman et al., 1981) while four polypeptides are required to form the acetylcholine receptor (Wu and Raftery, 1981).

Influence of Lipid on Protein Function

This topic is covered in detail in a later chapter and therefore only a brief outline on the use of reconstituted systems to study lipid-protein interactions is provided here.

During purification of a membrane protein any lipid that was present in the native bilayer can fairly readily be removed. When the protein is incorporated into a model membrane, therefore, its lipid environment can be absolutely defined. This allows studies on the requirements of an enzyme for particular lipids. Cytochrome oxidase and β-hydroxybutyrate dehydrogenase, for example, are believed to require the lipids cardiolipin and phosphatidylcholine respectively for optimal enzyme activity (Gazzotti et al., 1974; Vik et al., 1981). The ability to control the lipid environment of a protein also allows us to determine how the physical properties of the bilayer affect protein activity. While many enzymes are able to function normally in a wide range of lipid environments, bilayer characteristics such as acyl chain length or degree of saturation, or cholesterol content may influence the activity of certain enzymes (Carruthers and Melchoir, 1984; Cornea and Thomas, 1994).

In the area of lipid-protein interactions much interest has focused on the existence or otherwise of a tightly bound lipid layer at the interface between the bilayer and the membrane-spanning domain of intrinsic proteins. Many of these studies have used physical techniques such as electron spin resonance (ESR) or NMR spectroscopy to examine the influence of protein on the motional properties of specifically labeled lipids. These are either spin-labeled (for ESR spectroscopy) or, for example, deuterium labeled for NMR. Reconstituted systems are greatly favored for such studies not only because the labels are readily incorporated by including them in the lipid-detergent mixture but also because this procedure allows the final ratio of lipid-to-protein to be varied.

Protein-Protein Interactions

Reconstituted systems have proven very useful in determining both direct and indirect interactions between membranes proteins. As an example of a direct

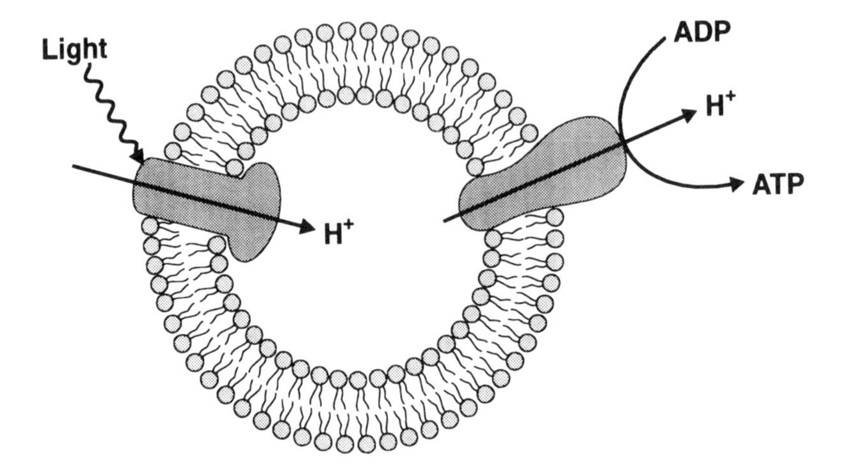

Figure 6. Schematic representation of reconstituted vesicles containing both bacteriorhodopsin and ATP synthase. Upon stimulation by light a proton gradient is generated by bacteriorhodopsin and this gradient is then utilized by the synthase to generate ATP.

interaction, the binding of the soluble proteins spectrin and actin to the erythrocyte anion channel, Band 3, was demonstrated in reconstituted vesicles. Using freeze-fracture electron microscopy the spectrin-actin-Band 3 complex could be shown to aggregate in a pH-dependent manner (Yu and Branton, 1976).

Reconstitution experiments also provided early support for Mitchell's chemiosmotic theory of energy transduction. In 1973 Racker and Kandrach prepared vesicles containing both the ATP synthase and cytochrome oxidase. They demonstrated that oxidation of cytochrome *c* by cytochrome oxidase resulted in generation of a transmembrane proton gradient and subsequent ATP synthesis. Additional evidence that ATP production was proton gradient-driven was provided by reconstituted vesicles containing ATP synthase and the light-driven proton pump, bacteriorhodopsin. As illustrated in Figure 6, when these vesicles are exposed to light the bacteriorhodopsin pumps protons into the vesicle lumen and this gradient is then utilized by the synthase to convert ADP to ATP (Racker and Stoeckenius, 1974).

Within the cell ATP is produced by the mitochondrial respiratory chain. The complexity of this system means, however, that the characteristics of individual components and quantification of the electrochemical potential that they can generate is often determined in reconstituted systems where no competing chain components are present (Kita et al., 1982). Increasingly complex reconstituted systems can also be prepared containing multiple components so that electron transfer through individual loops of the respiratory chain can be studied (Matsushita and Kaback, 1986).

SUMMARY

The complexity of most biological membranes makes it extremely difficult to define the properties and roles of individual components in the native structure. For this reason simpler membrane models have been developed. These may comprise either lipid alone or both lipid and protein and ideally they should resemble the biological membrane as closely as possible. Lipid model membranes include vesicular systems such as large multilamellar vesicles, small, large, and giant unilamellar vesicles, as well as planar bilayers and monolayers. Of these systems large unilamellar vesicles often represent the most appropriate membrane model. This is also the case with reconstituted protein-lipid systems where, in addition, the protein should preferably have a defined transmembrane orientation. Using such model systems, it is possible to define the roles of particular lipid species and to identify and characterize individual membrane enzymes. In addition, both lipid-protein and protein-protein interactions can be studied. As our basic understanding of membrane structure and function improves, the versatility of the model-building approach will permit the development of increasingly more sophisticated systems. These in turn should allow us a deeper insight into the complexities of cell biology.

REFERENCES

Bangham, A.D., Standish, M.M., & Watkins, J.D. (1965). Diffusion of univalent ions across the lamellae of swollen phospholipids. J. Mol. Biol. 13, 238–251.

Barenholtz, Y., Anselem, S., & Lichtenberg, D. (1979). A new method for preparation of phospholipid vesicles (liposomes)-French press. FEBS Lett. 99, 210–213.

Bayley, H., Höjeberg, B., Huang, K.-S., Khorana, H.G., Kiao, M.-J., Lind, C., & London, E. (1982). Delipidation, renaturation and reconstitution of bacteriorhodoposin. Meth. Enzym. 88, 74–81.

Blume, A. (1988). Applications of calorimetry to lipid model membranes. In: Physical Properties of Biological Membranes and their Functional Implications (Hidalgo, C., Ed.), pp. 71–121. Plenum Press, New York.

Boguslavsky, V., Rebecchi, M., Morris, A.J., Jhon, D.-Y., Rhee, S.G., & McLaughlin, S. (1994). Effect of monolayer surface pressure on the activities of phosphoinositide-specific phospholipase C. Biochemistry 33, 3032–3037.

Carruthers, A. & Melchoir, D.L. (1984). A rapid method of reconstituting human erythrocyte sugar transport proteins. Biochemistry 23, 2712–2718.

Comfurius, P. & Zwaal, R.F.A. (1977). The enzymic synthesis of phosphatidylserine and purification by CM cellulose column chromatography. Biochim. Biophys. Acta 488, 36–42.

Cornea, R.L. & Thomas, D.D. (1994). Effects of membrane thickness on the molecular dynamics and enzymatic activity of reconstituted Ca-ATPase. Biochemistry 33, 2912–2920.

Demel, R.A. & de Kruijff, B. (1976). The function of sterols in membranes. Biochim. Biophys. Acta 457, 109–132.

Evans, E. & Needham, D. (1987). Physical properties of surfactant bilayer membranes: Thermal transitions, elasticity, rigidity, cohesion and colloidal interactions. J. Phys. Chem. 91, 4219–4228.

Gazzotti, P., Bock, H.-G., & Fleischer, S. (1974). Role of lecithin in D-β-hydroxybutyrate dehydrogenase function. Biochem. Biophys. Res. Commun. 58, 309–315.

Hartshorne, R.P., Keller, B.V., Talvenheimo, J.A., Catterall, W.A., & Montal, M. (1985). Functional reconstitution of the purified brain sodium channel in planar lipid bilayers. Proc. Natl. Acad. Sci. USA 82, 240–244.

Helenius, A., Sarras, M., & Simons, K. (1981). Asymmetric and symmetric membrane reconstitution by detergent elimination. Studies with Semliki-Forest-virus spike glycoprotein and penicillinase from the membrane of *Bacillus licheniformis*. Eur. J. Biochem. 116, 27–35.

Herzlinger, D., Viitamen, P., Carrasco, N., & Kaback, H.R. (1984). Monoclonal antibodies against the *lac* carrier protein from *Escherichia coli* 2. Binding studies with membrane vesicles and proteoliposomes reconstituted with purified *lac* carrier protein. Biochemistry 23, 3688–3693.

Hope, M.J., Bally, M.B., Mayer, L.D., Janoff, A.S., & Cullis, P.R. (1986). Generation of multilamellar and unilamellar vesicles. Chem. Phys. Lipids 40, 89–107.

Hope, M.J., Redelmeier, T.E., Wong, K.F., Rodrigueza, W., & Cullis, P.R. (1989). Phospholipid asymmetry in large unilamellar vesicles induced by transmembrane pH gradients. Biochemistry 28, 4181–4187.

Huang, C. (1969). Studies on phosphatidylcholine vesicles. Formation and physical characteristics. Biochemistry 8, 344–349.

Kagawa, Y. & Racker, E. (1971). Partial resolution of the enzymes catalysing oxidative phosphorylation. XXV. Reconstitution of vesicles catalysing ^{32}Pi-adenosine triphosphate exchange. J. Biol. Chem. 246, 5477–5487.

Karlish, S.J.D. & Pick, U. (1981). Sidedness of the effect of sodium and potassium ions on the conformational state of the sodium-potassium pump. J. Physiol. Lond. 312, 505–529.

Kasahara, M. & Hinkle, P.C. (1977). Reconstitution and purification of the D-glucose transporter from human erythrocytes. J. Biol. Chem. 252, 7384–7390.

Kates, M. (1975). Techniques of lipidology. In: Isolation, Analysis and Identification of Lipids (Work, T.S. & Work, E., eds.). American Elsevier, New York.

Kita, K., Kasahara, M., & Anraku, Y. (1982). Formation of a membrane potential by reconstituted liposomes made with cytochrome b_{562}-O complex, a terminal oxidase of *Escherichia coli* K12. J. Biol. Chem. 257, 7933–7935.

Madden, T.D. & Cullis, P.R. (1985). Preparation of reconstituted cytochrome oxidase vesicles with defined transmembrane protein orientations employing a cytochrome *c* affinity column. Biochim. Biophys. Acta 808, 217–224.

Matsushita, K. & Kaback, H.R. (1986). D-lactate oxidation and generation of the proton electro-chemical gradient in membrane vesicles from *Escherichia coli* GR19N and in proteoliposomes reconstituted with purified D-lactate dehydrogenase and cytochrome *O* oxidase. Biochemistry 25, 2321–2327.

Mimms, L.T., Zampighi, G., Nozaki, Y., Tanford, C., & Reynolds, J.A. (1981). Formation of large unilamellar vesicles by detergent dialysis employing octylglucoside. Biochemistry 20, 833–840.

Newman, M.J., Foster, D.L., Wilson, T.H., & Kaback, H.R. (1981). Purification and reconstitution of functional lactose carrier from *Escherichia coli*. J. Biol. Chem. 256, 11804–11808.

Ostro, M.J. & Cullis, P.R. (1989). Use of liposomes as injectable drug-delivery systems. Am. J. Hosp. Pharm. 46, 1576–1587.

Racker, E. & Kandrack, A. (1973). Partial resolution of the enzymes catalyzing oxidative phosphorylation. Reconstitution of the third segment of oxidative phosphorylation. J. Biol. Chem. 248, 5841–5847.

Racker, E. & Stoeckenius, W. (1974). Reconstitution of purple membrane vesicles catalyzing light-driven proteon uptake and adenosine triphosphate formation. J. Biol. Chem. 249, 662–663.

Redelmeier, T.E., Mayer, L.D., Wong, K.F., Bally, M.B., & Cullis, P.R. (1989). Proton flux in large unilamellar vesicles in response to membrane potentials and pH gradients. Biophys. J. 56, 385–393.

Ross, A.H. & McConnell, H.M. (1978). Reconstitution of the erythrocyte anion channel. J. Biol. Chem. 253, 4777–4782.

Singer, S.J. & Nicolson, G.L. (1972). The fluid mosaic model of the structure of cell membranes. Science 175, 720–731.

Szoka, F. & Papahadjopoulos, D. (1980). Comparative properties and methods of preparation of lipid vesicles (liposomes). Ann. Rev. Biophys. Bioeng. 9, 467–508.

Vik, S.B., Georgevich, G., & Capaldi, R.A. (1981). Diphosphatidylglycerol is required for optimal activity of beef heart cytochrome *c* oxidase. Proc. Natl. Acad. Sci. USA 78, 1456–1460.

Wu, W.C.-S. & Raftery, M.A. (1981). Reconstitution of acetylcholine receptor function using purified receptor protein. Biochemistry 20, 694–701.

Yu, J. & Branton, D. (1976). Reconstitution of intramembrane particles in recombinants of erythrocyte protein Band 3 and lipid: Effects of spectrin-actin association. Proc. Natl. Acad. Sci. USA 73, 3891–3895.

RECOMMENDED READINGS

There is no textbook which deals solely with model systems; those cited below include such information together with a background to lipids and membranes. A review article covering the topic of protein reconstitution is also cited.

Gennis, R.B. (1989). Biomembranes: Molecular Structure and Function. Springer-Verlag, New York.

Houslay, M.D. & Stanley, K.K. (1982). Dynamics of Biological Membranes. John Wiley and Sons, New York.

Jain, M. (1988). Introduction to Biological Membranes. 2nd ed. John Wiley and Sons, New York.

Madden, T.D. (1986). Current concepts in membranes protein reconstitution. Chem. Phys. Lipids 40, 207–222.

Chapter 2

Lipid-Bilayer Heterogeneity

KENT JØRGENSEN, JOHN HJORT IPSEN, and OLE G. MOURITSEN[1]

Principles of Medical Biology, Volume 7A
Membranes and Cell Signaling, pages 19–38.

ISBN: 1-55938-812-9

BIOLOGICAL MEMBRANES: STRUCTURE AND FUNCTION

The biological membrane is a major structure in all living systems from one-celled prokaryotic microorganisms to complex many-celled eukaryotic organisms. The cell- or plasma membrane as shown in Figure 1 is a complex macromolecular assembly consisting of a lipid-bilayer with embedded trans-membrane proteins and superficially attached proteins, carbohydrates, and enzymes (Gennis, 1989). The membrane controls the trans-membrane flow of all material by acting as a physical barrier with a low permeability to organic molecules and ions. The membrane proteins are involved in active transport across the membrane in order to maintain steady-state conditions in the cell interior.

Recent research has shown that the fluid lipid-bilayer component of the biomembrane, apart from being a structural element of the whole cell, also displays a functional role important for various biochemical processes that take place in association with biomembranes (Kinnunen and Laggner, 1991). Examples of the many important biochemical processes associated with biomembranes include energy transduction, action of proteins as receptors for hormones, release of second messengers, immunological response, and activity of membrane-associated enzymes. Both experimental and theoretical results in the field of membranology have established that the lateral organization of the membrane components in the membrane, e.g., in terms of lipid domains (Jacobsen and Vaz, 1992; Bergelson et al., 1995), is of importance

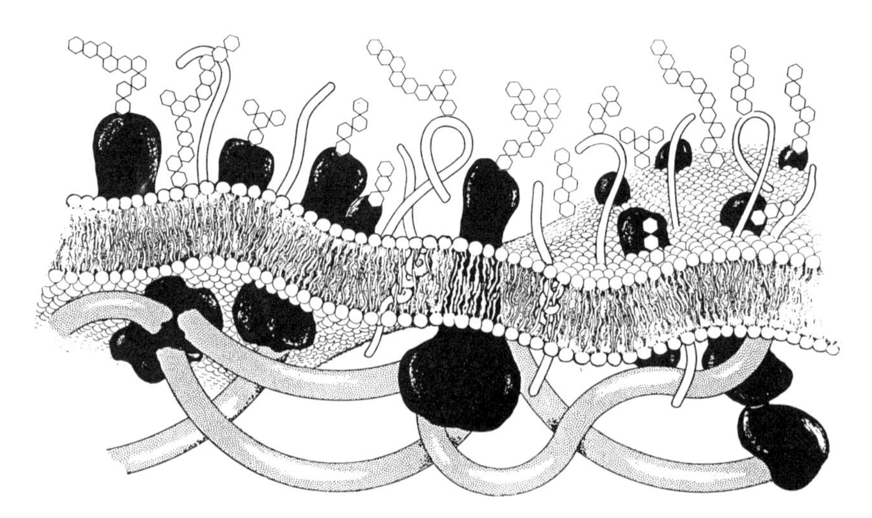

Figure 1. Schematic illustration of a eukaryotic cell membrane showing the fluid lipid bilayer composed of amphiphilic lipids constituting the two-dimensional membrane plane. Proteins and polypeptides are shown embedded in the lipid bilayer together with more superficially attached proteins and enzymes. The cytoskeleton attached to membrane proteins is also shown. (Courtesy of O. Broo Sørensen.)

for the biological functioning of the numerous membrane-associated processes. In order to understand how the many different molecular components of the membrane can influence the overall lateral heterogeneous structure of the membrane and its associated functions, it is of importance to understand the cooperative nature of the biomembrane as a molecular assembly (Sackmann, 1984; Mouritsen and Jorgensen, 1994).

The present chapter examines and emphasizes the various factors responsible for induction of dynamic membrane heterogeneity on microscopic and mesoscopic length scales (100–1000 Å) and how such heterogeneous membrane states may couple to active and passive membrane functions (Mouritsen and Jorgensen, 1994). Furthermore, the chapter presents a discussion of how various molecular agents interacting with membranes show effects that are related to the capacity of these agents to modulate membrane heterogeneity.

THE LIPID BILAYER AS A MANY-PARTICLE SYSTEM

The classical fluid-mosaic model of the lipid membrane proposed in 1972 by Singer and Nicolson pictures the membrane as a fluid two-dimensional lipid bilayer matrix in which transmembrane proteins are embedded and where peripheral proteins are superficially attached to either of the two sides of the bilayer (Singer and Nicolson, 1972). Central to the fluid-mosaic model is the concept of fluidity, which implies that the molecules in the membrane have a substantial degree of lateral mobility in the two-dimensional membrane plane. This allows for the lateral diffusion of both lipids and proteins in the membrane. The fluid-mosaic model regards the biomembrane in a simplistic way as "a sea of lipids with proteins floating around in it." Although this has provided the membrane biologist with a working model that has been extremely successful in understanding and rationalizing a large set of experimental observations, it has become clear that the fluid-mosaic model has several shortcomings (Israelachvili, 1977; Kinnunen and Mouritsen, 1994). By considering the membrane as laterally homogeneous, the fluid-mosaic model basically neglects membrane heterogeneity which seems to be an essential feature of the lateral order-function relationship that has been shown to be important for the performance of membrane proteins and enzymes.

The lateral organization of the membrane components is basically determined by the intramembrane interactions between these components (Bloom et al., 1991). It is therefore of major importance to understand the interaction between the membrane components in order to come to grips with which factors are responsible for the lateral organization and subsequently for the activity of membrane processes. In the years since the introduction of the fluid-mosaic model new insight has been gained into the importance of the cooperative behavior of the assembly of lipid molecules for the thermodynamic phase behavior and the associated physical properties of the lipids in the membrane (Cevc and Marsh, 1987; Bloom et al., 1991). A proper understanding of the relationship between lipid-lipid interactions and the heterogeneous membrane behavior may at the same time shed more light on lipid diversity and the functional role of the various lipids found in biological membranes. Examples of the many different lipids include phospholipids, sphingolipids, cardiolipids and cholesterol.

Model membranes consisting of simple one-component or few-component phospholipid bilayers, or bilayers reconstituted with proteins or amphiphilic polypeptides, have played a seminal role in the study of lipid-bilayer components. Such a simplistic model approach is deemed absolutely necessary as a basis in order to take full advantage of modern physical experimental techniques, including, for example, magnetic resonance spectroscopy (NMR and ESR), differential-scanning calorimetry, micromechanics, and scattering and diffraction techniques. Parallel to the experimental work, substantial progress has been made in the theoretical interpretation of the experimental data on physical membrane properties using molecular interaction models. The combined efforts have now brought the science of physical membranology to a stage where it is possible to outline some possible connections between membrane physical state and membrane function (Kinnunen and Mouritsen, 1994). It has become clear that the cooperative behavior of the many particle lipid-bilayer can give rise to dynamic heterogeneous membrane behavior which is much more complex than that embodied in the fluid-mosaic model.

DIFFERENT TYPES OF MEMBRANE HETEROGENEITY

Static Membrane Heterogeneity

Static membrane heterogeneity is generally caused by some static stabilizing conditions which can be of either mechanic or thermodynamic origin. Well-known examples of static heterogeneity in membranes include: (i) macroscopic morphologically different domains induced by, for example, coupling between different neighboring membranes in a stack, (ii) aggregation of proteins in single membranes due to thermodynamic phase separation or due to specific coupling to the cytoskeleton, and (iii) lateral phase separation in the bilayer driven by either intrinsic thermodynamic forces, by external fields, or by concentration gradients. The latter type of membrane heterogeneity is a plain consequence of the frequently overlooked fact that many-component membranes often tend to undergo phase-separation phenomena (Lee, 1977) which can lead to liquid domain formation on various length and timescales (Jørgensen and Mouritsen, 1995; Jørgensen et al., 1996).

Dynamic Membrane Heterogeneity

Dynamic membrane heterogeneity is a phenomenon that is a consequence of the many-particle nature of the bilayer molecular assembly. Large assemblies of mutually interacting molecules are subject to spontaneous fluctuations in lateral density or composition. Density fluctuations prevail near lipid phase transitions (Mouritsen, 1991), e.g., the gel-to-fluid transition of the lipid bilayer, which takes

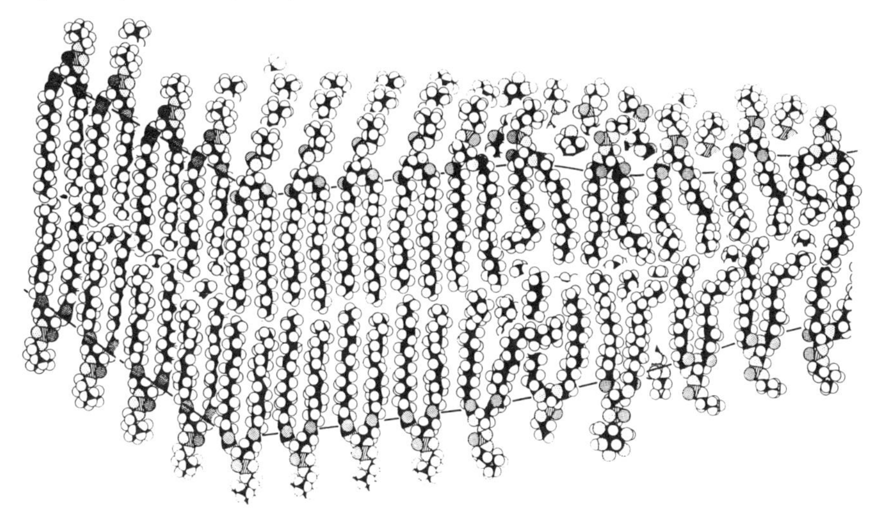

Figure 2. Phase transition of lipid-bilayers: Schematic illustration of the cooperative nature of the lipid membrane undergoing the main phase transition which takes the lipid-bilayer from the low-temperature gel phase (to the left) to the high-temperature fluid phase (to the right).

the bilayer from a phase with conformationally ordered lipid acyl chains (the gel phase) to a phase of considerable acyl-chain conformational disorder (the fluid phase), as shown in Figure 2. Close to the phase transition temperature the density fluctuations manifest themselves as domains of gel or fluid lipids (Mouritsen, 1991).

In lipid mixtures, compositional fluctuations manifested as domains of correlated lipids of the same type can give rise to a local concentration of the lipid species that is different from its global concentration. This type of dynamic heterogeneous membrane structure becomes more pronounced near phase boundaries or critical demixing points and it depends on the non-ideal mixing properties of the lipids constituting the lipid bilayer.

Microscopically, as well as mesoscopically, density- and compositional fluctuations manifest themselves in the dynamic formation of domains or clusters of correlated lipid molecules having properties and structures different from that of the bulk matrix. The fluctuation-induced dynamic lipid-bilayer heterogeneity is associated with a time scale characteristic of the lateral fluctuations in density or composition. Obviously, this time scale is strongly dependent on membrane composition and on the thermodynamic conditions of the membrane.

A similar dynamic heterogeneity can be defined in relation to lipid-protein interactions in membranes, where it has been found that on a certain time scale, the lipid environment adjacent to an integral membrane protein has a structure and composition that is different from that of the bulk lipid matrix, even though there is no specific chemical binding of the lipids to the proteins (Sperotto and Mouritsen, 1991).

Both in the case of static and dynamic lipid-bilayer heterogeneity, the membrane may display biologically differentiated regions which are considered important for a variety of functions associated with membranes (Biltonen, 1990; Hønger et al., 1996).

DYNAMIC MEMBRANE HETEROGENEITY INDUCED BY COOPERATIVE FLUCTUATIONS

The question as to which factors control dynamic membrane heterogeneity has received only scant attention. An exception is the much debated question of the possible existence of a lipid annulus around integral membrane proteins, a question which has now been solved as the result of proper time-scale considerations (Marsh, 1985). There are several reasons for the limited interest in dynamic lipid bilayer heterogeneity in membranes. In the first place, many membranologists consider the lipid bilayer as a mere anchoring place for more interesting membrane components such as proteins, receptors, and enzymes, and that the lipid properties have little influence on function. Secondly, it is extremely difficult experimentally to measure dynamic heterogeneity directly since most techniques involve implicit averaging procedures which do not allow a resolution of the lateral structure. This is true of thermodynamic and thermomechanic measurements, as well as of most spectroscopic techniques. One of the obstacles to obtaining direct information on dynamic heterogeneity of the lipid bilayer is that this type of heterogeneity which occurs on a short time scale is manifested at mesoscopic length scales on the order of 100–1000 Å which are not easily accessible using current experimental techniques (Bloom et al., 1991).

The most direct evidence of dynamic heterogeneity in lipid bilayers has been provided by fluorescence lifetime (Ruggerio and Hudson, 1989) and energy transfer measurements (Pedersen et al., 1996). However, dynamic heterogeneity may be inferred indirectly from measurements of membrane response functions, such as the specific heat or the lateral compressibility, which are bulk, integral measures of the thermal density fluctuations (Biltonen, 1990).

Evidence in favor of thermally induced dynamic membrane heterogeneity has also been obtained by theoretical calculations using specific molecular interaction models of lipid bilayers (Mouritsen, 1990). These calculations involve computer simulations on large arrays of lipid molecules taking into account the acyl-chain conformations and their mutual interactions.

The type of information which can be obtained from such computer simulations on a single-component phospholipid bilayer is illustrated in Figure 3 which shows the planar organization of the bilayer as a function of temperature. It is seen that the bilayer, in a wide range of temperatures, is subject to very strong lateral density fluctuations which manifest themselves in the formation of domains or clusters of correlated lipids having a structure and density which is different from that of the bulk equilibrium lipid matrix. These domains are dynamic and highly fluctuating

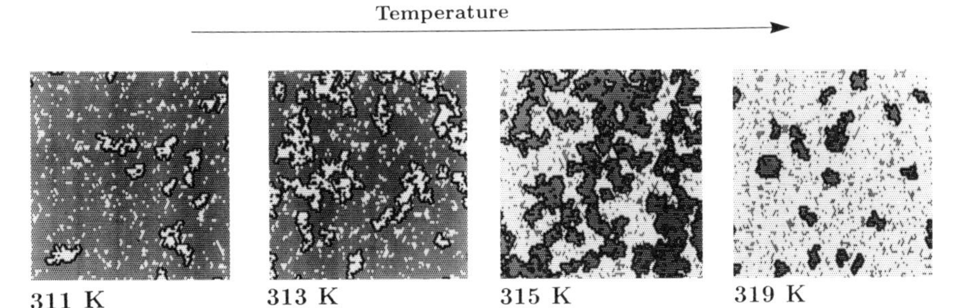

Figure 3. Dynamic membrane heterogeneity: Snapshots of microscopic configurations of a dipalmitoyl phosphatidylcholine bilayer characteristic of different temperatures near the gel-to-fluid phase transition at 314K. The configurations are obtained from computer simulations. Gray and light regions correspond to gel and fluid regions, respectively. The interfaces between the different regions are shown as black lines. Each snapshot shows a liquid bilayer area corresponding to 5000 liquid molecules ($\sim 550 \times 550$ Å^2).

entities which are consequences of the cooperative fluctuations of the membrane components.

The range over which these fluctuations are operative is described by a coherence length which is a measure of the average domain size. The coherence length depends on temperature and attains a maximum at the gel-to-fluid phase transition. For example, for dipalmitoyl phosphatidylcholine bilayers, the lipid domain size may become as large as several hundred lipid molecules. Hence, the dynamic membrane heterogeneity leads to membrane organization on the mesoscopic length scale. This average domain size depends on the lipid species in question and increases as the acyl-chain length decreases (Ipsen et al., 1990).

Associated with the lipid domains is a network of interfaces bounding the domains within the membrane. The interfaces, which are highlighted in Figure 3, have very special molecular packing properties; they are 'soft' and have a low interfacial tension. These interfaces may support specific passive and active membrane functions (Hønger et al., 1996).

Dynamic heterogeneity in lipid mixtures has been shown to be highly dependent on the mixing properties of the lipid species. Both theoretical and experimental studies have revealed that even the high-temperature mixed fluid phase is characterized by a dynamic heterogeneous membrane state (Knoll et al., 1991; Jørgensen et al., 1993). Hence, thermodynamic one-phase regions of mixtures may have considerable lateral structure and a dynamically heterogeneous organization of the lipid species. Examples of both static and dynamic membrane heterogeneity in a binary lipid mixture are shown in Figure 4. The snapshots of the lateral membrane organization of the membrane components derive from a model calculation of an equimolar mixture of dilauroyl phosphatidylcholine and distearoyl phosphatidylcholine ($DC_{12}PC$ and $DC_{18}PC$).

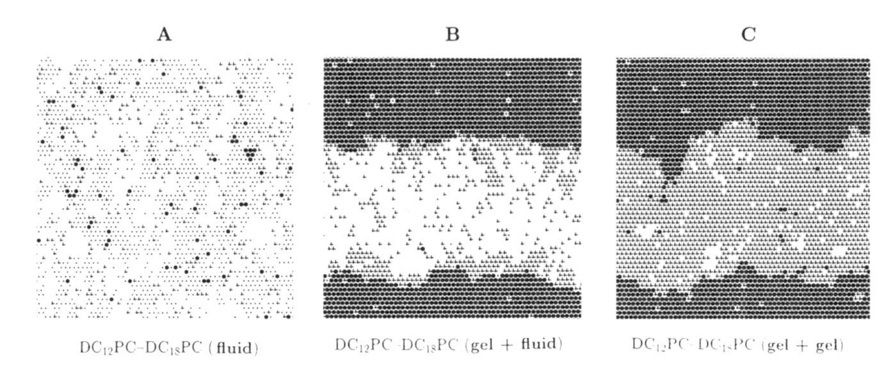

Figure 4. Static and dynamic membrane heterogeneity: Snapshots of microscopic configurations of an equimolar binary mixture of dilauroyl phosphatidylcholine and distearoyl phosphatidylcholine ($DC_{12}PC$ and $DC_{18}PC$) at different temperatures, corresponding to different regions in the phase diagram. (**A**) The high-temperature thermodynamic one-phase fluid region, (**B**) the gel-fluid coexistence region, and (**C**) the low-temperature gel-gel coexistence phase. Each snapshot shows a liquid bilayer area corresponding to 1800 liquid molecules (~330×330 Å^2).

The difference of six carbon atoms between the long hydrophobic acyl chains of the two lipid species is found to give rise to a highly non-ideal mixing behavior of the lipids manifested as immiscibility of the two lipid species in the gel phase, as seen in Figure 4C. At low temperatures the membrane is characterized by a static heterogeneity manifested as macroscopic phase coexistence of two gel phases. At higher temperature in the gel-fluid phase coexistence region, the snapshot in Figure 4B reveals a type of dynamic heterogeneity manifested as local interfacial ordering. Along the interfaces between the coexisting gel and fluid phases there is a pronounced enrichment of the interfaces by gel-state lipids of the lowest melting lipid. The enrichment of the interfaces is seen to involve many lipid layers, giving rise to differentiated regions of the membrane which have different properties from those of the bulk phases. Figure 4A shows that in the high temperature thermodynamic one-phase fluid region, the non-ideal mixing properties of the two lipids species can give rise to compositional fluctuations manifested as a local structure. The local ordering in the fluid phase is dynamically maintained as fluctuating domains of correlated lipids, characterized by a coherence length that depends on thermodynamic conditions, especially temperature and composition, as well as the distance from special critical demixing points in the phase diagram (Jørgensen et al., 1993).

PASSIVE MEMBRANE FUNCTIONS GOVERNED BY DYNAMIC MEMBRANE HETEROGENEITY

In a now classical experiment (Papahadjopoulos et al., 1973) it was discovered that the passive transmembrane permeability of alkali ions, e.g., Na^+, displayed an anomaly at the gel-to-fluid phase transition of phospholipid bilayers, as shown in Figure 5. At the transition the membranes become very leaky. Similar anomalies in the permeability have been observed for many other molecular species (Kanehisa and Tsong, 1978) including water.

It is possible to rationalize the full temperature dependence of the thermal anomaly of the passive permeability via a simple model (Cruzeiro-Hansson et al., 1988) which assigns a transmission coefficient to the interfacial region which is much larger than the transmission coefficients characterizing the clusters and the bulk membrane matrix. By using the temperature dependence of the interfacial region, as obtained by the computer simulations described above (cf. Figure 3) within this simple model, one obtains a theoretical prediction for the permeability, as shown in Figure 5. The correlation with the experimental data is astonishing. These results strongly support the hypothesis that dynamic membrane heterogeneity may control passive transmembrane functions.

Leakiness to ions and other molecular species is only one aspect of passive membrane phenomena mediated by membrane heterogeneity and interfaces. There are many other examples of phenomena which exploit the interfaces as easy membrane targets or as an appropriate environment for insertion of material. For example, it has been found that the spontaneous exchange of cholesterol between

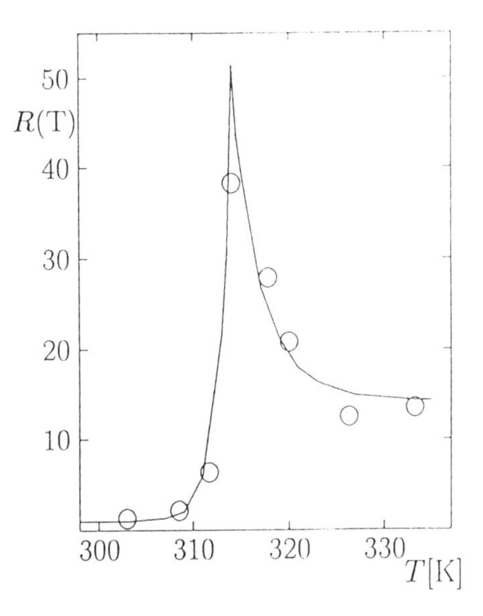

Figure 5. Dynamic membrane heterogeneity enhances the passive transmembrane permeability: Relative permeability, $R(T)$, of Na^+-ions in liposomes as measured by radioactive ^{22}Na-techniques (○) (Papahadjopoulos et al., 1973) and as calculated from a computer simulation (——) of dynamic heterogeneity in a dipalmitoyl phosphatidylcholine bilayer membrane.

small unilamellar phospholipid vesicles correlates with the interface formation (Bar et al., 1987). As another example, experiments have demonstrated that certain exchange proteins which facilitate phospholipid transfer between vesicles exhibit enhanced activity in the phase-transition region (Xü et al., 1983). Finally, calcium-induced fusion of lipid vesicles has been found to have a distinct maximum at the transition temperature (Leung and Ho, 1991).

ACTIVE MEMBRANE FUNCTIONS GOVERNED BY DYNAMIC MEMBRANE HETEROGENEITY

On a mesoscopic length scale, dynamic membrane heterogeneity provides a differentiated lipid environment for membrane-bound proteins, enzymes, and receptors. Specifically, the interfacial regions implied by this type of heterogeneity are regions with special packing properties which may attract proteins and possibly induce conformational changes in the proteins. Mechanisms along these lines for lipid-mediated protein functionality (Sackmann, 1984) have been suggested by several authors, e.g., for the activity of phospholipase A_2 (Grainger et al., 1989; Biltonen, 1990; Scott et al., 1990; Hønger et al., 1996) and for Ca^{2+}-ATPase (Asturias et al., 1990).

A particularly striking example is provided by the different phospholipases which, although of different structure, seem to be influenced and activated by similar factors. Phospholipase A_2 is an interfacially active enzyme: it acts most strongly on organized lipids in aggregates as in membranes. Despite the fact that the exact mechanism of the enzyme action is unclear, there is overwhelming experimental evidence that the rate of activation of the enzyme is controlled by the lipid fluctuations. Specifically, it has been found that the rate of activation may increase orders of magnitude in the membrane transition region and this increase correlates with the fluctuations (Biltonen, 1990). Supporting evidence for this result is provided in Figure 6, which reproduces the experimental data for the rate of activation for pancreatic phospholipase A_2 in the hydrolysis of lipid vesicles. This figure also shows the results of a computer-simulation calculation of the average size of the lipid domains reflecting the dynamic membrane heterogeneity in the transition region. There is an astonishing qualitative correlation between the two sets of data. It has been suggested that the heterogeneous lipid environment accompanying these fluctuations may provide a trigger mechanism for conformational changes in the enzyme which in turn controls the rate of activation (Biltonen, 1990; Mouritsen and Biltonen, 1993; Hønger et al., 1996). Furthermore, it can be conjectured that the phase separation processes, and then possibly the static heterogeneity, which follows from the enzymatic production of fatty acids, is also coupled to function (Burrack et al., 1993).

It has been suggested that the concept of hydrophobic matching between lipid-bilayer thickness and hydrophobic length of integral membrane-bound proteins (Mouritsen and Bloom, 1984; Mouritsen and Bloom, 1993) may provide a possible

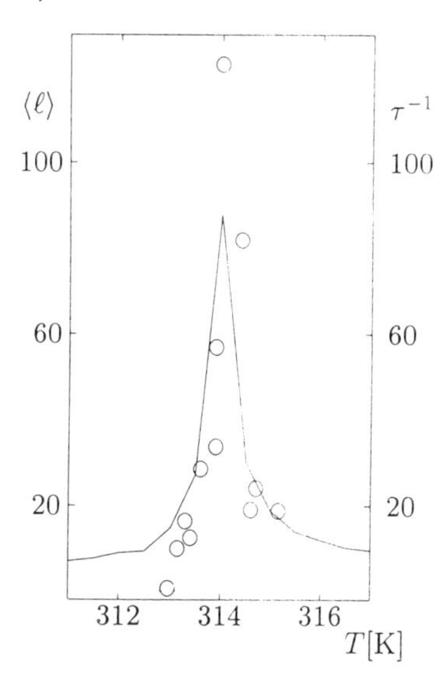

Figure 6. Dynamic membrane heterogeneity enhances active membrane functions: Rate of activation (○) (~ inverse time, τ^{-1} [$\text{min}^{-1} \times 10^{-3}$]), for hydrolysis of large unilamellar dipalmitoyl phosphatidylcholine vesicles in the neighborhood of the gel-to-fluid phase transition (314K) (adapted from Biltonen, 1990). τ is the time required to reach half-maximum activity of porcine pancreatic phospholipase A_2. Also shown are the results of a model calculation of the average lipid-domain size $\langle \ell \rangle$ (———), in a dipalmitoyl phosphatidylcholine bilayer (cf. Fig. 3). $\langle \ell \rangle$ is in units of number of lipid-acyl chains.

physical mechanism for nonspecific lipid-protein interactions in relation to static and dynamic membrane heterogeneity (Mouritsen and Biltonen, 1993). By changing the degree of hydrophobic matching, conformational changes in the protein may be induced which influence the activity of the protein (Johannsson et al., 1981). Changes in the hydrophobic-matching conditions can be visualized to occur at lipid-domain boundaries or in the immediate lipid environment of the protein by changing the local lipid composition or the lipid-chain order parameter. Such changes could be caused by active molecular agents interacting with membranes such as drugs. This question will be addressed further in the following section.

MANIPULATING MEMBRANE HETEROGENEITY

Obviously, it is not desirable for biological cell membranes to have high levels of passive permeability. Instead, cell function requires the regulation of transport and permeation of material across its membranes to take place in a selective and vectorial manner by means of specific pumps and channels. Hence it is important that biological membranes contain molecular compounds which seal the membrane by suppressing the dynamical fluctuations. Cholesterol is such a compound (Bloom and Mouritsen, 1988).

Cholesterol Influences Membrane Heterogeneity

The understanding of how cholesterol affects lipid membrane phase equilibria and determines the physical properties of membranes has proved to be a very elusive problem in the physical chemistry of biological membranes. Some clarification of the problem has only recently been achieved by novel experimental (Vist and Davis, 1990) and theoretical approaches (Ipsen et al., 1987). In large amounts ($\gtrsim$ 30 mol%), cholesterol seals the membrane by almost completely removing the density fluctuations (Corvera et al., 1992). However, it has been discovered that in smaller amounts ($\lesssim$ 10 mol%) cholesterol has the opposite effect and enhances the fluctuations and increases the degree of dynamic membrane heterogeneity (Cruzeiro-Hanson et al., 1989). This effect becomes apparent by comparing Figures 7A and 7B which show typical lateral membrane configurations as obtained from computer simulations on molecular interaction models. It is apparent that the presence of cholesterol induces more interfaces. Moreover, an analysis of the structure of these interfaces reveals that there is a distinct tendency for accumulation of cholesterol molecules at the interfaces. Hence, cholesterol exerts control of the dynamic heterogeneity of the membrane, both in terms of interface formation and in terms of inducing a non-random distribution of the two molecular species even within thermodynamic one-phase regions.

Drugs Affect Membrane Heterogeneity

Other molecular compounds interacting with membranes may lead to a similar change in dynamic membrane heterogeneity. A particularly prominent and immensely large class of such compounds are drugs, e.g., general and local anesthetics and a variety of chemical species which are pharmacologically active (Jørgensen et al., 1991a; Jørgensen et al., 1993). Figure 7 shows the effect on the dynamic membrane organization for a lipid bilayer model doped with foreign molecules which place themselves interstitially in the lipid matrix and which interact in a specific manner with excited lipid acyl-chain conformations (Jørgensen et al., 1991b). The concentration of the foreign molecules in the membrane is controlled by a membrane/water partitioning equilibrium. The figure shows that the foreign molecules lead to a dramatic enhancement of membrane heterogeneity inducing more interfacial regions. Moreover, the local concentration of the foreign molecules in the interfaces can easily be the global concentration, indicating a significant dynamic accumulation of the molecules in the interfaces. Hence, the foreign molecules modulate the dynamic membrane heterogeneity, leading in turn to a lateral organization of the membrane components which is very heterogeneous. The macroscopic consequences of these microscopic and mesoscopic events are several.

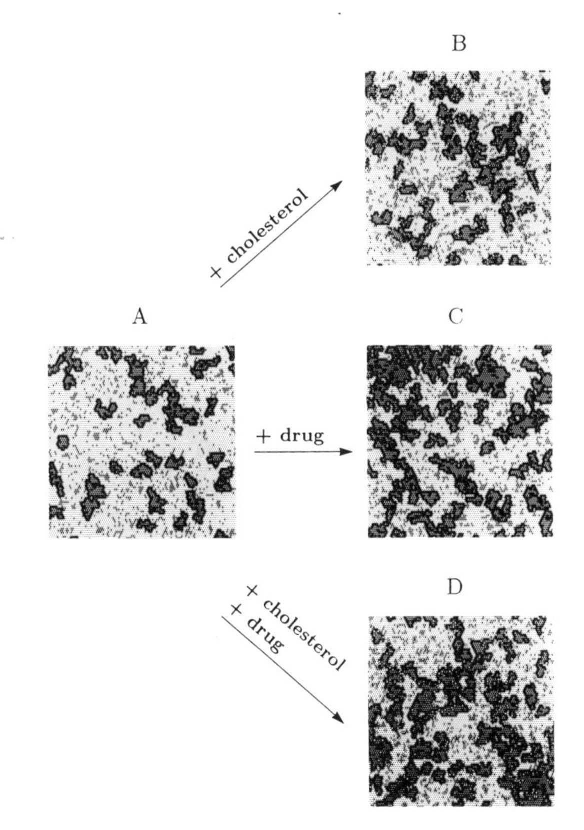

Figure 7. Dynamic membrane heterogeneity and how it is affected by various molecular compounds interacting with lipid bilayers: The configurations are obtained from computer simulations on a molecular interaction model of dipalmitoyl phosphatidylcholine bilayers at a temperature in the fluid phase. (**A**) The pure lipid bilayer. (**B**) The lipid bilayer in the presence of 9.5% cholesterol. (**C**) The lipid bilayer in the presence of 28% anesthetics. (**D**) The lipid bilayer in the presence of 9.5% cholesterol and 28% anesthetics. Each snapshot shows a liquid bilayer area corresponding to 5000 liquid molecules (~550 × 550 Å^2).

Firstly, the doping of the membrane leads to enhanced passive membrane permeability. Secondly, as shown in Figure 8, the partition coefficient (which is related to the concentration of the foreign molecules in the membrane) may display a peak in the gel-to-fluid transition region. This is a unique effect, known as enhanced adsorption in the field of heterogeneous catalysis, which is a consequence of the dynamic membrane fluctuations and how they are enhanced by the foreign molecules. It may be considered as a kind of synergistic effect by which the foreign molecules enhance the fluctuations, leading to more interfaces which in turn facilitate incorporation of more foreign molecules into the lipid bilayer. The

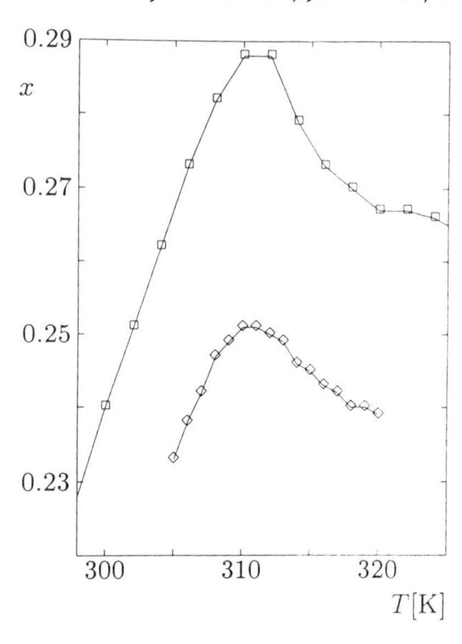

Figure 8. Dynamic membrane heterogeneity enhances the absorption of drugs: Partition coefficient x (□) for a drug compound (e.g., general anesthetics like halothane or local anesthetics like cocaine derivatives) interacting with a lipid bilayer. Data for the partition coefficient in the presence of 9.5% cholesterol (◇) are also shown. The data are derived from a computer simulation on a molecular interaction model of dipalmitoyl phosphatidylcholine bilayers.

heterogeneous membrane can be considered to consist of differentiated regions reflecting a differential solubility with regard to the drug molecules (Landau et al., 1979). It should be noted that this behavior is very different from the classical behavior (Jørgensen et al., 1991a) where the partitioning of an adsorbing species displays a simple jump-behavior in the transition region, reflecting the mere difference of solubility in the two different membrane phases.

Several pharmacological compounds have been found experimentally to lead to effects which are consistent with the behavior described above. Examples include volatile general anesthetics like halothane (Mountcastle et al., 1978), local anesthetics like cocaine derivatives (Singer and Jain, 1980) and non-steroidal anti-inflammatory agents like indomethacin (Lasoner and Weringa, 1990). Moreover, it is interesting to note that a completely different class of compounds interacting with membranes, the organochlorine insecticides lindane, DDT, parathion, and malathion (Antunes-Madeira and Madeira, 1987) show similar effects on the physical properties of lipid membranes (Jørgensen et al., 1991b; Saba et al., 1996).

In particular, the theoretical finding of the accumulation of drug molecules in specific regions in the membrane may serve to clarify and modify the notion of "clinical concentrations" of drugs. The accumulation of the drug molecules in the interfacial regions gives the drug more direct access to many proteins and receptors which also have a tendency to accumulate along lines of membrane defects (Sackmann, 1984). Hence it may be that the actual local concentration of drugs in certain regions of the heterogeneous membrane is more relevant in relation to defining the "clinical concentration" than the global concentration. These considerations may therefore have some bearing on the interpretation of studies on how drugs and pesticides influence protein function via modulation of the lipid environment in the neighborhood of the protein molecules (Fraser et al., 1990).

Anesthesia: A Membrane Perturbing Effect?

It is tempting to speculate that the ability of general and local anesthetics to couple to the dynamic membrane heterogeneity may provide some new clues to the important and mainly unsolved problem of the mechanism of anesthesia (Trudell, 1977). In order to elucidate the factors which control anesthesia it is of interest to study the effect of the solubility of drug molecules when other molecular species, like cholesterol, are present in the membrane. Figure 7D shows how the membrane heterogeneity appears when both interstitially intercalated molecules (anesthetics) and cholesterol are present in a model membrane. The membrane heterogeneity is changed and the local anesthetics concentration in the interfaces is diminished. The macroscopic consequence of these alterations is a suppression of the fluctuations and a lowering of the passive transmembrane permeability and a concomitant decrease in the anesthetics partition coefficient, as seen in Figure 8. Cholesterol simply squeezes out the drug. It would be interesting to explore whether these effects of cholesterol could provide an explanation for the varying effects of drugs on membranes of different composition (e.g., with respect to cholesterol) and whether such membrane "impurities" are related to the many side effects of drugs. In favor of the lipid membrane as a target for anesthetics is the pressure-reversal of anesthesia exerted by hydrostatic pressure. Apart from being the only antagonist of anesthesia, hydrostatic pressure, as a mere physical-chemical quantity, has the ability to influence the heterogeneous dynamic membrane behavior and the associated differentiated regions in a similar way as temperature (cf. Fig. 3).

The interesting question which now arises is whether there is control of active protein and enzyme function by coupling, via the physical state of the lipids, to the interfacial characteristics. As was pointed out above, many pharmacologically potent agents, as well as cholesterol in small concentrations, enhance interface formation via dynamic fluctuations. One would then predict that proteins, whose mechanism of action is coupled to the heterogeneity of the membrane and the associated formation of interfaces, could be manipulated by appropriate agents. Indeed it has been found experimentally for phospholipase A_2 that the rate of activation can be dramatically enhanced by cholate (Gheriani-Gruszaka et al., 1988).

Phospholipidosis

Phospholipidosis is characterized by an excessive accumulation of phospholipids in tissues leading to significant alterations in cellular membrane structure and function (Kodavanti and Mehendale, 1990). The storage disorder of phospholipids in tissues from liver, brain, kidney, and several other tissues occurs after long term administration of drugs from a diverse class of pharmacological agents including antipsychotics, antiarrhytmics, anesthetics, and antidepressants. Despite their diverse therapeutic applications, these compounds have several physiocochemical properties in common. Moreover, they have all been found to be membrane-perturbing agents. The effect of drugs from differnt pharmacological classes on the physical properties of the membrane is manifested in a decrease in transition temperature, a change in membrane composition, and an induced membrane fluidization. Extensive studies have revealed a close correlation between the influence of various drugs on the physical behavior of the membrane and the corresponding induction of a phospholipid storage disorder. The interesting question therefore arises as to what extent drug-phospholipid interaction and the induced changes in the heterogeneous membrane structure make phospholipids unsuitable substrates for phospholipases. Changes in the heterogenous membrane structure may affect the overall lateral-order function relationship that has been clearly demonstrated to exist between the heterogeneous membrane state and the related activity of phospholipase A_2 (Hønger et al., 1996). It seems conceivable that phospholipidosis induced by a variety of pharmacological agents might be mediated through a universal mechanism caused by a non-specific interaction with the heterogeneous membrane state, leading to a change in the membrane heterogeneity and alterations in the biologically differentiated regions in the two-dimensional membrane plane.

CONCLUSION AND FUTURE PERSPECTIVES

In the present chapter, results from studies on lipid-bilayer membranes have been presented which show that lipid bilayers sustain a substantial dynamic heterogeneity. This dynamic heterogeneity is a consequence of the often overlooked fact that the membrane molecular assembly is a many-particle system sustaining strong lateral density- and compositional fluctuations. It has been pointed out that the heterogeneity of the membrane varies with temperature and acyl-chain length of the phospholipid molecules in the membrane, as well as the mixing properties of the lipid species constituting the lipid-bilayer. Furthermore, it has been shown that foreign molecules interacting with the lipid membrane have a strong influence on the lateral density and compositional fluctuations and the associated dynamic heterogeneity of the lipid membrane state. Macroscopically, the dynamic membrane heterogeneity is manifested in enhanced response functions, such as specific heat, and an enhancement of the passive membrane permeability. The hypothesis put forward is that dynamic lipid-bilayer heterogeneity accompanied by the formation of lipid domains and their associated network of interfacial regions in the membrane plane support important membrane functions. This hypothesis is supported by the close correlation found between the calculated interface formation from lipid-bilayer model simulations and the experimentally measured passive permeability functions and the rate of activation of enzymes such as phospholipases. It is, moreover, proposed that by assuming that dynamic lipid-bilayer heterogeneity is a mesoscopic vehicle for membrane function, the interesting possibility arises of manipulating membrane function using molecular agents which modulate the dynamic membrane heterogeneity by changing the interfacial characteristics.

It has been pointed out that the physical effects on lipid-membrane properties as found experimentally for a large class of drugs interacting with membranes, including certain general and local anesthetics, could be rationalized via their ability to modulate the dynamic membrane heterogeneity. These findings lead to the suggestion that the action of various drugs may be lipid-mediated. This new standpoint may lead to a better understanding of the many side effects that follow long-term administration of a variety of drugs.

NOTE

1. Associate Fellow of the Canadian Institute of Advanced Research

REFERENCES

Antunes-Madeira, M.C. & Madeira, V.M.C. (1987). Partitioning of malathion in synthetic and native membranes. Biochim. Biophys. Acta 910, 61–66.

Asturias, F.J., Pascolini, D., & Blasie, J.K. (1990). Evidence that lipid lateral phase separation induces functionally significant structural changes in the Ca^{2+} ATPase of the sarcoplasmatic reticulum. Biophys. J. 58, 205–217.

Bar, L.K., Barenholz, Y., & Thompson, T.E. (1987). Dependence on phospholipid composition of the fraction of cholesterol undergoing spontaneous exchange between small unilamellar vesicles. Biochemistry 26, 5460–5465.

Bergelson, L.O., Gawrisch, K., Ferretti, J.A., & Blumenthal, R., (eds.) (1995). Special issue on domain organization in biological membranes. Mol. Memb. Biol. 12, 1–162.

Biltonen, R.L. (1990). A statistical-thermodynamic view of cooperative structural changes in phospholipid bilayer membranes: their potential role in biological function. J. Chem. Thermodyn. 22, 1–19.

Bloom, M. & Mouritsen, O.G. (1988). The evolution of membranes. Can. J. Chem. 66, 706–712.

Bloom, M., Evans, E., & Mouritsen, O.G. (1991). Physical properties of the fluid-bilayer component of cell membranes: a perspective. Q. Rev. Biophys. 24, 293–397.

Burrack, W.R., Yuan, Q., & Biltonen, R.L. (1993). Role of lateral phase separation in the modulation of phospholipase A_2 activity. Biochemistry 32, 583–589.

Cevc, G. & March, D. (1987). Phospholipid Bilayers. Physical Principles and Models. Wiley-Interscience, New York.

Corvera, E., Mouritsen, O.G., Singer, M.A., & Zuckermann, M.J. (1992). The permeability and the effect of acyl-chain length for phospholipid bilayers containing cholesterol: Theory and experiment. Biochim. Biophys. Acta 1107, 261–270.

Cruzeiro-Hansson, L. & Mouritsen, O.G. (1988). Passive ion permeability of lipid membranes modelled via lipid-domain interfacial area. Biochim. Biophys. Acta 944, 63–72.

Cruzeiro-Hansson, L., Ipsen, J.H., & Mouritsen, O.G. (1989). Intrinsic molecules in lipid membranes change the lipid-domain interfacial area: cholesterol at domain boundaries. Biochim. Biophys. Acta 979, 166–176.

Fraser, D.M., Louro, S.R.W., Horváth, L.I., Miller, K.W., & Watts, A. (1990). A study of the effect of general anesthetics on lipid-protein interactions in acetylcholine receptor enriched membranes form *Torpedo nobiliana* using nitroxide spin-labels. Biochemistry 29, 2664–2669.

Gennis, R.B. (1989). Biomembranes. Molecular Structure and Function. Springer-Verlag, London.

Gheriani-Gruszka, N., Almog, S., Biltonen, R.L., & Lichtenberg, D. (1988). Hydrolysis of phosphatidylcholine in phosphatidylcholine-cholate mixtures by porcine pancreatic phospholipase A_2. J. Biol. Chem. 263, 11808–11813.

Grainger, D.W., Reichert, A., Ringsdorf, H., & Salesse, C. (1989). An enzyme caught in action: direct imaging of hydrolytic function and domain formation of phospholipase A_2 in phosphatidylcholine monolayers. FEBS Lett. 252, 73–82.

Hønger, T., Jørgensen, K., Biltonen, R.L., & Mouritsen, O.G. (1996). Systematic relationship between phospholipase A_2 activity and dynamic lipid bilayer microheterogeneity. Biochemistry 35, 9003–9006.

Ipsen, J.H., Jørgensen, K., & Mouritsen, O.G. (1990). Density fluctuations in saturated phospholipid bilayers increase as the acyl-chain length decreases. Biophys. J. 58, 1099–1107.

Ipsen, J.H., Karlström, G., Mouritsen, O.G., Wennerström, H., & Zuckermann, M.J. (1987). Phase equilibria in the phosphatidylcholine-cholesterol system. Biochim. Biophys. Acta 905, 162–172.

Israelachvili, J.N. (1977). Refinement of the fluid-mosaic model of membrane structure. Biochim. Biophys. Acta 469, 221–225.

Jacobsen, K. & Vaz, W.L.C. (1992). Topical issue on domains in biological membranes. Comm. Mol. Cell. Biophys. 8, 1–114.

Johannsson, A., Smith, G.A., & Metcalfe, J.C. (1981). The effect of bilayer thickness on the activity of $(Na^+ -K^+)$-ATPase. Biochim. Biophys. Acta 641, 416–421.

Jørgensen, K., Ipsen, J.H., Mouritsen, O.G., Bennett, D., & Zuckermann, M.J. (1991a). A general model for the interaction of foreign molecules with lipid membranes: drugs and anesthetics. Biochim. Biophys. Acta 1062, 227–238.

Jørgensen, K., Ipsen, J.H., Mouritsen, O.G., Bennett, D., & Zuckermann, M.J. (1991b). The effects of density fluctuations on the partitioning of foreign molecules into lipid bilayers: Application to anesthetics and insecticides. Biochim. Biophys. Acta 1067, 241–253.

Jørgensen, K., Ipsen, J.H., Mouritsen, O.G., & Zuckermann, M.J. (1993). The effect of anesthetics on the dynamic heterogeneity of lipid membranes. Chem. Phys. Lipids.

Jørgensen, K., Sperotto, M.M., Mouritsen, O.G., Ipsen, J.H., & Zuckermann, M.J. (1993). Phase equilibria and local structure in binary lipid bilayers. Biochim. Biophys. Acta 1152, 135–145.

Jørgensen, K. & Mouritsen, O.G. (1995). Phase separation dynamics and lateral organization of two-component lipid membranes. Biophys. J. 69, 942–954.

Jørgensen, K., Klinger, A., Braiman, M., & Biltonen, R.L. (1996). Slow nonequilibrium dynamical rearrangement of the lateral structure of a lipid membrane. J. Phys. Chem. 100, 2766–2769.

Kanehisa, M.I. & Tsong, T.Y. (1978). Cluster model of lipid phase transition with application to passive permeation of molecules and structure relaxations in lipid bilayers. J. Amer. Chem. Soc. 100, 424–432.

Kinnunen, P. & Laggner, P., (eds.) (1991). Phospholipid Phase Transition, Topical Issue of Chem. Phys. Lipids 57, 109–408.

Kinnunen, P.K.J. & Mouritsen, O.G. (eds.) (1994). Functional dynamics of lipids in biomembranes. Special issue of Chem. Phys. Lipids 73, 1–236.

Knoll, W., Schmidt, G., Rötzer, H., Henkel, T., Pfeiffer, W., Sackmann, E., Mittler-Neher, S., & Spinke, J. (1991). Lateral order in binary lipid alloys and its coupling to membrane functions. Chem. Phys. Lipids 57, 363–374.

Kodavanti, P.U. & Mehendale, H.M. (1990). Cationic amphiphilic drugs and phospholipid storage disorder. Pharmacol. Rev. 42, 327–354.

Landau, M.L., Richter, J., & Cohen, S. (1979). Differential solubility in subregions of the membrane: A nonsteric mechanism of drug specificity. J. Med. Chem. 22, 325–327.

Lasoner, E. & Weringa, W.D. (1990). An NMR and DSC study of the interaction of phospholipid vesicles with some anti-inflammatory agents. J. Colloid Int. Sci. 139, 469–478.

Lee, A.G. (1977). Lipid phase transitions and phase diagrams. Biochim.Biophys. Acta 472, 285–344.

Leung, S.O. & Ho, J.T. (1991). Effect of membrane phase transition on long-time calcium-induced fusion of phosphatidylserine vesicles. Chem. Phys. Lipids 57, 103–107.

Marsh, D. (1985). ESR spin label studies of lipid-protein interactions. In: Progress in Protein-Lipid Interactions, vol. 1 (Watts, A. & J.J.H.H.M. De Pont, Eds.), pp.143–172. Elsevier, Amsterdam.

Mouritsen, O.G. (1990). Computer simulation of cooperative phenomena in lipid membranes. In: Molecular Description of Biological Membrane Components by Computer Aided Conformational Analysis, vol. 1 (R. Brasseur, Ed.), pp. 3–83. CRC Press, Boca Raton, Florida.

Mouritsen, O.G. (1991). Theoretical models of phospholipid phase transitions. Chem. Phys. Lipids 57, 179–194.

Mouritsen, O.G. & Biltonen, R.L. (1993). Protein-lipid interactions and membrane heterogeneity. In: Protein-Lipid Interactions. New Comprehensive Biochemistry 25, 1–39 (Watts, A., ed.).

Mouritsen, O.G. & Bloom, M. (1984). Mattress model of lipid-protein interactions in membranes. Biophys. J. 46, 141–153.

Mouritsen, O.G. & Bloom, M. (1992). Models of lipid-protein interactions in membranes. Ann. Rev. Biophys. Biomol. Struct. 22, 145–171.

Mouritsen, O. G. & Jørgensen, K. (1994). Dynamical order and disorder in lipid bilayers. Chem. Phys. Lipids 73, 3–26.

Mountcastle, D.B., Biltonen, R.L., & Halsey, M.J. (1978). Effect of anesthetics and pressure on the thermotropic behavior of multilamellar dipalmitoylphosphatidylcholine liposomes. Proc. Natl. Acad. Sci. USA 75, 4906–4910.

Papahadjopoulos, D., Jacobsen, K., Nir, S., & Isac, T. (1973). Phase transitions in phospholipid vesicles. Fluorescence polarization and permeability measurements concerning the effect of temperature and cholesterol. Biochim. Biophys. Acta 311, 330–348.

Pedersen, S., Jørgensen, K., Baekmark, T.R., & Mouritsen, O.G. (1996). Indirect evidence for domain formation in the transition region of phospholipid bilayers by two-probe fluorescence energy transfer. Biophys. J. 71, 554–560.

Ruggiero, A. & Hudson, B. (1989). Critical density fluctuations in lipid bilayers detected by fluorescence lifetime heterogeneity. Biophys. J. 55, 1111–1124.Sackmann, E. (1984). Physical basis of trigger processes and membrane structures. In: Biological Membranes, vol. 5 (D. Chapman, Ed.), pp. 105–143. Academic Press, London.

Sabra, M.C., Jørgensen, K., & Mouritsen, O.G. (1996). Lindane suppresses the lipid-bilayer permeability in the main transition region. Biochim. Biophys. Acta 1282, 85–92.

Scott, D.L., White, S.P., Otwinowski, Z., Yuan, W., Gelb, M.H., & Sigler, P.B. (1990). Interfacial catalysis: The mechanism of phospholipase A_2. Science 250, 1541–1546.

Singer, S.J. & Nicolson, G.L. (1972). The fluid mosaic model of the structure of cell membranes. Science 173, 720–731.

Singer, M.A. & Jain, M.K. (1980). Interaction of four local anesthetics with phospholipid bilayer membranes: Permeability effects and possible mechanisms. Can. J. Biochem. 58, 815–821.

Sperotto, M.M. & Mouritsen, O.G. (1991). Monte Carlo simulation studies of lipid order parameter profiles near integral membrane proteins. Biophys. J. 59, 261–270.

Trudell, J.R. (1977). A unitary theory of anesthesia based on lateral phase separation in nerve membranes. Anesthesiology 46, 5–10.
Vist, M.R. & Davis, J.M. (1990). Phase equilibria of cholesterol/DPPC mixtures: ^{2}H nuclear magnetic resonance and differential scanning calorimetry. Biochemistry 29, 451–464.
Xü, Y.-H., Gietzen, K., Galla, H.-J., & Sackmann, E. (1983). A simple assay to study protein-mediated lipid exchange by fluorescence polarization. Biochem. J. 209, 257–260.

RECOMMENDED READINGS

Gennis, R.B. (1989). Biomembranes. Molecular Structure and Function. Springer-Verlag, London.
Mouritsen, O.G. (1990). Computer simulation of cooperative phenomena in lipid membranes. In: Molecular Description of Biological Membrane Components by Computer Aided Conformational Analysis, vol. 1 (R. Brasseur, Ed.), pp. 3–83. CRC Press, Boca Raton, Florida.
Cevc, G. & March, D. (1987). Phospholipid Bilayers. Physical Principles and Models. Wiley-Interscience, New York.

Chapter 3

Lipid-Protein Interactions in Biological Membranes

PAUL BURN and BRUNO H. DALLE CARBONARE

Principles of Medical Biology, Volume 7A
Membranes and Cell Signaling, pages 39–66.

ISBN: 1-55938-812-9

INTRODUCTION

The two previous chapters dealt with model membrane systems and lipid-bilayer heterogeneity. We now focus our attention on biological membranes and their main components: lipids and proteins. To understand many aspects of the structure and function of biological membranes, it is crucial to know how the diverse components of these complex, highly active and dynamic structures are organized within the membrane lipid bilayer and how they interact with each other. Therefore, this chapter on lipid-protein interactions deals first with the composition and the basic principles governing organization of lipids and proteins in biological membranes. Then, it turns to the fluidity of membrane lipid bilayers and the mobility of proteins and lipids within biological membranes. In a following section, it concentrates on membrane-protein and lipid-protein interactions. Finally, in the last part of the chapter, the formation of domains in biological membranes, the dynamic and reversible association of amphitropic proteins with membranes, and the post-fluid mosaic model of biological membranes are considered.

BIOLOGICAL MEMBRANES

Biological membranes play a vital role in the structure and function of all living cells, prokaryotic and eukaryotic. They are very highly selective filters and devices for active transport and are involved in many complex processes of living cells such as endocytosis, exocytosis, cell adhesion, cell movement, cell-cell recognition, cell-cell communication and signal transduction. In addition, they permit the specialization of cellular functions within a cell by defining various compartments with different enzymatic activities which are a prerequisite for many complex biochemical processes.

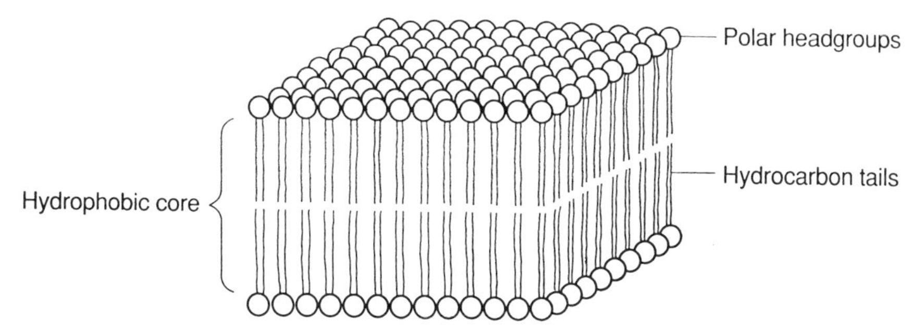

Figure 1. Three-dimensional view of a lipid bilayer. The polar head groups of the amphipathic lipid molecules face the aqueous environment whereas the hydrophobic hydrocarbon tails are sandwiched between the hydrophilic head groups and form the hydrophobic core of the lipid bilayer. Most membrane lipids spontaneously form bilayers in aqueous environments.

All biological membranes, including plasma membranes and internal membranes of eukaryotic cells are assemblies of lipid and protein molecules and in some cases of carbohydrates. Membranes are dynamic and fluid structures and their components are able to move freely within the plane of the membrane. Nevertheless, all membranes are structurally organized in a similar manner. As shown in Figure 1, lipid molecules are able to form sheet-like structures which provide the basic structure for the so-called lipid bilayer. This lipid bilayer is a relatively impermeable barrier and forms closed boundaries between compartments. The membrane proteins take over most of the membrane functions such as transport and transmembrane signalling or serve as a structural link to the extracellular matrix or to cytoskeletal elements.

Although all biological membranes are thought to be constructed on a common pattern, there exist noticeable diversities among different types of membranes, i.e., the composition and behavior of membranes from one cell type to another, and from one organelle to another, can vary remarkably. These variations give each kind of biological membrane its distinctive identity and specialized function. The diversity is primarily the result of the different functions of the proteins present in each membrane and the way in which they interact with lipids, with each other, or with cytoplasmic components. Although proteins clearly mediate the specific membrane functions, lipids are increasingly being recognized as active participants in membrane-associated processes.

COMPOSITION OF BIOLOGICAL MEMBRANES

All biological membranes, regardless of the source, contain lipids and proteins. In addition, in many membranes carbohydrates can be found up to a total membrane weight of 10%. The ratio of lipid to protein varies significantly from membrane to membrane, ranging from approximately 1:4 to 4:1 and is dependent on the specific function of the membrane. For example, myelin membranes which serve mainly to insulate nerve cell axons contain less than 25% of proteins whereas in membranes which are involved in energy transduction (e.g., internal membranes of mitochondria and chloroplasts) the protein content is approximately 75%.

Lipids

The most striking feature of lipids found in biological membranes is their enormous diversity. Any single membrane can contain well over 100 unique lipid species. The biological significance of this lipid heterogeneity is not known, but may be related to the recently recognized fact that lipids are active participants in many membrane-associated processes. Although the major role of lipids is to form the bilayer matrix, they are also involved in the proper organization of particular protein molecules within membranes. In addition, it has been revealed that they can act as second messengers, modulators, or activators of certain enzymes.

The three major types of lipids in biological membranes are phospholipids, glycolipids and cholesterol. All three are amphipathic molecules—that is, they have a hydrophilic ("water-loving" or polar) end and a hydrophobic ("water-hating" or non-polar) end. For example, phospholipids have polar head groups and two hydrophobic hydrocarbon tails. In addition, the polar head group of different phospholipids can have different net charges. The hydrocarbon tails usually contain an even number of carbon atoms, typically between 14 and 24. They are either saturated (no double bonds) or unsaturated (one or more double bonds). Double bonds are almost always in the *cis* configuration which leads to a bend of the otherwise straight hydrocarbon chains.

The lipid composition of biological membranes varies among different cells and among the various membranes of the same cell, as well as between the two leaflets (sides or faces of a lipid bilayer) of a membrane. Moreover, it has been recognized that the lipid composition of plasma membranes can change significantly during the activation or stimulation of a particular cell.

Proteins

All biological membranes contain proteins but in variable amounts ranging from 20% to about 80%. The types of protein associated with membranes vary from one cell type to another and from one organelle to another. The mitochondrial membrane protein composition differs significantly from that of the plasma membrane, and the plasma membrane protein components of a liver cell are strikingly different from those of an intestinal cell. The molecular weights of membrane proteins vary from approximately 10,000 to several hundred thousand. A few membrane proteins are present in high concentrations (10–50% of total membrane protein), but most are present at concentrations of only a fraction of a percent.

Carbohydrates

Carbohydrates are also important constituents of many biological membranes. They are especially abundant in the plasma membrane of eukaryotic cells but are, for example, absent from the mitochondrial membrane, the chloroplast lamellae, and several other intracellular membranes. Carbohydrates are associated either with proteins (glycoproteins) or with lipids (glycolipids) via covalent bonds. While most plasma membrane protein molecules exposed at the cell surface of eukaryotic cells are glycosylated, fewer than one in ten lipid molecules in the outer leaflet of most plasma membranes carries carbohydrates. A single glycoprotein can have many oligosaccharide side-chains whereas each glycolipid molecule has only one. In terms of weight, the proportion of carbohydrate in plasma membranes varies between 2–10%.

ORGANIZATION OF LIPIDS AND PROTEINS IN BIOLOGICAL MEMBRANES

The Lipid Bilayer

As already mentioned, lipid molecules have a hydrophilic or polar head group and a hydrophobic non-polar tail. It is exactly this amphipathic nature that causes lipid molecules to aggregate in water in such a way as to bury their hydrophobic tails and leave their hydrophilic heads exposed to water. They do this in one of two ways: (i) they form spherical micelles with the tails inward, or (ii) they arrange themselves into a bimolecular sheet, commonly described as lipid bilayer, with the hydrophobic tails sandwiched between the hydrophilic headgroups (Figure 1).

In a lipid bilayer the hydrocarbon tails of the lipids form a continuous hydrophobic interior (hydrophobic core) whereas the polar headgroups of the lipids face the surrounding watery surface (Figure 1). The full spectrum of forces (including van der Waals, electrostatic forces, and hydrogen-bonding interactions) that mediate noncovalent molecular interactions in biological systems stabilizes the lipid bilayer structure of biological membranes.

Membrane Proteins

Membrane proteins make contact with two environments: the lipid bilayer and its aqueous surrounding. Proteins which interact directly with the hydrophobic core of the lipid bilayer are called integral membrane proteins whereas proteins that are associated with the surface of the membrane only are called peripheral membrane proteins (Figure 2). The classification of membrane proteins into peripheral and integral is operational only and should not be mistaken for a molecular description of how these proteins are associated with the bilayer which, as a rule, is not known.

Integral or intrinsic membrane proteins interact directly with the hydrophobic core of the lipid bilayer. Like their lipid neighbors in the membrane, they are amphipathic molecules. Their amphipathic structure permits them to associate simultaneously with the hydrophobic hydrocarbon tails of the lipids in the interior of the membrane and the hydrophilic external environment. Those portions of the polypeptide which are exposed to the aqueous environment are usually enriched in amino acid residues with polar and ionizable side chains, whereas the residues in contact with the lipid hydrocarbon chains are primarily non-polar (Figure 2). The segments of a polypeptide chain which are embedded in the hydrophobic lipid tail environment are thought to be arranged mainly as α-helices or β-sheets. In these conformations the polar groups of the polypeptide backbone form hydrogen bonds with each other. These secondary structures, together with the mainly non-polar amino acid residues in these segments form a non-polar hydrophobic surface which can interact with the non-polar portions of the lipid bilayer. In addition, lipids covalently bound to the transmembrane segments of membrane proteins may facilitate the anchoring of the protein in the membrane.

Peripheral or extrinsic membrane proteins are thought to be attached to the membrane surface without having any direct contact with the hydrophobic interior of the lipid bilayer (Figure 2). They are usually bound to the membrane by electrostatic interactions either with the lipid polar headgroups or with other integral or peripheral membrane proteins.

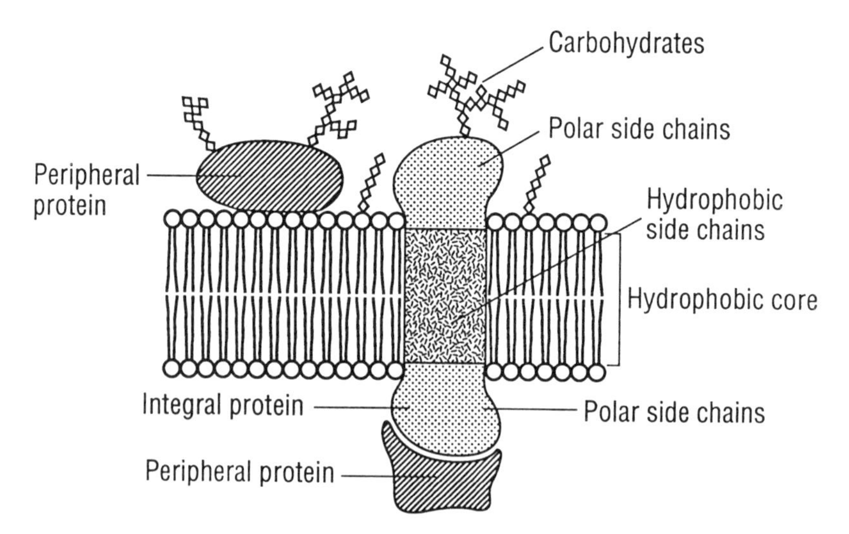

Figure 2. Schematic view of an integral membrane protein and two peripheral membrane proteins. The depicted integral protein, a transmembrane protein, contains polar side chains on both sides of the membrane and nonpolar side chains that are in contact with the hydrophobic interior of the membrane lipid bilayer. The peripheral proteins interact either with an exposed hydrophilic domain of the integral protein or with the polar head groups of the membrane lipids. Carbohydrates are bound covalently to lipids or proteins.

Asymmetry of Biological Membranes

All biological membranes are asymmetric. Many eukaryotic cell surfaces are highly polarized and exhibit distinct macroscopic domains. Inhomogeneities of this type are called lateral asymmetry of membranes. In addition, the inner (cytoplasmic) and outer (extra-cytoplasmic) surfaces (the two leaflets) of biological membranes differ in all their major components: proteins, lipids, and carbohydrates. This transverse asymmetry differentiates the two monolayer halves of the bilayer and is discussed below.

Each type of a membrane protein has a specific orientation within the phospholipid membrane, causing an asymmetry of the two membrane halves that gives them different characteristics and functional activities.

The transverse asymmetry of membranes can be easily recognized for glycoproteins, where in all cases the oligosaccharide residues are only found on the extracytoplasmic face of the membrane, and are totally absent from the inner half of the bilayer. This asymmetry comes about since all membrane proteins are inserted in a defined orientation from the cytoplasmic side of the membrane. Since sugars are highly hydrophilic, their residues tend to be located at the membrane surface rather than in the hydrocarbon core. Consequently, there is a high barrier for the rotation (flip-flop) of a glycoprotein from one side of the membrane to the other. Therefore, a spontaneous flip-flop event of a glycoprotein is rather unlikely to occur.

The lipid compositions of the two leaflets of biological membranes are strikingly different. For example, in the human red blood cell membrane all glycolipids and most of the phosphatidylcholine and sphingomyelin are found in the extracytoplasmic leaflet, whereas phosphatidylserine and phosphatidylethanolamine are preferentially located on the cytoplasmic face. Since the fatty acid tails of the former are usually more saturated than the latter, the asymmetry in the distribution of the head groups is furthermore accompanied by an asymmetry in the distribution of hydrocarbon tails as well. Consequently, there is a significant difference in charge between the two monolayer halves.

Some uncharged membrane lipids such as diacylglycerol, cholesterol, and fatty acids can spontaneously rotate between the two halves of a bilayer at an appreciable rate but most other membrane lipids do not. It is, therefore, not clear how the difference in the lipid composition in the two leaflets arises. It seems likely, however, that the asymmetry is generated during the biosynthesis of the bilayer in the endoplasmic reticulum, when enzymes are transferring lipid molecules from one monolayer to the other. Lipid-protein interactions between specific lipids and polypeptide domains of membrane-embedded proteins in one leaflet of the membrane may be important in maintaining this asymmetry. There is also some circumstantial evidence that cytoskeletal elements may be necessary to maintain the lipid asymmetry. Although the biological significance of the lipid asymmetry in membranes is not known at present, it may be involved in keeping membrane proteins properly oriented in the bilayer, a feature crucial for their function.

MEMBRANE FLUIDITY

Fluidity of Biological Membranes

Biological membranes are soft and flexible, two-dimensional structures with a fluid-like consistency and a viscosity like that of olive oil. The fluidity of a membrane depends on its composition, the temperature, and how the different components are packed and organized in the membrane. In general, membrane fluidity is decreased by lipids with long or saturated fatty acid chains, whereas lipids with short or unsaturated fatty acid chains tend to increase the fluidity of a membrane. Unsaturated fatty acid chains have kinks which interfere with the packing of adjacent fatty acids chains, thus resulting in an increase of membrane fluidity. Another determinant of membrane fluidity is cholesterol. Cholesterol tends to decrease the fluidity of biological membranes since it immobilizes lipids by fitting well into the kinked region of the fatty acid side chains.

In experiments with model lipid bilayers the fatty acyl chains of lipid molecules can exist in an ordered, rigid state or in a relatively disordered, fluid-like state, depending on the temperature. On the other hand, in biological membranes, ordered and disordered states coexist within the same membrane. Moreover, it seems that cells can adjust the lipid composition of their membranes to maintain an optimal fluidity in response to the surrounding environmental temperature. All animal and bacterial cells adapt to a decrease in temperature by increasing the proportion of unsaturated to saturated fatty acids in the membrane, which tends to maintain a fluid bilayer at reduced temperatures.

Mobility of Membrane Components in Biological Membranes

Membrane components, proteins and lipids, are constantly in motion. As mentioned above, the transition of a molecule from one membrane surface to the other is called transverse diffusion or flip-flop, whereas diffusion in the plane of a membrane is termed lateral diffusion. In contrast to lateral movement, the rotation of most proteins and lipids from one side of the membrane to the other is a very slow process. For example, it takes a phospholipid molecule about 10^9 times as long to flip-flop across a 50 Å membrane than it takes to diffuse a distance of 50 Å in the lateral direction.

From experimental evidence it seems that many membrane proteins do float quite freely in the lipid bilayer. However, the diffusion of membrane proteins is 100 to 10,000 times slower in membranes than in water. Apparently, the viscosity of the phospholipid core of biological membranes is 100 to 10,000 times higher than of water. A membrane protein can diffuse a distance of several microns within a

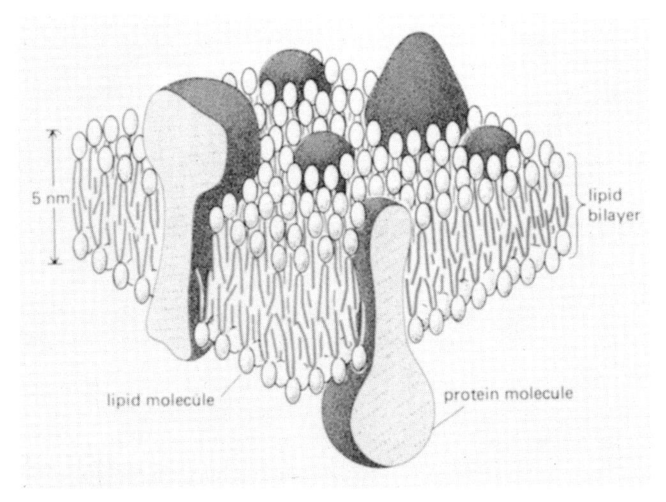

Figure 3. The fluid mosaic model of biological membranes. Membrane associated proteins are free to move laterally in the two-dimensional solution of a lipid bilayer. From Alberts et al. (1989). Molecular Biology of the Cell. Garland Publishing, New York.

membrane within approximately a minute. Phospholipid molecules diffuse 10 to 20 times more rapidly than proteins. This means that a lipid molecule can travel from one end of a bacterium to the other within a second.

The concept of proteins floating freely in a sea of lipid has been popularized by the fluid mosaic model of biological membranes (see Figure 3). One technique for demonstrating the mobility of proteins in the plane of the plasma membrane is the antibody-induced capping technique. It takes advantage of the fact that antibody molecules are bivalent, that is, they have two identical binding sites for the same antigen. Different antibodies directed toward a single protein species will often bind to different sites on the protein. When cells such as lymphocytes are exposed to antibodies specific for a particular cell surface antigen, the antibodies cross-link the surface protein molecules. Since the antigen molecules can diffuse laterally in the membrane, they accumulate in antibody-cross-linked networks called patches and caps on the cell surface (Figure 4).

Not all membrane components are freely mobile in the plane of the membrane. Cells can confine both membrane protein and lipid molecules to particular domains in a continuous lipid bilayer and they have several ways of immobilizing certain membrane molecules including: (i) self-assembly to large aggregates such as bacteriorhodopsin in the purple membrane of *Halobacterium* (large aggregates of this kind diffuse very slowly), (ii) interactions with the intracellular network of cytoskeletal elements, (iii) interactions with components of the extracellular matrix, and (iv) interactions with components on the surface of another cell.

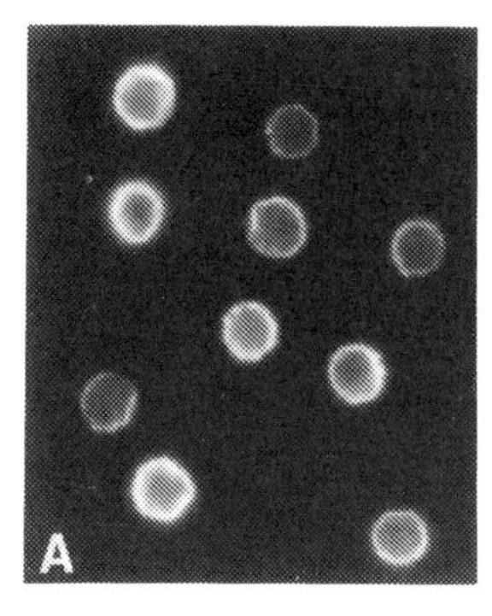

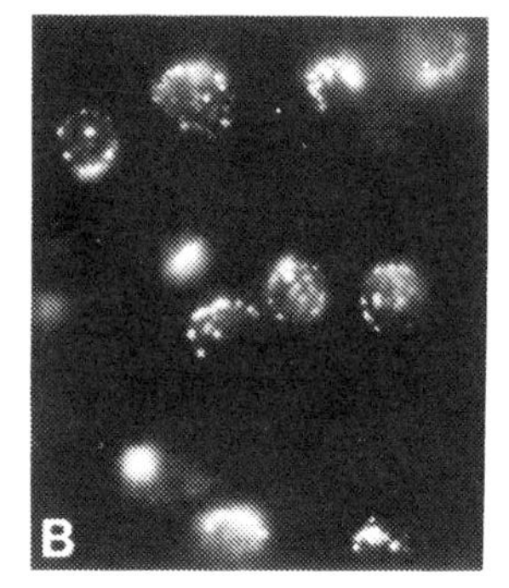

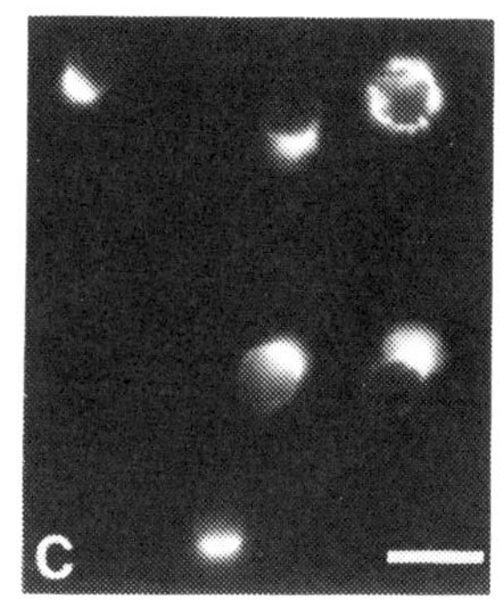

Figure 4. Distribution of a particular integral membrane protein (CD4) on the cell surface of lymphocytes before (A) and after antibody-induced patching (B) and capping (C). The mobility of CD4 in the plane of the membrane is visualized using antibody-mediated crosslinking together with indirect immunofluorescent microscopy techniques.

MEMBRANE-PROTEIN INTERACTIONS

Different Protein Attachment Modes to Membranes

In a previous section we described the operational classification of membrane proteins into peripheral and integral membrane proteins. At the molecular level, however, a great variety of protein attachments to membranes exists. Some of these are discussed briefly in this section and are schematically illustrated in Figure 5. The interactions between proteins and membranes are often very complex. Thus, a combination of several different modes of attachment may be responsible for the anchoring of an individual protein to a membrane:

1. *Transmembrane anchors* (Figures 5a and b): Some membrane proteins have polar regions on both sides of the membrane. Such proteins extend across the lipid bilayer and, thus, are called transmembrane proteins. They may cross the membrane only once (e.g., glycophorin) or many times (e.g., lactose permease or bacteriorhodopsin). Accordingly, they contain a single transmembrane segment or multiple transmembrane segments which interact directly with the hydrophobic part of the membrane. These transmembrane segments are usually identifiable from the primary acid sequence as a stretch of residues with non-polar character.
2. *Hydrophobic polypeptide anchors* (Figure 5c): Another class of membrane proteins is in direct contact with the hydrophobic part of the membrane without completely crossing the membrane. They are anchored in the membrane via a hydrophobic polypeptide segment and, therefore, their hydrophilic parts are exposed only on one side of the membrane.

3. *Lipid anchors* (Figures 5d, e, f, and g): Some other membrane proteins are covalently attached to lipids. Glycophospholipids (Figure 5e) as well as simple fatty acids (Figures 5d, f, and g) can serve as lipid anchors and may mediate or stabilize the anchoring of the protein to the membrane. In many cases however, it is not yet clear whether only the lipid tail is directly interacting with the interior of the membrane or whether also parts of the polypeptide backbone are embedded in the lipid bilayer.

4. *Hydrophobic domains* (Figures 5g and h): Some membrane proteins interact with the hydrophobic interior of the lipid bilayer without any substantial penetration of the membrane (e.g., *E. coli* pyruvate oxidase). Defined, secondary or tertiary folding patterns of the polypeptide chain may lead to the generation of hydrophobic domains exposed on the surface of the protein which may favor and mediate a direct interaction between these proteins and the hydrophobic part of the membrane.

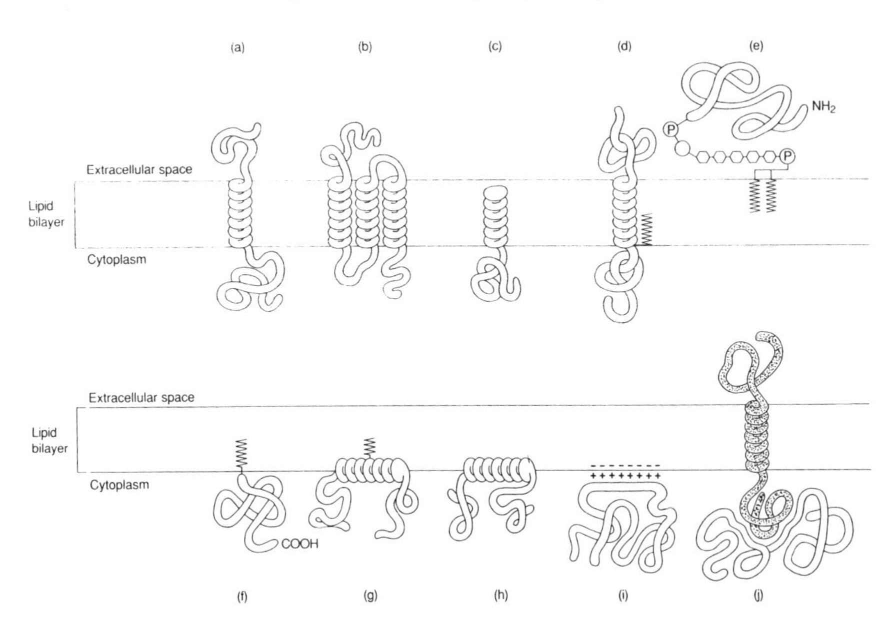

Figure 5. Different ways in which membrane proteins can be associated with membranes. Transmembrane proteins extend across the bilayer as a single α-helix (a) or as multiple α-helices (b). Other membrane proteins are attached to the bilayer by a hydrophobic polypeptide anchor (c), or solely by a covalently attached lipid such as a glycophospholipid (e) or a fatty acid (f). In some cases a combination of both hydrophobic segments of the polypeptide and covalently attached lipids are responsible for membrane attachment of the protein (d and g). Still other proteins contain hydrophobic domains on their surface which favor a direct protein-membrane association (g and h). Finally, many proteins are attached to the membrane only by noncovalent interactions with the polar headgroups of the membrane lipids (i) or, with other membrane proteins (j).

5. *Interaction with the surface of the bilayer* (Figure 5i): Some membrane proteins are just interacting with the surface of the bilayer without penetrating into the hydrophobic membrane interior. They usually interact electrostatically with the polar head groups of the membrane lipids.

6. *Interaction with other proteins* (Figure 5j): Some proteins interact only indirectly with the membrane by binding to other proteins which are either embedded in the bilayer or associated with the surface of the membrane. They bind to the exposed hydrophilic domains of these proteins without having any direct contact with the membrane lipids itself.

Structure Principles of Membrane Proteins

The special environment of the lipid bilayer imposes special constraints on the structure of membrane embedded segments of proteins. Hereof the constraints which keep hydrophilic groups and disordered polypeptide chains from the center of a water-soluble protein, equally exclude them from the interior of the lipid bilayer. Protein segments which are located within the hydrophobic environment of the lipids' hydrocarbon chains are thought to be arranged mainly as α-helices or β-sheets, formed from sequences of primarily nonpolar amino acids. The polar and hydrophilic amino acids, as well as the less regular structures, like turns and bends, with free hydrogen-bonding groups are mainly found in those parts of the protein exposed to the hydrophilic heads of lipids or to the aqueous environment. This, however, does not exclude charged or polar amino acid residues from being located within the hydrophobic bilayer (see below).

Examples of Membrane Proteins

At present high-resolution structure data are only available for one class of transmembrane proteins: the bacterial photosynthetic reaction center protein. Together with lower-resolution data of bacteriorhodopsin, these data provide the only models for most transmembrane proteins. Thus, the study of synthetic membrane-active peptides or of naturally occurring, small peptides such as melittin that bind to, or insert into, membranes is of particular interest in building and testing models of protein-lipid and protein-membrane interactions.

Bacterial Photosynthetic Reaction Center

Photosynthetic reaction centers are protein-pigment complexes localized in the photosynthetic membranes. They use captured light energy to synthesize ATP. In terms of the 3-dimensional structure, the photosynthetic reaction center of the purple nonsulfur bacteria *Rhodopseudomonas viridis* is the best characterized. It consists of a tetramer of four different polypeptides, called H (heavy), M (medium), L (light), and cytochrome. 13 prosthetic groups complement this complex to form

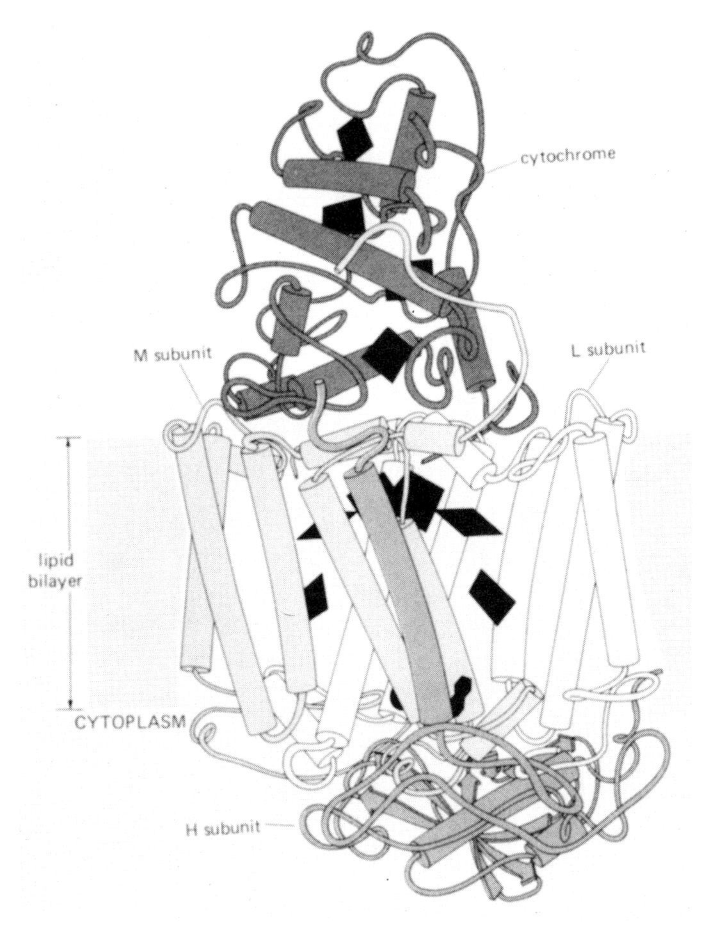

Figure 6. Schematic drawing of the polypeptide backbone of the four subunits (L, M, H, and cytochrome) of the photosynthetic reaction center of the bacterium *Rhodopseudomonas viridis.* The eleven membrane-spanning α-helices are indicated by cylinders. From Alberts et al. (1989). Molecular Biology of the Cell. Garland Publishing, New York.

the active photosynthetic reaction center in the plasma membrane of the bacterium. The reaction center has a total molecular weight of about 150,000.

The overall structure of the protein complex is somewhat like a sandwich as shown in Figure 6. The L and M subunits are folded quite similarly and are substantially integrated into the membrane along with a short segment of the H subunit. The L and M subunits each contain five membrane-spanning α-helical segments, and form together with the one helical segment of the H subunit the core of the reaction center. The cytochrome forms a cap on the extracytoplasmic surface of the bilayer, whereas the hydrophilic portion of the H subunit forms a similar cover on the cytoplasmic surface.

The structure of the complex as a whole is stabilized by: (i) interactions between the solvent-exposed portions of the subunits, (ii) favorable dipolar interactions between antiparallel α-helices which are closely packed together, (iii) van der Waals forces between closely packed helices, and (iv) shared ligation of the single iron atom by four histidine residues in the D and E helices of subunits L and M. The transmembranous complex is firmly anchored in the lipid bilayer by hydrophobic interaction between helices and the lipid core and by ionic interactions between the solvent-exposed, hydrophilic amino acid residues and the polar head groups of the lipids.

Several structural features of the protein complex are worth emphasizing. First, each of the eleven transmembrane segments are α-helical. Each helical segment is about 40 Å in length and sufficient to completely traverse the membrane. Second, the amino acids within these stretches are largely nonpolar. Each of the transmembrane helices in the L and M subunit has a stretch of at least 19 residues without an acidic or basic amino acid. Third, the portions of the L and M subunits connecting the transmembrane segments largely lie flat on each side of the membrane and form the contacts to the two hydrophilic subunits. Fourth, the eleven transmembrane helices are packed together as efficiently as are residues in the interior of water-soluble proteins. Those amino acid residues within the transmembrane region which are in contact with other proteins are generally hydrophobic, which is also the case for amino acids which are buried within soluble proteins. There are few polar interactions, such as hydrogen bonds or electrostatic salt bindings, between transmembrane helices.

Melittin

Melittin from bee venom is a water-soluble peptide. It is, however, a surface active peptide which forms monolayers at an air-water interface. The sequence of the 26-residue peptide includes many polar residues. Melittin can integrate into membranes and can lyse cells. The structure of melittin has been solved by X-ray crystallography, in two crystal forms. In both crystal forms the melittin polypeptide is a bent α-helical rod, with an inner surface consisting largely of hydrophobic side chains and an outer surface consisting of hydrophilic side chains. The asymmetry of hydrophobicity is often called amphiphilicity. The highly amphiphilic structure of the melittin helix suggests that it seeks a membrane surface because at such a phase boundary it can expose its hydrophilic face to the aqueous phase and still bury its largely hydrophobic face in the apolar phase. A model for melittin at the surface of a bilayer is shown in Figure 7. The melittin helix is oriented in such a way so that its hydrophilic side chains extend toward the aqueous phase, and its hydrophobic side chains are in contact with the hydrophobic interior of a lipid bilayer. Hypothetical interactions between two phosphatidylcholine molecules of

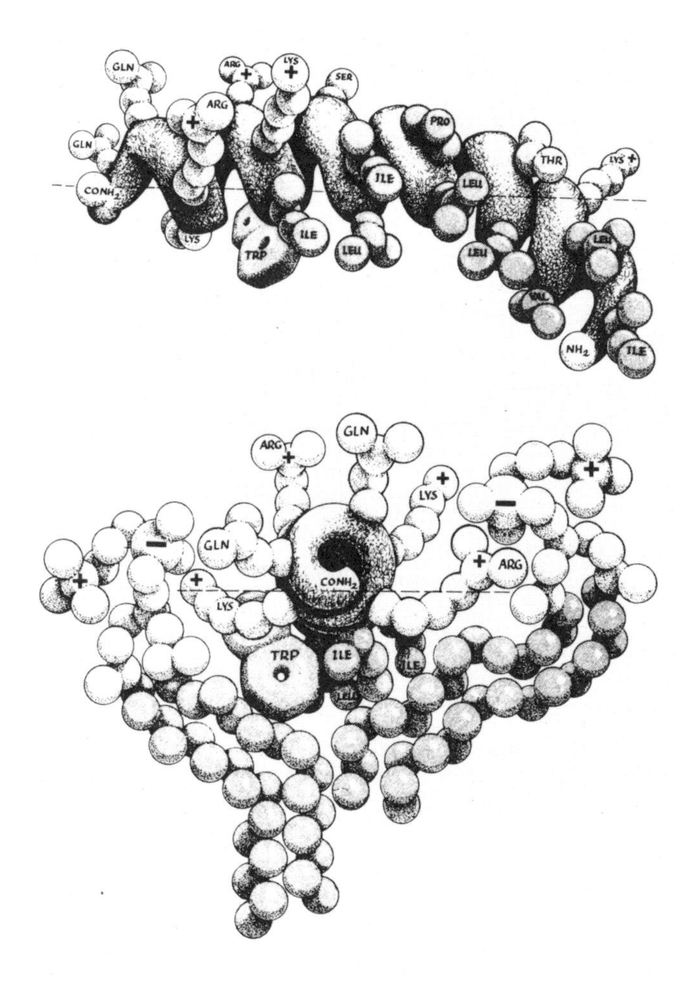

Figure 7. Schematic drawing of a melittin monomer binding to a membrane. The amphiphilic α-helix of melittin is shown at the membrane surface (dashed line) oriented so that hydrophilic side chains extend upwards towards the aqueous phase and hydrophobic side chains extend downwards into lipid. In the lower part of the figure, two phosphatidylcholine molecules have been added, showing polar contacts between their head groups with polar side chains on melittin and apolar contacts between the hydrocarbon tails and hydrophobic side chains of melittin. From Terwilliger, Weissmann, & Eisenberg (1982). The structure of melittin in the form I crystals and its implication for melittin's cytic and surface activities. Biophys. J. 37, 353–361.

the bilayer and the C-terminus of the melittin helix are shown as well. Similar amphiphilic structures on the surface of a folded protein may be the driving force for a protein to interact and insert into membranes.

Proteins with Covalently Attached Lipids

Proteins from a variety of sources (animals, bacteria, and viruses) contain covalently attached fatty acids or phospholipids. Most of these acylated proteins are not yet characterized, and the nature and linkage of the acylating group is not yet known. However, it is already apparent that acylated proteins are widespread in nature and that there is considerable diversity in the acylating group(s) and its linkage to these proteins. The different types of acylated proteins currently known suggest that a complex system of enzymes is modifying selected proteins with specific functional groups and thereby specifically and selectively alters their physical and biological properties.

Three main types of proteins containing covalently attached lipids exist in eukaryotic cells: (i) Proteins that are bound to palmitic acid, (ii) proteins covalently bound to myristic acid, and (iii) proteins which are linked to glycosyl-phosphatidylinositol. Each type is defined on the basis of the chemical structure of the acylating moiety and the nature of the covalent linkage to the modified protein. In the first two types, the fatty acid is directly attached to an amino acid residue, while in the third the fatty acid is part of a more complex structure.

Fatty acids can be attached to proteins directly via an ester bond or an amide linkage. Palmitic acid is most often attached to proteins by a thioester linkage to cysteine or by a hydroxyester bond to serine or threonine. These residues are usually located within the body of the polypeptide, and are near membrane-spanning segments, usually on the cytoplasmic side of the membrane. The addition of palmitic acid appears to occur posttranslationally. Myristic acid is most often attached to proteins through an amide linkage to an amino-terminal glycine. The N-terminal glycine appears to be an absolute requirement for amino-terminal myristilation. The myristilation appears to occur very early in the maturation of proteins and could be a true cotranslational event. Fatty acid acylated proteins appear to be primarily located at the cytoplasmic surface of the plasma membrane, but a significant fraction of myristilated proteins are soluble cytoplasmic proteins.

Another class of eukaryotic proteins contains a covalently attached glycophospholipid which is a derivative of phosphatidylinositol. It involves a complicated structure containing neutral and charged amino sugars, ethanolamine and phosphatidylinositol linked to the protein by an amide bond between the C-terminal amino acid residue and the ethanolamine moiety. The phospholipid tail is added posttranslationally after the proteolytic removal of 17–31 amino acids from the carboxyl-terminus of a precursor. Proteins attached to phosphatidylinositol appear to be located on the outer surface of the plasma membrane.

In several cases the sole mode of membrane attachment is via the lipid anchor. This is demonstrated by the action of phospholipase C, which can cleave the linkage between a protein and its phosphatidylinositol-anchor, resulting in the release of the protein from the membrane. In contrast, there are cases where the lipid moiety is bound to a transmembrane segment of a membrane-spanning protein. Here the protein is anchored in the membrane primarily by a hydrophobic stretch of its polypeptide chain. The lipid tail increases the hydrophobicity of this segment and strengthens the anchoring of the protein in the membrane by interacting with the hydrocarbon tails of the phospholipids in the interior of the membrane. Both lipid and polypeptide backbone are responsible for the proper attachment of this type of protein to the membrane. In most cases, however, it is not yet known to what extent lipid tails and polypeptide segments are involved in the interaction between the protein and the hydrophobic core of the membrane interior.

Although the functional significance of covalently attached fatty acids is not clear, several interesting proposals may be considered. These range from their role as hydrophobic anchors to possible functions, such as targeting moiety for proper localization, fusogen for uptake and secretion, or modulators for protein conformation that brings about a proper alignment of the protein in the bilayer for catalytic transport, and recognition functions. Other intriguing possibilities include involvement in the reversible association of proteins with membranes and a direct involvement of the released lipid tails in signal transduction events.

Proteins which Interact with the Surface of the Lipid Bilayer

Many membrane proteins are associated with the membrane primarily through interactions with integral membrane proteins. On the other hand, there also exists a diverse group of proteins which interact directly with the surface of the lipid bilayer. There, the membrane binding appears to be of two general kinds, which need not be mutually exclusive. First, binding may be mediated by an amphipathic secondary structural domain in the protein, usually an α-helix (Figure 8a). This secondary structure may be induced and stabilized by a lipid-protein interaction. Second, binding may be primarily electrostatic, involving a positively charged region of the protein and acidic membrane phospholipids (Figure 8b). This kind of interaction is also likely to involve a substantial hydrophobic component depending on how much the protein penetrates the surface of the bilayer.

A considerable number of soluble proteins can bind to the bilayer surface transiently or under specific conditions (see below). In addition, there are several known enzymes which depend on the membrane binding in order to maintain their enzymatic activity. Other examples of polypeptides which interact with the surface of the bilayer are provided by amphipathic peptide hormones or by blood clotting factors. These various types of protein-membrane association all involve the noncovalent interaction between particular membrane lipids and polypeptide chains. In many cases, specific lipids and specific lipid-protein interactions are

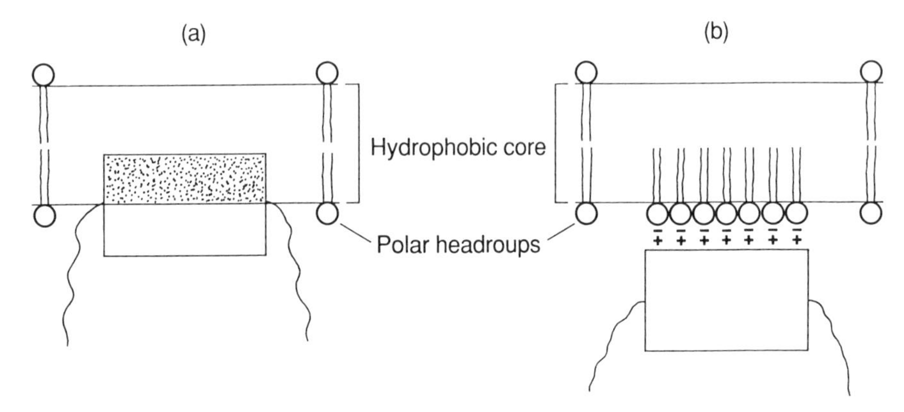

Figure 8. Two alternative ways in which proteins interact with the surface of a membrane lipid bilayer. (**a**) Hydrophobic interactions between secondary amphiphilic protein structures and the hydrophobic core of the lipid bilayer, or (**b**) electrostatic interactions with the charged head groups of the membrane lipids. Both ways involve noncovalent interactions between particular structures of the folded protein and membrane lipids.

important for the association as well as the optimal function of particular membrane proteins.

LIPID-PROTEIN INTERACTIONS

Biological membranes form a two-dimensional matrix in or on which proteins are present. These proteins have been found to alter the order and mobility of the lipid acyl chains, the translational mobility of the whole lipid, and the lateral organization of the bulk lipid. On the other hand, lipids are able to modulate the structure and function of proteins.

To understand the effect of proteins on lipids and vice versa it is important to realize that, in most cellular membranes, the membrane surface area occupied by proteins is at least as extensive as the surface area occupied by lipids. Using an average protein molecular weight of 50,000, this would mean a molar lipid/protein ratio of about 60:1 if only phospholipids were present. If one takes into account that proteins represent a cylinder extending to both sides of the bilayer, about 50% of the lipids at any time would be adjacent to a protein molecule (Figure 9). In other words, it can be estimated that in many cellular membranes only about three layers of lipid separate the proteins at the point of closest approach of nonaggregated membrane proteins. Most biological membranes are, therefore, not well described by the commonly used illustrations to represent membrane structure.

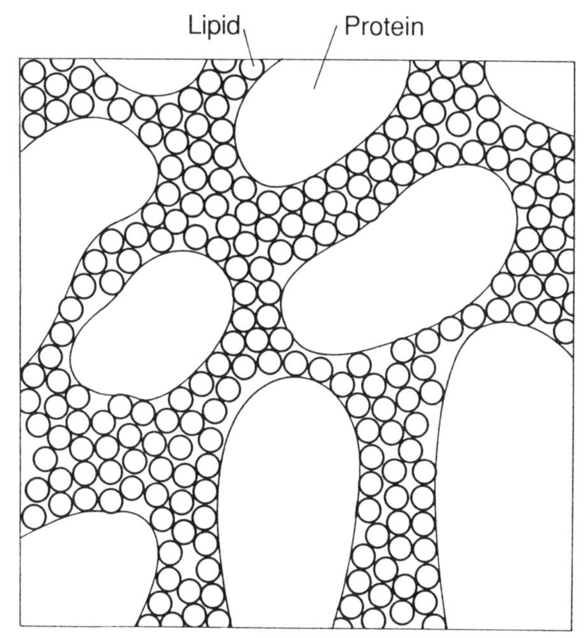

Figure 9. Schematic representation of the relationship between the amount of lipid and the amount of protein in a typical cellular membrane. About 50% of the lipids are adjacent to a protein molecule. The inner mitochondrial membrane would have an even higher protein content and, thus, less lipid than this figure represents. The myelin membrane would have less protein and more lipid.

Effect of Proteins on Lipids

As soon as the existence of integral membrane proteins was postulated, questions arose as to how such proteins were sealed into the membrane, what degree of perturbation occurs at the interface of the semirigid polypeptide chain with the fluid bilayer, and whether there are any specific protein-lipid associations. Lipid molecules in the environment near the interface with a protein molecule could experience unique organizational and conformational constraints. Indeed, many investigations indicate that integral proteins are surrounded by a layer of boundary (or annular) lipids (Figure 10). The protein in contact with the membrane, therefore, appears to immobilize a fraction of the lipids in the bilayer. It should be pointed out, however, that these protein-lipid interactions are thought to be temporary, reversible, and dynamic in nature. Annular lipids are expected to interact with protein through van der Waals interactions and also perhaps through electrostatic interactions at the polar head group region. The lipid annulus may act to seal the protein in the bilayer, thus preventing nonspecific leakage over the protein-lipid interface. Furthermore, such boundary lipid may act as a buffer to insulate integral proteins from the effects

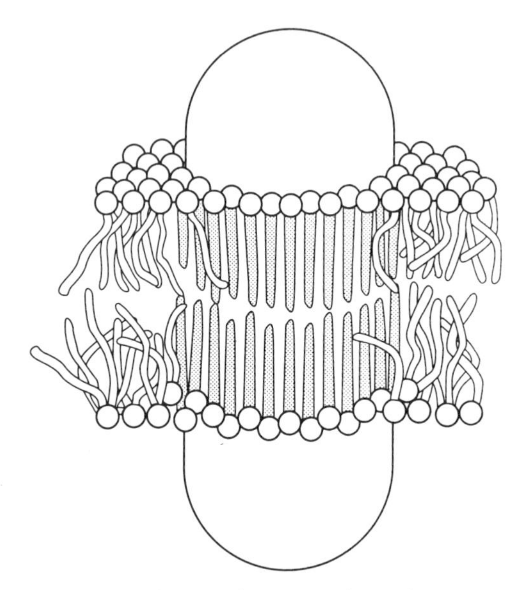

Figure 10. Three-dimensional view of an integral membrane protein surrounded by a layer of boundary or annular lipids. The lipids interact with the protein through van der Waals interactions and electrostatic interactions at the polar head group region. These interactions immobilize a fraction of the lipids in the bilayer. Nevertheless, the protein-lipid interactions appear to be reversible and rather dynamic.

of the bulk lipid properties of biological membranes, or from other membrane proteins.

In line with the fluid mosaic model of biological membranes, it was postulated that specific lipids are tightly coupled to integral membrane proteins. If the annulus were found to select for specific lipid species, integral membrane proteins might be expected to produce a lateral phase separation of lipids in biological membranes. Indeed some proteins have been identified to interact preferentially with a certain class of membrane lipids or rather with specific types of lipids. Thus, specific protein-lipid interactions contribute significantly to lateral compartmentation in biological membranes. Moreover, specific interactions between proteins and membrane lipids are responsible for lateral heterogeneity in biological membranes. They may be one of the driving forces for the formation of microdomains within membranes (see below). Compared to the bulk composition of a membrane these microdomains would be enriched in certain lipids and proteins.

Effect of Lipids on Proteins

The role of lipids in the function of membrane proteins is probably most dramatically demonstrated by the fact that most of the membrane functions are lost when lipids or part of them are removed or degraded by appropriate phospholipases. In addition, depletion of lipids from many membrane proteins during isolation procedures leads to a loss of their function. Reconstitution of such lipid bilayers often does not completely restore the function of the isolated protein. In many cases, however, the presence of specific lipids in the reconstituted system can overcome the loss of function.

Lipids seem to have also a dramatic effect on the catalytic properties of proteins. Some membrane proteins require the presence of specific lipids just as many enzymes in aqueous solution require a particular ion for activity. Enzymes localized in a membrane display all the modes of regulation described for soluble enzymes: induction, allosteric modulation by ligands, and post-translational modification. While some of these modes are further modified by solubility and accessibility of the substrates and products as regulated by the membrane, the amphiphilic environment of the bilayer offers additional possibilities: anisotropy of organization, regulation of conformational change, aggregation, and desegregation. Regulation or modulation of membrane proteins, for example, can occur by alterations in the microenvironment (lipid composition or lipid organization) of the proteins. Such changes can be induced by external perturbations under certain pathological conditions, and probably by genetic regulation.

The precise nature of lipid-protein interaction is not known in many systems. However, it is clear that the lipid molecules in membranes provide a specific microenvironment for many proteins. It appears that lipids are responsible for the proper structure and function of membrane proteins, and that they are involved in the organization, modulation and regulation of many proteins in biological membranes.

DOMAINS IN BIOLOGICAL MEMBRANES

The fluid mosaic model of biological membranes presumes a homogeneous distribution of protein and lipid components in the plane of the bilayer. There is, however, evidence showing that there are lateral inhomogeneities in biological membranes i.e., that there are domains or regions within some membranes that are physically separated from each other or from other portions of the membrane. These regions have distinct compositions and display different functions. Whether as large specialized regions, or microdomains of proteins or lipids, they are maintained by: (i) specific protein-protein interactions between membranes, (ii) specific structures within the membrane, (iii) interactions with cytoskeletal elements, or (iv) protein and lipid aggregation in the plane of the membrane.

Several categories of membrane domains have been recognized, including:

1. *Macroscopic domains.* Examples are the apical and basolateral domains of polarized epithelial cells or the stacked and stroma-exposed regions of photosynthetic thylakoid membranes. Macroscopic domains are large, morphologically

distinct regions of the cell membrane which are separated by barriers. They have different lipid and protein compositions, as well as different functions.

2. *Domains of aggregated proteins.* Aggregation of proteins within the plane of a membrane can result in relatively large patches, domains or caps that are enriched in a particular protein, accompanied by whatever other components (e.g., lipids or proteins) are favored by that environment. Examples are the purple membrane patches in *Halobacterium halobium* containing bacteriorhodopsin or the gap junctions containing connexin.

3. *Domains defined by the cytoskeleton.* Cytoskeletal elements (microtubules, microfilaments, and intermediate filaments) have been recognized to be involved in lateral compartmentation of many membrane proteins and lipids by organizing specific membrane components via interactions with cytoskeletal proteins. Patching and capping of cell surface antigens and the concentration of specific receptors in coated pits prior to endocytosis involves interactions with cytoskeletal elements that lead to the formation of lateral inhomogeneities and to domain formation in biological membranes.

4. *Lipid microdomains.* From studies with lipid model systems it is well known that, under appropriate conditions, lipids will undergo lateral phase separation, leading to stable coexisting lamellar domains. These lateral phase separations can be induced by temperature changes, pressure or ionic strength or by addition of divalent cations or proteins. Electron microscopic and biophysical techniques have provided evidence that similar lipid and lipid-protein domains also exist in biological membranes.

The concept of microdomains in biological membranes is attractive. Microdomains could provide unique lipid environments for proteins sequestered in the same membrane. Specific lipid-protein interactions within these microdomains may promote the functional organization of proteins and lipids within the bilayer and may optimize or modulate the activity of enzymes associated with biological membranes.

REVERSIBLE ASSOCIATION OF PROTEINS WITH MEMBRANES

Amphitropic Proteins

A growing number of cellular proteins including enzymes, oncogene products and cytoskeleton-associated proteins appear to interact reversibly with membranes. These proteins cannot be classified easily into cytoplasmic, peripheral or integral membrane proteins. They belong to a newly recognized class of membrane proteins termed amphitropic proteins. It seems that these proteins can exist in different states and in different compartments of the same cell. Amphitropic proteins are found in either a soluble form in the cytoplasm or in contact with the hydrophobic part of

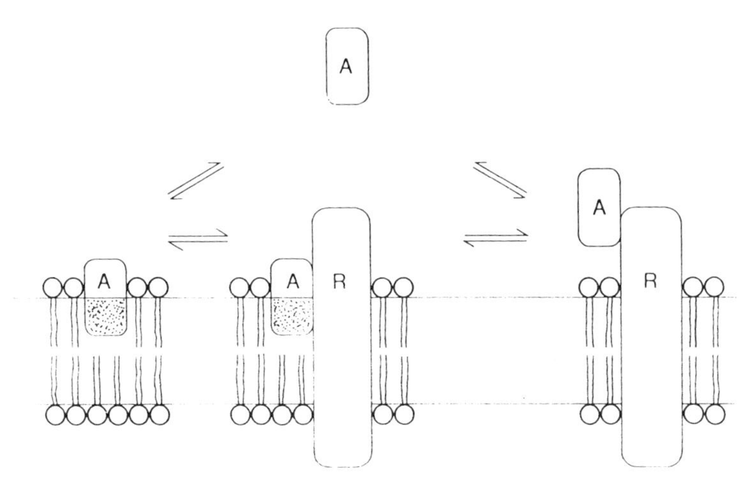

Figure 11. Two possible ways in which amphitropic proteins (A) interact with membranes. Dependent on particular lipids, cytosolic amphitropic proteins may insert directly into a lipid bilayer (*left side*). Secondary interactions with integral receptor molecules (R) may stabilize the membrane-inserted amphitropic protein in the lipid bilayer (*center*). Alternatively, cytosolic amphitropic proteins may first interact with an integral receptor protein (*right side*) and only thereafter establish stabilizing interactions with the hydrophobic membrane interior (*center*).

bilayers in a membrane-embedded form (Figure 11). Depending on the conditions within an individual cell, they exhibit properties of cytoplasmic, peripheral or integral membrane proteins, and, therefore, can be isolated in a soluble cytoplasmic form or associated with membranes.

How these amphitropic proteins interact with membranes is not always known in detail. In particular, the question of how deeply their polypeptide segments penetrate the hydrophobic core of membranes has yet to be answered in many cases. However, many of the amphitropic proteins interact with specific lipids in a noncovalent manner or contain a covalently bound lipid moiety, suggesting that specific protein-lipid interactions are particularly important in the reversible association of amphitropic proteins with membranes.

Reversible Association of Amphitropic Proteins with Membranes

A crucial question concerning the dynamic association of amphitropic proteins with membranes is what might trigger the translocation of these proteins from the cytoplasmic compartment into the membrane compartment? In the case of proteins containing covalently bound lipid, a mechanism involving post-translational modification of those proteins with lipids might be important (Figure 12a). The addition of a hydrophobic lipid tail to a cytoplasmic, water-soluble protein has been shown to result in its membrane insertion. Removal of this lipid moiety from the protein

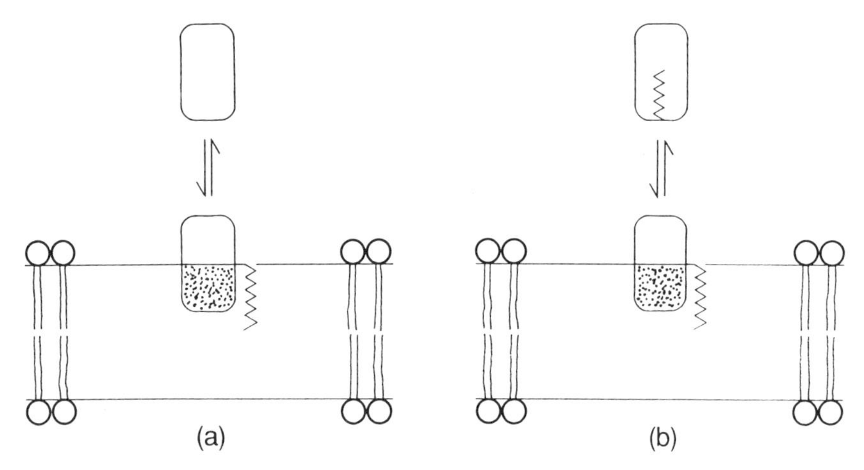

Figure 12. Two possible mechanisms for the reversible association of amphitropic proteins with membranes involving covalently bound lipids. (a) Acylation-deacylation cycles of amphitropic proteins, or (b) induced conformational changes of lipid containing amphitropic proteins, may mediate reversible membrane insertion.

could reverse this effect. Such a mechanism of acylation and deacylation could, therefore, be involved in the dynamic and reversible process of anchoring amphitropic proteins to membranes in living cells. Indeed, acylating and deacylating enzyme activities have been identified and appear to be involved in acylation-deacylation cycles. Moreover, a post-translational turnover of covalently bound fatty acids has been reported for several acylated proteins. In addition, enzymes capable of cleaving the linkage between protein and lipid moieties of phosphatidylinositol-anchored membrane proteins have been identified. Phospholipase C, for example, releases the protein moiety of certain phosphatidylinositol-anchored membrane proteins from the membrane and at the same time generates a free diacylglycerol molecule in the plane of the membrane (Figure 13b). A particularly intriguing possibility is that the diacylglycerol portion itself can act as a second messenger and may be involved in signal transduction across the membrane and/or in the generation of lipid domains within particular membrane areas.

Instead of a deacylation-reacylation cycle in the lifetime of a single protein, the presence or absence of covalently bound fatty acid can be regulated by other mechanisms. Differential splicing to produce messages encoding two proteins from the same gene, of which only one is modified by specific lipids, would produce a similar effect, although in a less instantaneous way. Alternatively, amphitropic proteins may contain a covalently bound lipid that is buried in the hydrophobic interior of the folded protein. A reversible conformational induced change, for example, by phosphorylation, could lead to the exposure of this hydrophobic region

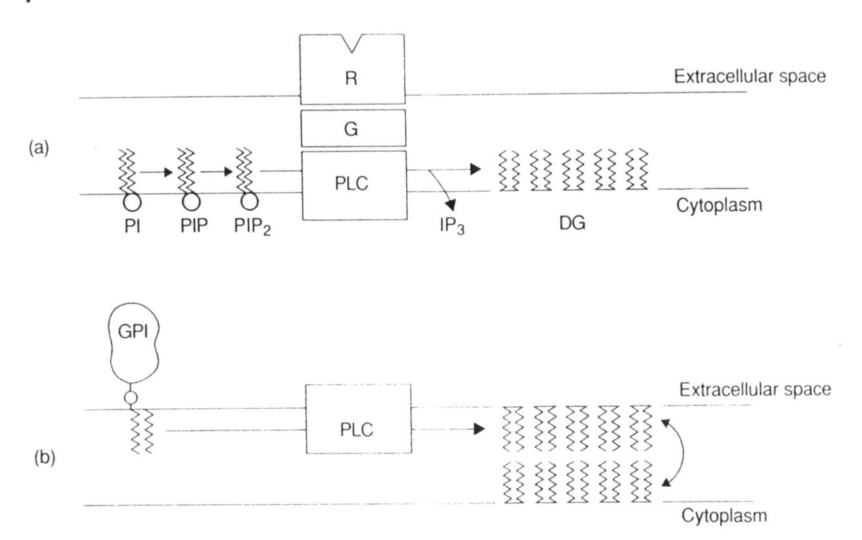

Figure 13. Two different possibilities of transient alterations in the lipid composition of biological membranes induced by phospholipase C (PLC) action. (a) Receptor-mediated activation of the phosphatidylinositol-cycle and PLC leads to the generation of the second messenger diacylglycerol (DG). (b) Release of glycosyl-phosphatidylinositol-anchored proteins (GPI) leads to the production of DG within the plane of the membrane by the action of PLC. *Note:* (i) DG generation may lead to microdomain formation within the membrane (a,b), and (ii) DG produced within one leaflet of the bilayer may rotate between the two halves of the bilayer as indicated in (b), and thus may be involved in transmembrane signaling. PI, phosphatidylinositol; PIP, phosphatidylinositol 4-phosphate; PIP_2, phosphatidylinositol 4,5-bisphosphate; IP_3, inositol 1,4,5-trisphosphate; R, receptor; G, G-protein.

on the protein surface and provide the driving force for membrane insertion (Figure 12b).

To understand the exact nature of the reversible interaction between an amphitropic protein containing a covalently attached lipid and membranes, many questions have to be addressed, such as: (i) Does an amphitropic protein containing a covalently bound lipid moiety bind to the membrane only via a hydrophobic lipid tail or do parts of the polypeptide chain insert into the hydrophobic core of the membrane as well? (ii) Are there specific proteins in membranes that have a selective affinity for proteins containing covalently bound lipids? (iii) What enzymes are involved in possible modification steps?

A modification of the amphitropic protein is not absolutely necessary to trigger a membrane association. The induction of changes in the lipid composition within the membrane could have a similar effect. Ligand-receptor interactions, for example, which in many different cell systems lead to an increased activity in the phosphatidylinositol-cycle, are accompanied by an increased formation of diacylglycerol and lead to transient changes in the lipid composition of membranes (Figure 13a). The release of diacylglycerol from phosphatidylinositol-anchored membrane proteins by phospholipase C action may lead to similar alterations in the membrane lipid composition (Figure 13b). These alterations could be generated only locally in restricted membrane areas. Microdomain formation within the membrane would in this case lead to a directed insertion of an amphitropic protein into that particular site. This non-random, directed insertion would have the advantage that no lateral movement of the protein within the plane of the membrane would be needed to bring the protein to its place of function.

THE POST-FLUID MOSAIC MODEL OF BIOLOGICAL MEMBRANES

Despite the great success of the fluid mosaic model in interpreting a wide variety of experimental findings, more recent evidence suggests that the simple concept of the membrane as a two-dimensional sea of lipid with "icebergs" of integral proteins (Figure 3) must be modified. The new model of biological membranes is called the post-fluid mosaic or plate-tectonics model of biological membranes (Figure 14) and takes into account the existence of discrete membrane domains composed of proteins or lipids or protein-lipid complexes. In this model, patches of different composition segregate in the plane of the membrane to form separate lipid and lipid-protein domains. The continuum of the biological membrane is broken up into

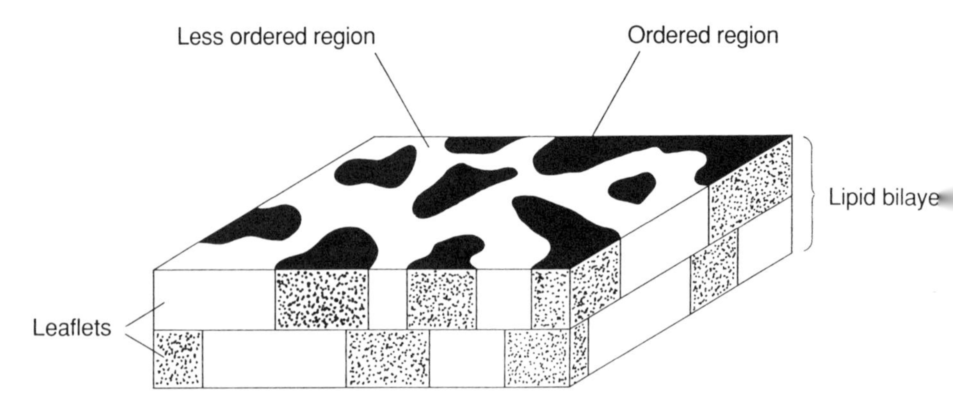

Figure 14. The plate-tectonics model of biological membranes. The nonrandom topological view of a membrane is shown schematically with regions of differing composition and organization. Ordered regions that are in motion with respect to one another are depicted to coexist along with intervening regions of relative disorder. Note that the composition and organization of lipids and proteins is not necessarily identical in both leaflets of the membrane.

a number of ordered regions. These regions are in motion with respect to one another. They are separated from each other by relatively disorganized regions containing membrane lipids in bilayer, hexagonal, or micellar form. The whole is a dynamic structure, and amphitropic proteins are among the dynamic components. They may be involved in the generation of new membrane domains that are constantly formed within the plane of the membrane in response to a variety of signals.

SUMMARY

Lipids and proteins are the main components of any biological membrane. Whereas the fluid-like lipid bilayer determines the basic structure of membranes, proteins are responsible for most membrane functions. Like the lipid molecules, most membrane proteins are mobile in the plane of the membrane, and both are distributed asymmetrically within the bilayer.

Membrane proteins interact with membranes by a variety of different means. Many membrane proteins extend across the lipid bilayer. Others do not span the bilayer but instead are attached to the one or the other side of the membrane. Some membrane proteins have nonpolar domains which interact with the hydrophobic core of the bilayer. Others are associated with the surface of the membrane by a combination of electrostatic and hydrophobic noncovalent interactions. Amphitropic proteins are considered to bind reversibly to membranes in response to a physiological signal.

Lipids provide a specific microenvironment for many membrane proteins. They are involved in the organization, activation, modulation and regulation of membrane proteins. At any instant in time, a substantial fraction of the lipids in a biological membrane are adjacent to proteins. The layer of lipids surrounding a protein is called the boundary layer or the lipid annulus. Some membrane proteins function only when embedded in specific lipid domains. The lipids surrounding a particular membrane protein may not necessarily have the same composition as the bulk membrane. Proteins as well as lipids can be immobilized and confined to particular domains in a continuous lipid bilayer. The formation of microdomains within membranes, which involves specific lipid-lipid, lipid-protein, and protein-protein interactions, is particularly crucial for understanding the structure and function of biological membranes. At present, biological membranes are best described as a dynamic complex of discrete, interacting microdomains composed of ordered lipid and protein structures, which are in motion with respect to one another within the less organized, fluid-like, bulk lipid bilayer.

RECOMMENDED READINGS

Aloia, R.C., Curtain, C.C., & Gordon, L.M. (1988). Advances in Membrane Fluidity, Vol. 2. Lipid Domains and the Relationship to Membrane Function. Alan R. Liss, Inc., New York.

Bittar, E.E. (Ed.) (1992). Fundamentals of Medical Cell Biology, Vol. 4, Membranology and Subcellular Organelles. JAI Press, Inc., Greenwich, CT.

Burn, P. (1988). Amphitropic proteins: a new class of membrane proteins. Trends Biochem Sci. 13, 79–83.

Datta, D.B. (1987). Comprehensive Introduction to Membrane Biochemistry. Floral, Madison, WI.

Edidin, M. (1992). Trends Cell Biol., Vol. 2, 376–380.

Finean, J.B., Coleman, R., & Michell, R.H. (1984). Membranes and Their Cellular Functions. 3rd ed. Blackwell Scientific Publications, Oxford.

Gennis, R.B. (1989). Biomembranes Molecular Structure and Function. Springer-Verlag, New York.

Jain, M.K. (1988). Introduction to Biological Membranes. 2nd edn. John Wiley & Sons, New York.

Quinn, P.J. (1976). The Molecular Biology of Cell Membranes. The Macmillan Press Ltd., London.

Singer, S.J. & Nicholson, G.L., (1972). The fluid mosaic model of the structure of cell membranes. Science 175, 720–731.

Yeagle, P. (1987). The Membranes of Cells. Academic Press, Orlando, FL.

Chapter 4

Lipid Modifications of Proteins and Their Relevance to Protein Targeting

PHILIPPE ZLATKINE and ANTHONY I. MAGEE

Principles of Medical Biology, Volume 7A
Membranes and Cell Signaling, pages 67–91.

ISBN: 1-55938-812-9

INTRODUCTION

The last fifteen years have seen the recognition of covalent attachment of lipids to proteins as a major form of post-translational modification (see Schlesinger, 1993). A variety of lipids (fatty acids, isoprenoids, glycosylphosphatidylinositol) can be attached in a number of different chemical linkages (amide, thioester, thioether). In some cases these modifications occur on *bona fide* transmembrane proteins and in these instances the function of the lipid modification is unclear and may be subtly to affect local protein-membrane or protein-protein interactions. However, in many cases the modified proteins are otherwise hydrophilic and cytosolic and the lipid modifications clearly have a role in membrane localization and intracellular targeting. This chapter deals with this function of lipid modifications. We have not attempted to give an exhaustive review of the field but rather to select a few examples where functional significance is better proven. A number of reviews are cited where the reader can gain a more detailed description of various aspects.

GLYCOSYLPHOSPHATIDYLINOSITOL (GPI) ANCHORS

GPI anchors are found exclusively on proteins localized to the cell surface, or destined to end up there eventually. The structure and biosynthesis of this class of lipid modification has been reviewed recently (Lisanti et al., 1990; Tartakoff and Singh, 1992; McConville and Ferguson, 1993). A preformed lipid precursor is attached to the protein rapidly after its translation and translocation into the lumen of the endoplasmic reticulum. Attachment occurs to the C-terminal hydrophobic sequence. The primary structural signal for GPI addition is tripartite, involving a cleavage/attachment domain requiring small amino acids in the first and third position, a spacer domain of 8–12 residues and a hydrophobic domain of 11 or more residues (Coyne et al., 1993; Kodukula et al., 1993). Several GPI anchor structures have now been determined but they all contain an identical core structure consisting of phosphatidylinositol linked α1-6 to glucosamine followed by three mannose residues linked via a phosphodiester bond to ethanolamine and thence in amide linkage to the C-terminus of the protein (McConville and Ferguson, 1993). There is considerable variation in this structure due to additional substituents. These can include palmitoylation of the inositol ring, additional ethanolamine phosphate residues on the mannoses, extra sugar residues including mannose, glucose, N-acetylgalactosamine, galactose, N-acetylglucosamine, and sialic acid. In addition, the glycerol moiety of the phosphatidylinositol can be substituted in the 1 and 2 positions with acyl or alkyl groups of varying chain lengths and unsaturation.

Because of the phospholipid nature of the anchoring group, GPI-linked proteins can only penetrate one leaflet of cellular membranes. The high lateral mobility of lipids relative to proteins in biological membranes has led to the suggestion that GPI-anchored proteins may also have high diffusion rates. Initial measurements favored this hypothesis but data gathered over several years have shown that many

GPI-anchored proteins have lateral mobilities comparable to those of transmembrane proteins and that the situation in real biological systems is complex. Protein-protein interactions mediated by the extracellular domain of GPI-linked proteins are a major factor in determining lateral mobility (Zhang et al., 1993). It is clear, however, that GPI-anchored proteins can access different lipid microdomains in cellular membranes from those occupied by transmembrane proteins (Edidin and Stroynowski, 1991). GPI-anchored proteins are often associated with a non-ionic detergent-insoluble complex which is also enriched in glycosphingolipids (Brown and Rose, 1992; Fiedler et al., 1993) and whose integrity depends on cholesterol (Rothberg et al., 1990a,b; Cerneus et al., 1993). These complexes have been interpreted as being "rafts" formed due to the colligative properties of glycolipids (Simons and Wandinger-Ness, 1990). Such clusters contain not only GPI-anchored proteins but also non-receptor tyrosine kinases (Štefanová et al., 1991) as well as other proteins, although it is not clear to what extent the interactions between molecules in the cluster are merely physical or are due to specific protein-protein or protein-lipid interactions.

GPI Anchors and Protein Targeting

GPI anchors are involved in protein targeting in several ways. Firstly, cleavage of the C-terminal hydrophobic signal is required for transport out of the endoplasmic reticulum (Moran and Caras, 1992). Newly synthesized GPI-linked proteins are initially non-ionic detergent soluble during transport (Cerneus et al., 1993) and are highly mobile in the membrane (Hannan et al., 1993). Protein localization studies in polarized cells have revealed that GPI-linked proteins show a polarized distribution. In epithelial cells they are usually found in the apical membrane (Wilson et al., 1990; Lisanti et al., 1990a,b; Fukuoka et al., 1992) although in Fischer rat thyroid epithelial cells they are preferentially basolateral (Zurzolo et al., 1993). GPI-linked proteins have also been reported to be selectively localized to axons of hippocampal neurons in culture (Dotti et al., 1991) although this may not be the case *in vivo* (Morris, 1992). The selective delivery of GPI-anchored proteins to particular membrane domains in polarized cells has been postulated to be due to the formation of the lipid rafts mentioned above, which occurs after transfer to the Golgi apparatus (Brown and Rose, 1992). How these lipid rafts are then targeted to specific membrane domains is as yet unclear.

As well as specifying selective exocytic trafficking, GPI anchors also contribute to the selectivity of endocytosis. GPI-anchored proteins are excluded from clathrin coated pits and enter the cell via non-coated plasma membrane invaginations called caveolae (Rothberg et al., 1990a,b; Keller et al., 1992). They are associated with a detergent-insoluble complex also containing the protein caveolin (Sargiacomo et al., 1993; Zurzolo et al., 1994). Endocytosis via caveolae of GPI-linked CD4 (a T cell surface marker which is a coreceptor for antigen stimulation of the T cell receptor) was much slower than that of the transmembrane form of the protein, and recycling occurred via a compartment distinguished from that involved in recycling of the transmembrane protein by its insensitivity to the weak base primaquine which disturbs intravesicular pH (Keller et al., 1992).

Finally, targeting of GPI-linked proteins can be achieved by selective enzymatic cleavage of the GPI anchor by phospholipases. Thus release of cell surface GPI-linked proteins into the extracellular milieu can be achieved in a regulated fashion. Although this attractive mechanism has been proposed for a number of GPI-linked proteins (see Lisanti et al., 1990a,b; McConville and Ferguson, 1993) it has only been observed in a few cases. For example, the GPI-linked Fc receptor III on neutrophils is released on stimulation of the cells with fMetLeuPhe, and may then play a role in inflammatory responses (Huizinga et al., 1988).

FATTY ACYLATION

Palmitic and myristic acid are the predominant fatty acids found attached to proteins in eukaryotic cells (Magee and Courtneidge, 1985; McIlhinney et al., 1987; Schmidt, 1989). Palmitoylation is a post-translational process, in which palmitic acid (16:0) is attached to proteins via a thioester linkage to cysteine residues (Magee and Courtneidge, 1985; McIlhinney et al., 1985; Olson et al., 1985). Palmitoyl-CoA seems to be the acyl donor and the palmitoylation site usually occurs near the transmembrane region of integral membrane proteins, on the cytoplasmic side, but can also occur on cysteines of otherwise soluble cytoplasmic proteins, e.g., ras (Buss and Sefton, 1986). For many palmitoylated proteins this linkage is biologically labile and the fatty acid can turn over faster than the protein (Omary and Trowbridge, 1981; Magee et al., 1987; Staufenbiel, 1987; Skene and Virag, 1989). Myristoylation in contrast is a cotranslational and irreversible process, although some exceptions exist (da Silva and Klein, 1990; McIlhinney and McGlone, 1990; Manenti et al., 1993). The myristate fatty acid (14:0) is attached via the more stable amide linkage to the amino terminal glycine (Towler et al., 1988; James and Olson, 1990) or less commonly a lysine residue (Stevenson et al., 1993). A N-myristoyl transferase catalyses the myristoylation of the substrate using myristoyl-CoA as a co-substrate. An increasing number of studies have shown in the last decade the importance of fatty acylation in intracellular protein targeting.

Involvement of Fatty Acylation in Protein Anchoring and Targeting to Membranes

Whereas palmitoylation does not appear to have a significant effect on targeting to membrane of transmembrane palmitoylated proteins such as the acetylcholine receptor (Olson et al., 1984), the β_2 adrenergic receptor (O'Dowd et al., 1989) or CD4 (Crise and Rose, 1992), many peripheral membrane proteins seem to require palmitoylation for their membrane anchorage and targeting.

GAP 43, also called Neuromodulin, is a neuro-specific calmodulin-binding protein which is implicated in neuronal growth and regeneration. GAP 43 is palmitoylated *in vivo* on two cysteines close to the N-terminus, and this modification seems to provide a hydrophobic moiety for anchoring the protein to the membrane (Skene and Virag, 1989). The addition of the N-terminal domain of GAP 43 to chloramphenicol acetyltransferase (CAT), normally a cytosolic protein when expressed in eukaryotic cells, directs CAT to sites in growth cones normally occupied by GAP 43 (Zuber, 1989). Palmitoylation at Cys 3 and 4 of the N-terminal domain of GAP 43 is both necessary and sufficient for its membrane association. As shown by site direct mutagenesis, there is a strong correlation between the extent of the protein palmitoylation and targeting to membranes (Liu et al., 1993).

Another example is the mechanism by which the keratinocyte transglutaminase (TGK) anchors to membranes. TGK is attached on the cytosolic side of the plasma membrane by palmitoyl fatty acids (Chakravarty and Rice, 1989). Membrane anchorage occurs near the amino terminus of the enzyme. The importance of a cluster of five cysteines, where palmitoylation was presumed to occur was demonstrated using TGK mutants. TGK in which the cluster was deleted or the cysteines were all converted to alanine or serine were cytosolic. Moreover, attachment of a 32 residue segment containing the cysteine cluster to involucrin, a highly soluble keratinocyte protein, resulted in membrane anchorage of the hybrid protein (Phillips et al., 1993).

Lck, a T lymphocyte specific member of the Src family of tyrosine kinase is both palmitoylated and myristoylated (Paige et al., 1993; Koegl et al., 1994). Removal of cysteines at positions 3 and 5 abolishes palmitoylation and results in a protein that is only weakly associated with the plasma membrane (Turner et al., 1990). It is therefore possible that palmitoylation plays an important role in directing the subcellular localization of Lck (Paige et al., 1993).

Myristoylation of Lck has been shown through mutational analysis to affect membrane localization of the protein (Abraham and Veillette, 1990). A non-myristoylated mutant was unable stably to interact with the plasma membrane and exhibited an impaired tyrosine kinase activity. Moreover, inhibition of myristoylation with 2-hydroxymyristic acid was shown to be an effective method for reducing the amount of Lck available at the plasma membrane (Nadler et al., 1993). Thus N-myristoylation of Lck seems to be essential for a proper targeting of the protein to the plasma membrane.

Another member of the src tyrosine kinase family, $p60^{src}$, requires myristoylation for its membrane association. The myristate acts in concert with a cluster of amino terminal lysine residues as a membrane targeting motif (Silverman and Resh, 1992). Chimeric proteins consisting of the first few amino acids of $p60^{src}$ fused to the cytosolic proteins α globulin (Pellman *et al.*, 1985), v-fps polypeptide and pyruvate kinase (Kaplan et al., 1990) directed myristoylation and membrane association. Furthermore, *in vitro* binding of $p60^{src}$ to membrane can be inhibited with myristoylated peptides corresponding to the N-terminus of the protein (Resh, 1989;

Silverman and Resh, 1992). The saturable and specific nature of $p60^{src}$ binding to membrane *in vitro* suggested that $p60^{src}$ is targeted to a membrane myristoyl src receptor (Resh, 1989; Resh and Ling, 1990). However the mitochondrial ADP/ATP carrier was identified as the 32 kD protein which bound the N-terminal myristoylated peptide, and was shown not to bind full length $p60^{src}$ (Sigal and Resh, 1993). To date, no specific myristoyl-Src receptor has been characterized. The N-terminal membrane binding motif of $p60^{src}$ may confer a more general membrane binding affinity than previously thought. This motif may be involved in the targeting of the $p60^{src}$ via a protein-protein interaction in the membrane or an interaction with a specific domain in the lipid bilayer. Possibly both type of interaction may be involved as it is known that some proteins are surrounded by or even select a specific lipid domain (Warren et al., 1975; Metcalf and Warren, 1977; Simmonds et al., 1982, Gordon and Mobley, 1985).

The Marcks protein, a specific protein kinase C substrate, is targeted to the membrane by its amino terminal myristoylated membrane binding domain (Aderem et al., 1988). Two modes of interaction of Marcks with membrane phospholipids have been proposed. *In vitro* Marcks interacts with neutral phosphatidylcholine containing membranes through its N-terminal myristoyl moiety, and the phosphorylation of the proteins does not change this interaction. On the other hand, the protein interacts with the negatively charged membranes through both the myristoyl motif and phosphorylation sites (Taniguchi and Manenti, 1993). The stimulation dependent translocation of Marcks between the cytoplasmic and the membrane fraction, which involves both myristoylation and a phosphorylation-dephosphorylation cycle, suggested also the presence of a specific receptor (Thelen et al., 1991). In this case also a receptor-specific amino-terminal myristate recognition and targeting is under debate (George and Blackshear, 1992).

An important recent example of membrane targeting by myristoylation is the endothelial cell nitric oxide synthase (EC-NOS). This enzyme is responsible for nitric oxide production and is a member of a family containing at least two other members (neural and macrophage). EC-NOS associates with the particulate fraction of cells whereas the other two isoforms are cytosolic. Recent cDNA cloning has revealed that EC-NOS contains a potential N-terminal myristoylation site but the cytosolic isoform does not (Sessa et al., 1992). The wild type protein incorporates ^{3}H-myristic acid and mutation of the myristoylation site abolishes membrane binding (Busconi and Michel, 1993; Sessa et al., 1993).

Modulation of myristoylation-dependent membrane binding is observed for the hippocampus-specific recoverin family member hippocalcin (Kobayashi et al., 1993). This myristoylated protein contains three Ca^{2+}-binding EF hand motifs and exhibits Ca^{2+}-dependent membrane binding. A non-myristoylated mutant was unable to bind to membranes even in the presence of Ca^{2+}. Thus conformational changes distant from the myristoylation site can influence the exposure of the fatty acid for membrane interaction.

From this series of examples emerge two major features of protein fatty acylation and intracellular targeting. Firstly, the energetic contribution of myristate and palmitate hydrophobicity (14 or 16 carbon) to membrane binding seems to be barely enough to partition a protein into the plasma membrane (Black and Mould, 1991; Peitzsh and McLaughlin, 1993). Thus, most of the time fatty acylation is required but needs to be coupled with a specific sequence of amino acids for an efficient membrane targeting (James and Olson, 1989; Kaplan et al., 1990; Liu et al, 1991). Secondly, the membrane targeting of these proteins is probably regulated either through an acylation-deacylation cycle of palmitoylated proteins such as GAP 43 (Sudo et al., 1992), ankyrin (Staufenbiel, 1987) and $p21^{ras}$ (Magee et al., 1987), or through an active reversible binding of myristoylated proteins to the membrane. Membrane binding of Marcks can be reversed by PKC-catalyzed phosphorylation (Thelen et al., 1991), binding of recoverin or hippocalcin to membranes requires calcium ions (Dizhoor et al., 1993; Kobayashi et al., 1993) and binding of ADP ribosylation factor depends on a GTP-GDP exchange cycle resulting in a conformational change and exposure of the buried myristate (Randazzaro et al., 1993). These regulatory processes may permit modulation of protein-membrane interaction and may alter the degree of membrane protein anchorage or targeting in response to physiological conditions.

Involvement of Protein Fatty Acylation in Membrane Protein-Protein Interactions

There is increasing evidence that fatty acylation of proteins may mediate protein-protein interactions or targeting in the membrane rather than protein-membrane interaction.

The α subunits of several members of the heterotrimeric G proteins ($\alpha_{i1},\alpha_{i2},\alpha_{i3},\alpha_o,\alpha_z,\alpha_t$) involved in cellular signaling are myristoylated (Buss et al., 1987; Jones et al., 1990; Mumby et al., 1990; Neuber et al., 1992). N-terminal myristoylation is necessary for the association of $G_o\alpha$ and $G_i\alpha l$ with the plasma membrane of transfected COS cells (Mumby et al., 1990). Reconstitution experiments with phospholipid vesicles show that the presence of $\beta\gamma$ is required for the association of the α_o subunit (Sternweis, 1986). Myristoylation of the α subunit of G_o increases its affinity for the $\beta\gamma$ subunits (Linder et al., 1991). Thus myristate is an important determinant of the association of $G_o\alpha$ with $\beta\gamma$ and the failure of non-myristoylated α subunit to target to membranes may be due to the inability to form a high affinity complex with $\beta\gamma$. Furthermore, signal transduction and regulation of adenylyl cyclase require the addition of myristic acid to the α_{i2} subunit (Gallego et al., 1992). These findings can probably be extended to the other myristoylated G protein α subunits.

More recently, palmitoylation of several G protein α subunits ($\alpha_{i1},\alpha_{i2},\alpha_{i3},\alpha_o,\alpha_s,\alpha_q,\alpha_z$) has been demonstrated (Parenti et al., 1993; Linder et al., 1993; Degtyarev et al., 1993a; Wedegaertner et al., 1993), including many that are myristoylated (α_i,α_o,α_z). The palmitate was shown to be attached onto a cysteine residue near the amino termini of the protein. Palmitoylation is required for membrane attachment and signaling function of $G_s\alpha$ and $G_q\alpha$ (Wedegaertner et al., 1993). Reversible palmitoylation of $G\alpha$ subunits may provide an additional mechanism for regulation of signal transduction. Indeed the palmitoylation of α_s is a dynamic process; an increased palmitoylation of α_s was established after activation via the β adrenergic receptor or cholera toxin (Degtyarev et al., 1993b). Increasing incorporation of palmitate suggests that some changes in the α subunit conformation may regulate protein-protein interaction in the signal transduction cascade.

Palmitoylation of the human β2-adrenergic receptor may play a crucial role in the normal coupling of the receptor to adenylyl cyclase signal transduction (O'Dowd et al., 1989). Mutation of the palmitoylated cysteine to a non-acylated glycine prevented the coupling of G_s to the receptor which correlated with a drastically reduced ability to mediate stimulation of adenylyl cyclase. As was proposed for the serotonin 1β receptor (Ng et al., 1993), the anchorage to the membrane of a palmitate residue attached to a cysteine creates an intracellular loop of the receptor terminal cytoplasmic tail. Proper recognition, and protein-protein targeting in intracellular signaling could involve secondary or tertiary information.

A study of fatty acylation of protein tyrosine kinases (PTKs) Lck and Fyn has shown that palmitoylation of the amino terminal cysteine residues together with myristoylation of the amino terminal glycine residue define an important motif for the association of PTKs with GPI-anchored proteins (Shenoy-Scaria et al., 1993). Moreover, native $p60^{src}$, a myristoylated PTK, failed to associate with GPI anchored DAF (decay accelerating factor). Replacement of the 10 amino terminal residues of $p60^{src}$ by the corresponding motif of Lck or Fyn, or a mutation of serine to cysteine at position 3 and 6 in $p60^{src}$ conferred on this protein the ability to be palmitoylated and to associate with the GPI-anchored DAF. This result underlines the importance of a variety of acylated motifs for protein to membrane or protein to protein targeting. GPI-anchored membrane proteins associate preferentially with glycolipid rich membrane subdomains (Brown and Rose, 1992). It is possible that palmitoylation together with myristoylation is needed to localize the PTKs to the same lipid subdomain. However, the mediation of transmembrane proteins is also possible. Moreover, the preferential association of GPI-anchored proteins with glycolipid might explain the cosegregation of these components to the apical surface of epithelial cells (Lisanti et al., 1988; Simons and Van Meer, 1988). Similarly, protein fatty acylation by interacting with lipid subdomains or by promoting protein-protein interactions might play an important role in protein intracellular sorting.

Role of Protein Fatty Acylation in Vesicular Targeting to Membranes

Fatty acylation of protein seems to play a crucial role in intracellular protein transport. Fatty acyl CoA was shown to be required for vesicular transport between cisternae of the Golgi stack in a cell free system (Glick and Rothman, 1987). The transfer of palmitate from palmitoyl-CoA is required for the budding of non-clathrin coated transport vesicles from Golgi cisternae (Pfanner et al., 1989). Fusion of transport vesicles also requires fatty acyl CoA and is inhibited by a non-hydrolyzable analog of the co-enzyme (Pfanner et al., 1990). An important question is whether fatty acylation occurs on the sorting machinery or on passenger protein targeted to specific destinations. Recent studies concerning myristoylation of proteins in the yeast secretory pathway suggest that each round of vesicular budding, targeting and fusion is accompanied by a round of fatty acid acylation and deacylation of a component of the vesicular machinery rather than secretory proteins (Simon and Aderem, 1992). It appears from these results that several transport components are covalently modified by a fatty acid such as palmitate or myristate and thereby, are activated to perform their role in the formation and/or the fusion of transport vesicles. The hydrophobic lipid moiety of the component might directly take part in the budding or fusion process by perturbing the lipid bilayer. On the other hand, fatty acylation might serve as a membrane anchor to promote stable assembly of acylated components. Finally, the presence of additional fatty acid might influence protein-protein interaction in the trafficking machinery. The reversible palmitoylation of one or more proteins involved in the trafficking machinery might be a mechanism that allows rapid recycling of transport components. The turnover rate of the palmitate moiety can be considerably higher than the turnover rate of the protein itself (Magee et al., 1987, 1988). The release of the fatty acid would allow recycling of the acylated transport component, setting it free for new rounds of transport.

One approach used to identify possible targets for protein palmitoylation in the trafficking process was to exploit the inhibition of intracellular transport during mitosis. By comparing the palmitoylation patterns between interphase and mitotic cells a major palmitoylated protein p62 was identified. Acylation and deacylation of p62 could be important in vesicular transport, and was found to be regulated during mitosis (Mundy and Warren, 1992). The function of p62 is unknown but it might be a component of the transport mechanism which is thought to be shared by all vesicle mediated steps from the ER to the trans Golgi network. Interestingly, another reversibly palmitoylated protein with a molecular mass of 63 kD has been identified as a resident of a membrane network between rough ER and Golgi apparatus (Schweizer et al., 1993). These two proteins share the property of a dramatically increased acylation during mitosis and when intracellular transport is inhibited by brefeldin A. This type of protein might be a component of the coat complex that is thought to be assembled from dispersed cytosolic subunits, the coatomers. The presence of the coat on the surface of a Golgi cisterna may lead to membrane deformation into a vesicle. Once the vesicle is formed the coat is removed and the vesicle can enter the fusion process (Rothman and Orci, 1990). An acylation and deacylation cycle may attach and release the coatomers by reversible addition of a palmitoyl group. The increased palmitoylation during mitosis or treatment with brefeldin A might be due to a block prior to the deacylation step and an accumulation of fatty acylated vesicular protein.

A myristoylated GTP binding protein, the ADP ribosylation factor (ARF), seems to have a central role in Golgi coat formation. Mammalian ARFs have a conserved glycine at amino acid residue 2 (Tsuchiya et al., 1991) which is the site of attachment of myristate. Myristoylated ARF, but not a non-myristoylated mutant, bound to Golgi membranes and allowed subsequent coatomer binding (Palmer et al., 1993). Myristoylation was found to be necessary for GTP-binding of ARF to Golgi membranes (Haun et al., 1993), to endosome membrane (Lenhard et al., 1992) and to nuclear vesicles (Borman et al., 1992). Whether myristoylation is required (Franco et al., 1993) or not (Helms et al., 1993) for the interaction of ARF with the nucleotide exchange protein remains under debate, but it seems nevertheless clear that functionally important and specific protein-protein interactions can occur only after myristoylated ARF binds to the membrane (Randazzo et al., 1993; Helms et al., 1993; Franco et al., 1993). Once activated by a guanine nucleotide exchange protein and associated with the plasma membrane, myristoylated ARF might diffuse freely to associate stably with a target protein, namely the coatomer subunits. Binding of GTP might allow ARF incorporation in the coat and hydrolysis of the bound GTP on encounter with the target membrane might disassemble the coat and recycle ARF and the coatomers to the cytosol (Rothman and Orci, 1992). As several different ARF proteins exist and have been identified in different subcellular locations (Lenhard et al., 1992; Balch et al., 1992; Taylor et al., 1992) a specificity may exist in their targeting. Each subtype of coated vesicles may have a different ARF protein carrying vesicle targeting information to specify its site of attachment. One region of divergence between ARF proteins is the N-terminal domain adjacent to the myristoylation site (Tsuchiya et al., 1991). A cooperative interaction between the fatty acyl chain and the N-terminal motif might create distinct structures recognized by or for protein-protein interaction at the target membrane (Magee and Newman, 1992).

Such a cooperation between the acyl chain and a protein sequence is common in the family of ras-like proteins. A number of small 21–24 kD GTP binding ras-like proteins such as rabs, sec, and YPT, which are involved in the secretory pathway, cycle between cytosolic and membrane-bound forms. Some have been reported to be acylated (Molenaar et al., 1988; Newman et al., 1992) and fatty acylation might transiently associate these proteins with a budding vesicle. Upon targeting these proteins might be deacylated and return to the cytosolic pool. Moreover, GTP binding and hydrolysis might be involved in anchoring these molecules to and releasing them from the vesicles (Bourne, 1988; 1990) by coupling with the acylation-deacylation cycle.

In polarized epithelial cells, different proteins accumulate in the apical and the basolateral plasma membranes. One proposed mechanism is that proteins contain a targeting motif similar to the signal that directs traffic of membrane and secreted proteins (Wickner and Lodish, 1985; Matlin, 1986; Verner and Schatz, 1988).

Different proteins bearing a specific targeting motif might interact with the sorting machinery (via protein-protein interaction or protein-lipid interaction) of polarized epithelia to direct proteins to appropriate membrane domains (apical or basolateral). The Xlcaax-1 protein found in the plasma membrane of *Xenopus leavis* oocytes and XTC cells (Kloc et al., 1991) belongs to the group of membrane associated proteins with C-terminal prenylation signals as described later in this chapter. As shown by mutation experiments, the association of this protein to the plasma membrane of oocytes is dependent on a CXXX box and two cysteine palmitoylation sites providing a secondary membrane targeting signal (Kloc and Etkin, 1993). Interestingly, Xlcaax-1 is present in the basolateral membrane of kidney tubules and other polarized epithelia (Cornish et al., 1992). It is tempting to speculate that in polarized cells the targeting motif is attached and/or decoded by the sorting machinery leading to the basolateral segregation of the protein. A signal for one class of basolateral membrane proteins in Madin-Darby canine kidney (MDCK) cells and hepatocytes is characterized by the amino acid residue located close to the transmembrane domain (Yokode et al., 1992; Mostov et al., 1992). The structural characteristic of this basolateral sorting signal suggests that it may be recognized by cytosolic proteins related to the family of adaptor proteins that regulate signal mediated endocytosis of receptor at the cell surface (for review see Nelson, 1992). Similarly, the GPI-anchored protein DAF, sorted to the apical membrane in MDCK cells, possesses a 37 amino acid motif that may convey apical targeting information (Lisanti et al., 1989). The elucidation of the relevance of these complex and varied lipid modifications of proteins in their membrane targeting, together with the understanding of protein-lipid interactions may be the key in the understanding of protein and lipid sorting in epithelial cells.

PROTEIN PRENYLATION

Prenylation is a newly discovered lipid modification of proteins which involves the addition of isoprenoids, precursors in the steroid biosynthetic pathway, to cysteine residues in thioether linkage (reviewed in Cox and Channing, 1992; Sinensky and Lutz, 1992; Clarke, 1992; Giannakouros and Magee, 1993). The modified cysteine residues occur in C-terminal motifs of three kinds: CXXX, CXC, and CC, where C is cysteine and X is any amino acid. The latter two motifs are modified exclusively by the twenty carbon isoprenoid geranylgeranyl, which can be added to both of the cysteines in the motif. $CX_1X_2X_3$ sequences can be modified with either geranylgeranyl (if the X_3 residue is large and hydrophobic, e.g., leucine) or the fifteen carbon isoprenoid farnesol (if the X_3 residue is small and relatively hydrophobic, e.g., alanine, serine, methionine). Not all CXXX-like sequences become prenylated: residues in the X_1 and X_2 position can modify substrate activity resulting in a non-functional motif e.g., in G protein α-subunits (Jones and Spiegel, 1990). However, the context of the CXXX motif can play a permissive role (Cox et al., 1993).

Recognition and selective prenylation of these sequences is achieved by a family of related soluble heterodimeric enzymes, the prenyltransferases. Farnesyl transferase (FT) is responsible for farnesylation of CXXX motifs using farnesyl diphosphate as the cosubstrate. Geranylgeranyl transferase I (GGTI) modifies the alternative CXXX motifs utilizing geranylgeranyl diphosphate. Geranylgeranyl transferase II (GGTII) is active on the CXC and CC motifs, and contains a third subunit called Rab Escort Protein (REP), which is required for multiple rounds of catalysis, probably by removing doubly prenylated product from the enzyme active site. REP is related to Rab3A GDI (Andres et al., 1993; see below) and the product of the human choroideraemia gene (an X-linked retinal degenerative disease; Seabra et al., 1992). In general, FT and GGTI require only a CXXX peptide of appropriate sequence and are active on small synthetic peptides. GGTII, on the other hand, requires considerable upstream sequence information and is inactive on peptides.

Prenylation is often followed by further modification (Gutierrez et al., 1989). Prenylated CXXX motifs are subjected to proteolytic removal of the X residues probably by a soluble protease activity (see Clarke, 1992; Giannakouros and Magee, 1993). Subsequently, the now C-terminal prenylated cysteine residue becomes carboxyl-methylated by a membrane-bound methyltransferase using S-adenosyl methionine as the methyl donor. Prenylated CXC and CC motifs are not proteolyzed but CXC sequences do become carboxyl-methylated whereas CC sequences do not (Newman et al., 1992; Wei et al., 1992).

Following prenylation and further modification of C-terminal sequences, additional cysteine residues within a few amino acids upstream of the prenylation site may be present which can be further processed. If the cysteine occurs immediately adjacent to the CXXX sequence, this can involve a second prenylation. For example, the RhoB protein has a C-terminal sequence CCKVL which becomes doubly modified with geranylgeranyl and farnesyl (Adamson et al., 1992). This may be due to the most C-terminal cysteine in the CXXX motif acting as a farnesylation site (despite the presence of C-terminal leucine), followed by proteolytic processing and carboxyl-methylation. The resulting adduct CC(SFar)OMe resembles a GGTII recognition sequence and can be geranylgeranylated on the upstream cysteine by that enzyme. More commonly, however, upstream cysteines are palmitoylated (Hancock et al., 1989; 1990). Palmitoylation is dependent on prior prenylation, probably due to the need to attach the hydrophobic prenoid modification in order that the protein can acquire some membrane binding affinity and be presented to a membrane-associated palmitoyltransferase (Gutierrez and Magee, 1991).

Prenylation is a widespread modification of eukaryotic proteins, occurring in yeast (Newman et al., 1992; Giannakouros et al., 1992; 1993), plants (Zhu et al., 1993; Sweizewska et al., 1993) and mammals (reviewed in Giannakouros and Magee, 1992; Clarke, 1992). In all known cases prenylated proteins are otherwise hydrophilic in nature, with no hydrophobic spanning peptide regions. The addition of one or more prenoids and the associated modifications would be expected to confer hydrophobic membrane-binding properties (Black, 1992) and this has been borne out in the cases where it has been studied (Epand et al., 1993). However, as well as general membrane affinity, the C-terminal modifications somehow contribute to the specificity of intracellular targeting such that prenylated proteins usually reside in particular membrane compartments.

Prenylation and Targeting of ras-related Proteins

The membrane targeting role of prenylation has been thoroughly studied in the ras family of proteins (reviewed in Newman and Magee, 1993). C-terminal prenylation of ras proteins is necessary but not sufficient for plasma membrane localization of the protein (Hancock et al. 1990; 1991a). In some cases (e.g., N-, H-ras) upstream palmitoylation sites are also required, presumably providing a cooperative hydrophobic signal. In the case of the K(B)-ras protein, palmitoylation sites are absent but are replaced by a highly basic stretch of lysine residues which contribute in an additive way to localization. These have been presumed to interact electrostatically with negatively charged phospholipid head groups (e.g., phosphatidylserine) in the cytoplasmic leaflet of the plasma membrane (Hancock et al., 1991b). The principle of cooperation between prenylated C-termini and upstream palmitoylated or basic regions holds true in many cases. This type of membrane binding has the potential to be regulated, for example, by phosphorylation of polybasic regions, thus reducing their net charge. Indeed a potential serine phosphorylation site occurs within the polybasic region of K(B)-ras. Alternatively, reversible palmitoylation of upstream cysteines may play a role. Rapid turnover of acyl groups has been observed on many proteins including ras (Magee et al., 1987). The enzymology of this process is poorly understood, although a thioesterase which deacylates ras proteins has been purified (Camp and Hoffman, 1993). No palmitoyltransferase has yet been purified to homogeneity although some progress has been made in this area (Schmidt, 1989; Gutierrez and Magee, 1991).

The post-translational modifications of ras proteins are crucial for function, as well as localization. For example, the ability of ras proteins to cause cell transformation (Hancock et al., 1990; 1991a) and interact with downstream signaling systems (Horiuchi et al., 1992; Kuroda et al., 1992; Itoh et al., 1993) is dependent on prenylation, but palmitoylation is not absolutely essential for function although it does considerably potentiate activity (Hancock et al., 1991a). The mechanism by which isoprenoids contribute to specific membrane localization is unclear. However, subtle differences can affect protein function as exemplified by "tail switching" experiments with ras (Cox et al., 1992). Placing a CXXX sequence signaling geranylgeranylation onto oncogenic ras which is normally farnesylated has no observable phenotypic effect. In contrast, the same change on a normal non-transforming ras allele expressed at moderate levels reversed its activity, making it growth inhibitory. Thus small differences in the C-terminal modifications may alter function by subtly changing the interaction of the ras protein with the bilayer, with membrane subdomains or with other proteins.

The rab subfamily of the ras superfamily are implicated in regulating membrane trafficking events between specific subcellular compartments (recently reviewed in Simons and Zerial, 1993; Novick and Brennwald, 1993). Models for their role in this process usually envisage recycling of vesicle-associated rab proteins via a soluble intermediate. The rab proteins contain C-terminal prenylation signals usually of the CXC and CC type, but sometimes have CXXX motifs (Magee and Newman, 1992). They are thus singly or doubly modified with geranylgeranyl moieties. The modified C-termini of rab proteins are largely responsible for their specific subcellular localization, although considerable sequence upstream of the prenylation sites is required for this targeting (Chavrier et al., 1991; Stenmark et al., 1994; Beranger et al., 1994). In contrast to ras proteins, where C-terminal modification seems to be independent of the rest of the molecule, there is strong evidence for intramolecular interactions between the prenylation sites of rab proteins and the rest of the molecule. Thus mutations in the nucleotide binding site of rab 5 (Sanford et al., 1993) or the effector domain of rab 1b (Wilson and Maltese, 1993) block prenylation. Similarly, the prenylation status of rab 6 affects its nucleotide binding kinetics and GTPase activity (Yang et al., 1993).

In order to permit the removal of the highly hydrophobic tails of rabs and other ras-related proteins from membranes, a series of "transport" proteins has evolved which can bind the lipid anchor and form a soluble 1:1 complex with them. In many cases these proteins also influence the nucleotide exchange rate of the ras-related protein in a positive or negative way and have thus been called guanine nucleotide dissociation stimulators (GDS) or inhibitors (GDI) (Takai et al., 1992). A number of such proteins have now been identified which can have broad (e.g., smgGDS and rabGDI; Ullrich et al., 1993), or narrow (e.g., rhoGDS) substrate specificities. C-terminal modification of ras-related proteins is absolutely required for the binding and activity of GDS and GDI proteins. There is accumulating evidence that GDS/GDI proteins functionally regulate ras-related proteins by modulating their intracellular localization (Mizuno et al., 1992; Abo et al., 1994; Elazar et al., 1994; Ullrich et al., 1994). In cells, GDS and GDI proteins can have opposing effects on the same ras-related protein (Kikuchi et al., 1992) which results in a balanced system which is therefore exquisitely sensitive to perturbation and highly regulatable. The recognition of rab proteins by rabGDI seems to depend on geranylgeranylation, since farnesylated rabs are poorly recognized (Ullrich et al., 1993). However, apart from this, rabGDI-rab binding does not appear to be highly constrained by the exact nature of the C-terminal modifications since a wide range of differently modified rab proteins are recognized (Ullrich et al., 1993) and bovine brain rabGDI will even recognize the evolutionarily distant rab protein Sec4 from *S. cerevisiae* (Garrett et al., 1993).

The mechanism of regulation of GDS/GDI binding to ras-related proteins can involve multiple factors. Clearly the nucleotide bound to the ras-related protein is important for the binding of some of these regulatory proteins. For example, rabGDI and rhoGDI bind only to the GDP form of their partners whereas smgGDS binds to both GDP and GTP forms (Takai et al., 1992). Phosphorylation of either rabs (Bailly et al., 1991; Van der Sluijs et al., 1992) or GDIs (Steele-Mortimer et al., 1993) can affect the interaction between these proteins suggesting mechanisms by which membrane trafficking and other processes dependent on ras-related proteins can be integrated with other cellular events such as the cell cycle.

Prenylation and Targeting of Other Proteins

G proteins of the heterotrimeric class also undergo prenylation of their γ subunits at CXXX sequences (Spiegel et al., 1991). Different γ subunits can be modified with farnesyl or geranylgeranyl moieties and prenylation is absolutely required for membrane association of βγ dimers, but not for formation of the dimer (Muntz et al., 1992). Lipid modification of trimeric G proteins is thus complex, since α-subunits can also be modified by myristoylation or palmitoylation or both. The relative contributions of all these modifications to the ultimate subcellular location of trimeric G proteins is as yet unclear, as is the mechanism.

Prenylated βγ subunits can have an additional targeting function. The β-adrenergic receptor kinase family (βARK 1 and 2 also known as GRK 2 and 3) are responsible for post-stimulation phosphorylation of active G-protein coupled seven-transmembrane receptors, which results in their desensitization. This process involves translocation of the kinases from the cytosol to the membrane, which is mediated by geranylgeranylated βγ dimers. Farnesylated βγ dimers are much less effective (Pitcher et al., 1992; Inglese et al., 1992a,b; Boekhoff et al., 1994). Sequences in the C-terminal half of the kinases bind to the βγ complex. This system provides an elegant regulatory loop, since free βγ subunits are themselves produced by the agonist-dependent activation of G protein by receptor, causing dissociation of α from βγ. These can in turn cause translocation of the negative regulatory kinase resulting in desensitization of the system.

Rhodopsin kinase, which inactivates the retinal light receptor rhodopsin, is itself farnesylated at a C-terminal CXXX motif, thus removing the need for a βγ-mediated translocation mechanism (Inglese et al., 1992a,b). The alternative mode of membrane association for rhodopsin kinase may reflect the different kinetics required for desensitization of the retinal light transducing system. Another class of signal regulating proteins which is bound to membranes by C-terminal prenylation is the cyclic nucleotide phosphodiesterases (Braun et al., 1991; Qin et al., 1992).

The nuclear lamins are major components of the nuclear lamina, the meshwork of filaments which is localized to the nucleoplasmic face of the inner nuclear membrane. These proteins are members of the intermediate filament family and must be targeted into the nucleus from their site of synthesis in the cytoplasm, and then to the nuclear membrane. This is again achieved by the cooperation of two signals. Newly synthesized lamins carry CXXX sequences at their C-termini which undergo farnesylation, proteolysis and carboxyl-methylation (Vorburger et al., 1989; Farnsworth et al., 1989; Kitten and Nigg, 1991). In conjunction with a peptide-based nuclear localization signal these modifications specify transport into the nucleus and targeting to the nuclear membrane (Holtz et al., 1989). After nuclear import the A type lamins are further processed by proteolytic cleavage upstream of the C-terminus, thus removing the prenylation site (Weber et al., 1989; Beck et al., 1990; Lutz et al., 1992). This could explain why A-type lamins become soluble when the nuclear membrane depolymerizes at mitosis, while prenylated B type lamins remain associated with nuclear membrane elements (Stick et al., 1989).

SUMMARY AND CONCLUSIONS

All of the diverse types of lipid modification of proteins discovered in recent years can contribute in varying degree to protein targeting within cells. Frequently, multiple lipid modifications on a single protein act in concert with one another, or with primary amino acid sequences, to specify membrane binding and intracellular targeting. Outstanding questions which must be addressed in the coming years are as follows: (i) What are the enzymes involved in lipid modification of proteins and how are they regulated? (ii) How do lipid moieties contribute to membrane binding and targeting; is this via lipid-lipid or lipid-protein interactions or both? (iii) Are there other as yet unidentified lipid modifications? These questions will require the full range of modern molecular cell biological and biochemical approaches for their answer.

ACKNOWLEDGMENTS

Work in the authors' laboratory is supported by the U.K. Medical Research Council and the Leukaemia Research Fund grant 93/13. We thank our many colleagues in the field for useful comments and the provision of pre-publication information.

REFERENCES

Abo, A., Webb, M.R., Grogan, A., & Segal, A.W. (1994). Activation of NADPH oxidase involves the dissociation of p21^{rac} from its inhibitory GDP/GTP exchange protein (rhoGDI) followed by its translocation to the plasma membrane. Biochem. J. 298, 585–591.

Abraham, N. & Veillette, A. (1990). Activation of p56^{lck} through mutation of a regulatory carboxy-terminal tyrosine residue requires intact sites of autophosphorylation and myristylation. Mol. Cell. Biol. 10, 5197–5206.

Adamson, P., Marshall, C.J., Hall, A., & Tilbrook, P.A. (1992). Post-translational modifications of p21rho proteins. J. Biol. Chem. 267, 20033–20038.

Aderem, A.A., Albert, K.A., Keum, M.M., Wang, J.K.T., Greengard, P., & Cohn, Z.A. (1988). Stimulus-dependent myristoylation of a major substrate for protein kinase C. Nature, 332, 362–364.

Andres, D.A., Seabra, M.C., Brown, M.S., Armstrong, S.A., Smeland, T.E., Cremers, F.P.M., & Goldstein, J.L. (1993). cDNA cloning of component A of Rab geranylgeranyl transferase and demonstration of its role as a rab escort protein. Cell 73, 1091–1099.

Bailly, E., McCaffrey, M., Touchot, N., Zahraoui, A., Goud, B., & Bornens, M. (1991). Phosphorylation of two small GTP-binding proteins of the Rab family by p34^{cdc2}. Nature 350, 715–718.

Balch, W.E., Kahn, R.A., & Schwaninger, R. (1992). ADP-ribosylation factor is required for vesicular trafficking between the endoplasmic reticulum and the *cis*-Golgi compartment. J. Biol. Chem. 267, 13053–13061.

Beck, L.A., Hosick, T.J., & Sinensky, M. (1990). Isoprenylation is required for the processing of the lamin A precursor. J. Cell Biol. 110, 1489–1499.

Beranger, F., Paterson, H., Powers, S., de Gunzburg, J., & Hancock, J.F. (1994). The effector domain of Rab6, plus a highly hydrophobic C terminus, is required for Golgi apparatus localization. Mol. Cell. Biol. 14, 744–758.

Black, S.D. & Mould, D.R. (1991). Development of hydrophobicity parameters to analyze proteins which bear post- or cotranslational modifications. Analyt. Biochem. 193, 72–82.

Black, S.D. (1992). Development of hydrophobicity parameters for prenylated proteins. Biochem. Biophys. Res. Comm. 186, 1437–1442.

Boekhoff, I., Inglese, J., Schleicher, S., Koch, W.J., Lefkowitz, R.J., & Breer, H. (1994). Olfactory desensitization requires membrane targeting of receptor kinase mediated by $\beta\gamma$-subunits of heterotrimeric G proteins. J. Biol. Chem. 269, 37–40.

Borman, A.L., Taylor, T.C., Melançon., & Wilson, K.L. (1992). A role for ADP-ribosylation factor in nuclear vesicle dynamics. Nature 358, 512–514.

Bourne, H.R. (1988). Do GTPases direct membrane traffic in secretion? Cell 53, 669–671.

Bourne, H.R., Sanders, D.A., & McCormick, F. (1990). The GTPase superfamily: A conserved switch for diverse cell functions. Nature 348, 125–131.

Braun, P.E., De Angelis, D., Shtybel, W.W., & Bernier, L. (1991). Isoprenoid modification permits 2′,3′-cyclic nucleotide 3′-phosphodiesterase to bind to membranes. J. Neuro. Res. 30, 540–544.

Brown, D.A. & Rose, J.K. (1992). Sorting of GPI-anchored proteins to glycolipid-enriched membrane subdomains during transport to the apical cell surface. Cell 68, 533–544.

Busconi, L. & Michel, T. (1993). Endothelial nitric oxide synthase. J. Biol. Chem. 268, 8410–8413.

Buss, J.E. & Sefton, B.M. (1986). Direct identification of palmitic acid as the lipid attached to p21^{ras}. Mol. Cell. Biol. 6, 116–122.

Buss, J.E., Mumby, S.M., Casey, P.J., Gilman, A.G., & Sefton, B.M. (1987). Myristoylated α subunit of guanine nucleotide-binding regulatory proteins. Proc. Natl. Acad. Sci. USA 84, 7493–7497.

Camp, L.A. & Hofmann, S.L. (1993). Purification and properties of a palmitoyl-protein thioesterase that cleaves palmitate from H-ras. J. Biol. Chem. 268, 22566–22574.

Cerneus, D.P., Ueffing, E., Posthuma, G., Strous, G.J., & van der Ende, A. (1993). Detergent insolubility of alkaline phosphatase during biosynthetic transport and endocytosis. J. Biol. Chem. 268, 3150–3155.

Chakravarty, R. & Rice, R.H. (1989). Acylation of keratinocyte transglutaminase by palmitic and myristic acids in the membrane anchorage region. J. Biol. Chem. 264, 625–629.

Chavrier, P., Gorvel, J.-P., Stelzer, E., Simons, K., Gruenberg, J., & Zerial, M. (1991). Hypervariable C-terminal domain of rab proteins acts as a targeting signal. Nature 353, 769–772.

Clarke, S. (1992). Protein isoprenylation and methylation at carboxyl-terminal cysteine residues. Ann. Rev. Biochem. 61, 355–386.

Cornish, J.A., Kloc, M., Decker, G.L., Reddy, B.A., & Etkin, L.D. (1992). Xlcaax-1 is localized to the basolateral membrane of kidney tubule and other polarized epithelia during *Xenopus* development. Dev. Biol. 150, 108–120.

Cox, A.D. & Der Channing, J. (1992). Protein prenylation: More than just glue? Curr. Opin. Cell Biol. 4, 1008–1016.

Cox, A.D., Hisaka, M.M., Buss, J.E., & Der Channing, J. (1992). Specific isoprenoid modification is required for function of normal, but not oncogenic, ras protein. Mol. Cell. Biol. 12, 2606–2615.

Cox, A.D., Graham, S.M., Solski, P.A., Buss, J.E., & Der Channing, J. (1993). The carboxyl-terminal CXXX sequence of $G_i\alpha$, but not Rab5 or Rab 11, supports ras processing and transforming activity. J. Biol. Chem. 268, 11548–11552.

Coyne, K.E., Crisci, A., & Lublin, D.M. (1993). Construction of synthetic signals for glycosyl-phosphatidylinositol anchor attachment. J. Biol. Chem. 268, 6689–6693.

Crise, B. & Rose, J.K. (1992). Identification of palmitoylation sites on CD4, the human immunodeficiency virus receptor. J. Biol. Chem. 267, 13593–13597.

Degtyarev, M.Y., Spiegel, A.M., & Jones, T.L.Z. (1993a). The G protein α_s subunit incorporates [^{3}H]palmitic acid and mutation of cysteine-3 prevents this modification. Biochemistry 32, 8057–8061.

Degtyarev, M.Y., Spiegel, A.M., & Jones, T.L.Z. (1993b). Increased palmitoylation of the G_s protein α subunit after activation by the β-adrenergic receptor or cholera toxin. J. Biol. Chem. 32, 23679–23772.

Dizhoor, A.M., Chen, C.-K., Olshevskaya, E., Sinelnikova, V.V., Phillipov, P., & Hurley, J.B. (1993). Role of the acylated amino terminus of recoverin in Ca^{2+}-dependent membrane interaction. Science 259, 829–832.

Dotti, C.G., Parton, R.G., & Simons, K. (1991). Polarized sorting of glypiated proteins in hippocampal neurons. Nature, 349, 158–161.

Edidin, M. & Stroynowski, I. (1991). Differences between the lateral organization of conventional and inositol phospholipid-anchored membrane proteins. A further definition of micrometer scale membrane domains. J. Cell Biol. 112, 1143–1150.

Elazar, Z., Mayer, T., & Rothman, J.E. (1994). Removal of Rab GTP-binding proteins from Golgi membranes by GDP dissociation inhibitor inhibits inter-cisternal transport in the Golgi stacks. J. Biol. Chem. 269, 794–797.

Epand, R.F., Xue, C.B., Wang, S.-H., Naider, F., Becker, J.M., & Epand, R.M. (1993). Role of prenylation in the interaction of the a-factor mating pheromone with phospholipid bilayers. Biochemistry 32, 8368–8373.

Farnsworth, C.C., Wolda, S.L., Gelb, M.H., & Glomset, J.A. (1989). Human lamin B contains a farnesylated cysteine residue. J. Biol. Chem. 264, 20422–20429.

Fiedler, K., Kobayashi, T., Kurzchalia, V., & Simons, K. (1993). Glycosphingolipid-enriched, detergent-insoluble complexes in protein sorting in epithelial cells. Biochemistry 32, 6365–6373.

Franco, M., Chardin, P., Chabre, M., & Paris, S. (1993). Myristoylation is not required for GTP-dependent binding of ADP-ribosylation factor ARF1 to phospholipids. J. Biol. Chem. 268, 24531–24534.

Fukuoka, S.-I., Freedman, S.D., Yu, H., Sukhatme, V.P., & Scheele, G.A. (1992). GP-2/THP gene family encodes self-binding glycosylphosphatidylinositol-anchored proteins in apical secretory compartments of pancreas and kidney. Proc. Natl. Acad. Sci. USA 89, 1189–1193.

Gallego, C., Gupta, S.K., Winitz, S., Eisfelder, B.J., & Johnson, G.L. (1992). Myristoylation of $G\alpha_{i2}$ polypeptide, a G protein α subunit, is required for its signalling and transformational functions. Proc. Natl. Acad. Sci. USA 89, 9695–9699.

Garrett, M.D., Kabcenell, A.K., Zahner, J.E., Kaibuchi, K., Sasaki, T., Taki, Y., Cheney, C.M., & Novick, P.J. (1993). Interaction of Sec4 with GDI proteins from bovine brain, *Drosophila melanogaster* and *Saccharomyces cerevisiae*. FEBS Letts. 331, 223–238.

George, D.J. & Blackshear, P.J. (1992). Membrane association of the myristoylated alanine-rich C kinase substrate (MARCKS) protein appears to involve myristate-dependent binding in the absence of a myristoyl protein receptor. J. Biol. Chem. 267, 24879–24885.

Giannakouros, T., Armstrong, J., & Magee, A.I. (1992). Protein prenylation in *Schizosaccharomyces pombe*. FEBS Letts. 297, 103–106.

Giannakouros, T. & Magee, A.I. (1993). Protein prenylation and associated modifications. In: Lipid Modifications of Proteins. (M.J. Schlesinger, ed.) pp. 135–162. CRC Press Inc., Boca Raton, FL.

Giannakouros, T., Newman, C.M.H., Craighead, M.W., Armstrong, J., & Magee, A.I. (1993). Post-translational processing of *S.pombe* YPT5 protein: *In vitro* and *in vivo* analysis of processing mutants. J. Biol. Chem. 268, 24467–24474.
Glick, B.S. & Rothman. J.E. (1987). Possible role for fatty acyl-coenzyme A in intracellular protein transport. Nature 326, 309–312.
Gordon, J.I. & Mobley, P.W. (1985). Membrane Fluidity in Biology, Cellular Aspects. Academic Press, Inc., New York.
Gutierrez, L., Magee, A.I., Marshall, C.J., & Hancock, J.F. (1989). Post-translational processing of $p21^{ras}$ is two-step and involves carboxyl-methylation and carboxy-terminal proteolysis. EMBO J. 8, 1093–1098.
Gutierrez, L. & Magee, A.I. (1991). Characterization of an acyl transferase acting on $p21^{N\text{-}ras}$ in a cell-free system. Biochim. Biophys. Acta. 1078, 147–154.
Hancock, J.F., Magee, A.I., Childs, J.E., & Marshall, C.J. (1989). All ras proteins are polyisoprenylated but only some are palmitoylated. Cell 57, 1167–1177.
Hancock, J.F., Paterson, H., & Marshall, C.J. (1990). A polybasic domain or palmitoylation is required in addition to the CAAX motif to localize $p21^{ras}$ to the plasma membrane. Cell 63, 133–139.
Hancock, J.F., Cadwallader, K., & Marshall, C.J. (1991a). Methylation and proteolysis are essential for efficient membrane binding of prenylated $p21^{K\text{-}ras(B)}$. EMBO J. 10, 641–646.
Hancock, J.F., Cadwallader, K., Paterson, H., & Marshall, C.J. (1991b). A CAAX or a CAAL motif and a second signal are sufficient for plasma membrane targeting of ras proteins. EMBO J. 10, 4033–4039.
Hannan, L.A., Lisanti, M.P., Rodriguez-Boulan, E., & Edidin, M. (1993). Correctly sorted molecules of a GPI-anchored protein are clustered and immobile when they arrive at the apical surface of MDCK cells. J. Cell Biol. 120, 353–358.
Haun, R.S., Tsai, S.-C., Adamik, R., Moss, J., & Vaughan, M. (1993). Effect of myristoylation on GTP-dependent binding of ADP-ribosylation factor to Golgi. J. Biol. Chem. 268, 7064–7068.
Helms, J.B., Palmer, D.J., & Rothman, J.E. (1993). Two distinct populations of ARF bound to Golgi membranes. J. Cell Biol. 121, 751–760.
Holtz, D., Tanaka, R.A., Hartwig, J., & McKeon, F. (1989). The CaaX motif of lamin A functions in conjunction with the nuclear localization signal to target assembly to the nuclear envelope. Cell 59, 969–977.
Horiuchi, H., Kaibuchi, K., Kawamura, M., Matsuura, Y., Suzuki, N., Kuroda, Y., Kataoka, T., & Takai, Y. (1992). The post-translational processing of *ras* p21 is critical for its stimulation of yeast adenylate cyclase. Mol. Cell. Biol. 12, 4515–4520.
Huizinga, T.W.J., van der Schoot, C.E., Jost, C., Klaassen, R., Kleijer, M., von dem Borne, A.E.G.K., Roos, D., & Tetteroo, P.A.T. (1988). The PI-linked receptor FcRIII is released on stimulation of neutrophils. Nature 333, 667–669.
Inglese, J., Glickman, J.F., Lorenz, W., Caron, M.C., & Lefkowitz, R.J. (1992a). Isoprenylation of a protein kinase. J. Biol. Chem. 267, 1422–1425.
Inglese, J., Koch, W.J., Caron, M.G., & Lefkowitz, R.J. (1992b). Isoprenylation in regulation of signal transduction by G-protein-coupled receptor kinases. Nature 359, 147–150.
Itoh, T., Kaibuchi, K., Tadayuki, M., Yamamoto, T., Matsuura, Y., Maeda, A., Shimizu, K., & Takai, Y. (1993). The post-translational processing of *ras* p21 is critical for its stimulation of mitogen-activated protein kinase. J. Biol. Chem. 268, 3025–3028.
James, G. & Olson, E.N. (1989). Myristoylation, phosphorylation and subcellular distribution of the 80-kD protein kinase C substrate in BC_3H1 myocytes. J. Biol. Chem. 264, 20928–20933.
James, G. & Olson, E.N. (1990). Fatty acylated proteins as components of intracellular signaling pathways. Biochemistry 29, 2623–2684.
Jones, T.L.Z., Simonds, W.F., Merendino, Jr., J.J., Brann, M.R., & Spiegel, A.M. (1990). Myristoylation of an inhibitory GTP-binding protein α subunit is essential for its membrane attachment. Proc. Natl. Acad. Sci. USA 87, 568–572.

Jones, T.L.Z. & Spiegel, A.M. (1990). Isoprenylation of an inhibitory G protein α subunit occurs only upon mutagenesis of the carboxyl terminus. J. Biol. Chem. 265, 19389–19392.

Kaplan, J.M., Varmus, H.E., & Bishop, J.M. (1990). The *src* protein contains multiple domains for specific attachment to membranes. Mol. Cell. Biol. 10, 1000–1009.

Keller, G.-A., Siegel, M.W., & Caras, I.W. (1992). Endocytosis of glycophospholipid-anchored and transmembrane forms of CD4 by different endocytic pathways. EMBO J. 11, 863–874.

Kikuchi, A., Kuroda, S., Sasaki, T., Kotani, K., Hirata, K.-I., Katayama, M., & Takai, Y. (1992). Functional interactions of stimulatory and inhibitory GDP-GTP exchange proteins and their common substrate small GTP-binding protein. J. Biol. Chem. 267, 14611–14615.

Kitten, G.T. & Nigg, E.A. (1991). The CaaX motif is required for isoprenylation, carboxyl methylation, and nuclear membrane association of lamin B_2. J. Cell Biol. 113, 13–23.

Kloc, M., Reddy, B., Crawford, S., & Etkin, L.D. (1991). A novel 110-kDa maternal CAAX box-containing protein from *Xenopus* is palmitoylated and isoprenylated when expressed in baculovirus. J. Biol. Chem. 266, 8206–8212.

Kloc, M., Li, X.X., & Etkin, L.D. (1993). Two upstream cysteines and the CAAX motif but not the polybasic domain are required for membrane association of Xlcaax in *Xenopus* oocytes. Biochemistry 32, 8207–8212.

Kobayashi, M., Takamatsu, K., Saitoh, S., & Noguchi, T. (1993). Myristoylation of hippocalcin is linked to its calcium-dependent membrane association properties. J. Biol. Chem. 268, 18898–18904.

Kodukula, K., Gerber, L.D., Amthauer, R., Brink, L., & Udenfriend, S. (1993). Biosynthesis of glycosylphosphatidylinositol (GPI)-anchored membrane proteins in intact cells: Specific amino acid requirements adjacent to the site of cleavage and GPI attachment. J. Cell Biol. 120, 657–664.

Koegl, M., Zlatkine, P., Ley, S.C., Courtneidge, S.A., & Magee, A.I. (1994). Palmitoylation of multiple Src-family kinases at a homologous N-terminal motif. Biochem. J. 303, 749–753.

Kuroda, Y., Suzuki, N., & Kataoka, T. (1992). The effect of post-translational modifications on the interaction of Ras2 with adenylyl cyclase. Science 259, 683–686.

Lenhard, J.M., Kahn, R.A., & Stahl, P.D. (1992). Evidence for ADP-ribosylation factor (ARF) as a regulator of *in vitro* endosome-endosome fusion. J. Biol. Chem. 267, 13047–13052.

Linder, M.E., Pang, I.-H., Duronio, R.J., Gordon, J.I., Sternweis, P.C., & Gilman, A.G. (1991). Lipid modifications of G protein subunits. J. Biol. Chem. 266, 4654–4659.

Linder, M.E., Middleton, P., Hepler, J.R., Taussig, R., Gilman, A.G., & Mumby, S.M. (1993). Lipid modifications of G proteins: α subunits are palmitoylated. Proc. Natl. Acad. Sci. USA 90, 3675–3679.

Lisanti, M.P., Sargiacomo, M., Graeve, L., Saltiel, A.R., & Rodriguez-Boulan, E. (1988). Polarized apical distribution of glycosyl-phosphatidylinositol-anchored proteins in a renal epithelial cell line. Proc. Natl. Acad. Sci. USA 85, 9557–9561.

Lisanti, M.P., Caras, I.W., Davitz, M.A., & Rodriguez-Boulan, E. (1989). A glycophospholipid membrane anchor acts as an apical targeting signal in polarized epithelial cells. J. Cell Biol. 109, 2145–2156.

Lisanti, M.P., Caras, I.W., Gilbert, T., Hanzel, D., & Rodriguez-Boulan, E. (1990a). Vectorial apical delivery and slow endocytosis of a glycolipid-anchored fusion protein in transfected MDCK cells. Proc. Natl. Acad. Sci. USA 87, 7419–7423.

Lisanti, M.P., Le Bivic, A., Saltiel, A.R., & Rodriguez-Boulan, E. (1990b). Preferred apical distribution of glycosyl-phosphatidylinositol (GPI) anchored proteins: A highly conserved feature of the polarized epithelial cell phenotype. J. Membrane Biol. 113, 155–167.

Liu, Y., Chapman, E.R., & Storm, D.R. (1991). Targeting of neuromodulin (GAP-43) fusion proteins to growth cones in cultured rat embryonic neurons. Neuron, 6, 411–420.

Liu, Y., Fisher, D.A., & Storm, D.R. (1993). Analysis of the palmitoylation and membrane targeting domain of neuromodulin (GAP-43) by site-specific mutagenesis. Biochemistry 32, 10714–10719.

Lutz, R.J., Trujillo, M.A., Denham, K.S., Wenger, L., & Sinensky, M. (1992). Nucleoplasmic localization of prelamin A: Implications for prenylation-dependent lamin A assembly into the nuclear lamina. Proc. Natl. Acad. Sci. USA 89, 3000–3004.

Magee, A.I. & Courtneidge, S.A. (1985). Two classes of fatty acid acylated proteins exist in eukaryotic cells. EMBO J. 4, 1137–1144.

Magee, A.I., Gutierrez, L., McKay, I.A., Marshall, C.J., & Hall, A. (1987). Dynamic fatty acylation of p21$^{N\text{-}ras}$. EMBO J. 6, 3353–3357.

Magee, A.I. & Hanley, M. (1988). Sticky fingers and CAAX boxes. Nature 335, 114–115.

Magee, A.I. & Newman, C. (1992). The role of lipid anchors for small G proteins in membrane trafficking. Trends Cell Biol. 2, 318–323.

Manenti, S., Sorokine, O., Van Dorsselaer, A., & Taniguchi, H. (1993). Isolation of the non-myristoylated form of a major substrate of protein kinase C (MARCKS) from bovine brain. J. Biol. Chem. 268, 6878–6881.

Matlin, K.S. (1986). The sorting of proteins to the plasma membrane in epithelial cells. J. Cell Biol. 103, 2565–2568.

McIlhinney, R.A.J., Pelly, S.J., Chadwick, J.K., & Cowley, G.P. (1985). Studies on the attachment of myristic and palmitic acid to cell proteins in human squamous carcinoma cell lines: Evidence for two pathways. EMBO J. 4, 1145–1152.

McIlhinney, R.A.J., Chadwick, J.K., & Pelly, S.K. (1987). Studies on the cellular location, physical properties and endogenously attached lipids of acylated proteins in human squamous-carcinoma cell lines. Biochem. J. 244, 109–115.

McIlhinney, R.A.J. & McGlone, K. (1990). Evidence for a non-myristoylated pool of the 80 kDa protein kinase C substrate of rat brain. Biochem. J. 271, 681–685.

McConville, M.J. & Ferguson, M.A.J. (1993). The structure, biosynthesis and function of glycosylated phosphatidylinositols in the parasitic protozoa and higher eukaryotes. Biochem. J. 294, 305–324.

Metcalfe, J.C. & Warren, G.B. (1977). Lipid-protein interactions in a reconstituted calcium pump. In: International Cell Biology (R. Brinkley & K. Porter, eds.) pp. 15–23, Rockefeller University Press, New York.

Mizuno, T., Kaibuchi, K., Ando, S., Musha, T., Hiraoka, K., Takaishi, K., Asada, M., Nunoi, H., Matsuda, I., & Takai, Y. (1992). Regulation of the superoxide-generating NADPH oxidase by a small GTP-binding protein and its stimulatory and inhibitory GDP-GTP exchange proteins. J. Biol. Chem. 267, 10215–10218.

Molenaar, C.M.T., Prange, R., & Gallwitz, D. (1988). A carboxyl-terminal cysteine residue is required for palmitic acid binding and biological activity of the *ras*-related yeast *YPT1* protein. EMBO J. 7, 971–976.

Moran, P. & Caras, I.W. (1992). Proteins containing an uncleaved signal for glycophospholipidinositol membrane anchor attachment are retained in a post-ER compartment. J. Cell Biol. 119, 763–772.

Morris, R. (1992). Thy-1, the enigmatic extrovert on the neuronal surface. BioEssays 14, 715–722.

Mostov, K., Apodaca, G., Aroeti, B., & Okamoto, C. (1992). Plasma membrane protein sorting in polarized epithelial cells. J. Cell Biol. 116, 577–583.

Mumby, S.M., Heukeroth, R.O., Gordon, J.I., & Gilman, A.G. (1990). G-protein α-subunit expression, myristoylation, and membrane association in COS cells. Proc. Natl. Acad. Sci. USA 87, 728–732.

Mundy, D.I. & Warren, G. (1992). Mitosis and inhibition of intracellular transport stimulate palmitoylation of a 62-kD protein. J. Cell Biol. 116, 135–146.

Muntz, K.H., Sternweis, P.C., Gilman, A.G., & Mumby, S.M. (1992). Influence of γ subunit prenylation on association of guanine nucleotide-binding regulatory proteins with membranes. Mol. Biol. Cell. 3, 49–61.

Nadler, M.J.S., Harrison, M.L., Ashendel, C.L., Cassady, J.M., & Geahlen, R.L. (1993). Treatment of T cells with 2-hydroxymyristic acid inhibits the myristoylation and alters the stability of p56^{lck}. Biochemistry 32, 9250–9255.

Nelson, W.J. (1992). Regulation of cell surface polarity from bacteria to mammals. Science 258, 948–955.

Neubert, T.A., Johnson, R.S., Hurley, J.B., & Walsh, K.A. (1992). The rod transducin α subunit amino terminus is heterogeneously fatty acylated. J. Biol. Chem. 267, 18247–18277.

Newman, C.M.H., Giannakouros, T., Hancock, J.F., Fawell, E.H., Armstrong, J., & Magee, A.I. (1992). Post-translational processing of *Schizosaccharomyces pombe* YPT proteins. J. Biol. Chem. 267, 11329–11336.

Newman, C.M.H. & Magee, A.I. (1993). Post-translational processing of the ras superfamily of small GTP-binding proteins. BBA Reviews on Cancer 1155, 79–96.

Ng, G.Y.K., George, S.R., Zastawny, R.L., Caron, M., Bouvier, M., Dennis, M., & O'Dowd, B.F. (1993). Human serotonin$_{1B}$ receptor expression in Sf9 cells: Phosphorylation, palmitoylation, and adenyl cyclase inhibition. Biochemistry 32, 11727–11733.

Novick, P. & Brennwald, P. (1993). Friends and family: The role of the Rab GTPases in vesicular traffic. Cell 75, 597–601.

O'Dowd, B.F., Hnatowich, M., Caron, M.G., Lefkowitz, R.J., & Bouvier, M. (1989). Palmitoylation of the human β_2-adrenergic receptor. J. Biol. Chem. 264, 7564–7569.

Olson, E.N., Glaser, L., Merlie, J.P., & Lindstrom, J. (1984). Expression of acetylcholine receptor α-subunit mRNA during differentiation of the BC_3H1 muscle cell line. J. Biol. Chem. 259, 3330–3336.

Olson, E.N., Towler, D.A., & Glaser, L. (1985). Specificity of fatty acid acylation of cellular proteins. J. Biol. Chem. 260, 3784–3790.

Omary, M.B. & Trowbridge, I.S. (1981). Covalent binding of fatty acid to the transferrin receptor in cultured human cells. J. Biol. Chem. 256, 4715–4718.

Omary, M.B. & Trowbridge, I.S. (1981). Biosynthesis of the human transferrin receptor in cultured cells. J. Biol. Chem. 256, 12888–12892.

Paige, L.A., Nadler, M.J.S., Harrison, M.L., Cassidy, J.M., & Geahlen, R.L. (1993). Reversible palmitoylation of the protein-tyrosine kinase $p56^{lck}$. J. Biol. Chem. 268, 8669–8674.

Palmer, D.J., Helms, J.B., Beckers, C.J.M., Orci, L., & Rothman, J.E. (1993). Binding of coatomer to Golgi membranes requires ADP-ribosylation factor. J. Biol. Chem. 268, 12083–12089.

Parenti, M., Viganó, M.A., Newman, C.M.H., Milligan, G., & Magee, A.I. (1993). A novel N-terminal motif for palmitoylation of G-protein α subunits. Biochem. J. 291, 349–353.

Peitzsch, R.M. & McLaughlin, S. (1993). Binding of acylated peptides and fatty acids to phospholipid vesicles: Pertinence to myristoylated proteins. Biochemistry 32, 10436–10443.

Pellman, D., Garber, E.A., Cross, F.R., & Hanafusa, H. (1985). An N-terminal peptide from $p60^{src}$ can direct myristylation and plasma membrane localization when fused to heterologous proteins. Nature 314, 374–377.

Pfanner, N., Orci, L., Glick, B.S., Amherdt, M., Arden, S.R., Malhotra, V., & Rothman, J.E. (1989). Fatty acyl-coenzyme A is required for budding of transport vesicles from Golgi cisternae. Cell 59, 95–102.

Pfanner, N., Glick, B.S., Arden, S.R., & Rothman, J.E. (1990). Fatty acylation promotes fusion of transport vesicles with Golgi cisternae. J. Cell Biol. 110, 955–961.

Phillips, M.A., Qin, Q., Mehrpouyan, M., & Rice, R.H. (1993). Keratinocyte transglutaminase membrane anchorage: Analysis of site-directed mutants. Biochemistry 32, 11057–11063.

Pitcher, J.A., Inglese, J., Higgins, J.B., Arriza, J.L., Casey, P.J., Kim, C., Benovic, J.L., Kwatra, M.M., Caron, M.G., & Lefkowitz, R.J. (1992). Role of $\beta\gamma$ subunits of G proteins in targeting the β-adrenergic receptor kinase to membrane-bound receptors. Science 257, 1264–1267.

Qin, N., Pittler, S.J., & Baehr, W. (1992). *In vitro* isoprenylation and membrane association of mouse rod photoreceptor cGMP phosphodiestersase α and β subunits expressed in bacteria. J. Biol. Chem. 267, 8458–8463.

Randazzo, P.A., Yang, Y.C., Rulka, C., & Kahn, R.A. (1993). Activation of ADP-ribosylation factor by Golgi membranes. J. Biol. Chem. 268, 9555–9563.

Resh, M.D. (1989). Specific and saturable binding of pp60$^{v\text{-}src}$ to plasma membrane: Evidence for a myristyl-*src* receptor. Cell 58, 281–286.

Resh, M.D. & Ling, H.-P. (1990). Identification of a 32K plasma membrane protein that binds to the myristylated amino-terminal sequence of p60*v-src*. Nature 346, 84–86.

Rothberg, K.G., Ying, Y.-S., Kolhouse, J.F., Kamen, B.A., & Anderson, R.G.W. (1990a). The glycophospholipid-linked folate receptor internalizes folate without entering the clathrin-coated pit endocytic pathway. J. Cell Biol. 110, 637–649.

Rothberg, K.G., Ying, Y.-S., Kamen, B.A., & Anderson, R.G.W. (1990b). Cholesterol controls the clustering of the glycosylphospholipid-anchored membrane receptor for 5-methyltetrahydrofolate. J. Cell Biol. 111, 2931–2938.

Rothman, J.E. & Orci, L. (1990). Movement of proteins through the Golgi stack: A molecular dissection of vesicular transport. FASEB J. 4, 1460–1468.

Rothman, J.E. & Orci, L. (1992). Molecular dissection of the secretory pathway. Nature 355, 409–415.

Sanford, J.C., Pan, Y., & Wessling-Resnick, M. (1993). Prenylation of Rab5 is dependent on guanine nucleotide binding. J. Biol. Chem. 268, 23773–23776.

Sargiacomo, M., Sudol, M., Tang, Z., & Lisanti, M.P. (1993). Signal transducing molecules and glycosyl-phosphatidylinositol-linked proteins form a caveolin-rich insoluble complex in MDCK cells. J. Cell Biol. 122, 789–807.

Schlesinger, M.J. (1992). Lipid Modifications of Proteins. CRC Press, Boca Raton, Florida.

Schmidt, M.F.G. (1989). Fatty acylation of proteins. Biochim. Biophys. Acta 988, 411–426.

Schweizer, A., Rohrer, J., Jenö, P., DeMaio, A., Buchman, T.G., & Hauri, H.-P. (1993). A reversibly palmitoylated resident protein (p63) of an ER-Golgi intermediate compartment is related to a circulatory shock resuscitation protein. J. Cell Sci. 104, 685–694.

Seabra, M.C., Brown, M.S., Slaughter, C.A., Südhof, T.C., & Goldstein, J.L. (1992). Purification of component A of rab geranylgeranyl transferase: Possible identity with the choroideremia gene product. Cell, 70, 1049–1057.

Sessa, W.C., Harrison, J.K., Barber, C.M., Zeng, D., Durieux, M.E., D'Angelo, D.D., Lynch, K.R., & Peach, M.J. (1992). Molecular cloning and expression of a cDNA encoding endothelial cell nitric oxide synthase. J. Biol. Chem. 267, 15274–15276.

Sessa, W.C., Barber, C.M., Zeng, D., & Lynch, K.R. (1993). Mutation of *N*-myristoylation site converts endothelial cell nitric oxide synthase from a membrane to a cytosolic protein. Circulation Res. 72, 921–924.

Shenoy-Scaria, A.M., Timson Gauen, L.K., Kwong, J., Shaw, A.S., & Lublin, D.M. (1993). Palmitoylation of an amino-terminal cysteine motif of protein tyrosine kinases p56^{lck} and p59^{fyn} mediates interaction with glycosyl-phosphatidylinositol-anchored protein. Mol. Cell. Biol. 13, 6385–6392.

Sigal, C.T. & Resh, M.D. (1993). The ADP/ATP carrier is the 32-kilodalton receptor for an NH_2-terminally myristylated Src peptide but not for pp60^{src} polypeptide. Mol. Cell. Biol. 13, 3084–3092.

da Silva, A.M. & Klein, C. (1990). A rapid post-translational myristoylation of a 68-kD protein in *D. discoideum*. J. Cell Biol. 111, 401–407.

Silverman, L. & Resh, M.D. (1992). Lysine residues form an integral component of a novel NH_2-terminal membrane targeting motif for myristoylated pp60$^{v\text{-}src}$. J. Cell Biol. 119, 415–425.

Simmonds, A.C., East, J.M., Jones, O.T., Rooney, E.K., McWhirter, J., & Lee, A.G. (1982). Annular and non-annular binding sites on the (Ca^{2+} + Mg^{2+})-ATPase. Biochim. Biophys. Acta 693, 398–406.

Simon, S.M. & Aderem, A. (1992). Myristoylation of proteins in the yeast secretory pathway. J. Biol. Chem. 267, 3922–3931.

Simons, K. & van Meer, G. (1988). Lipid sorting in epithelial cells. Biochemistry 27, 6197–6202.

Simons, K. & Wandinger-Ness, A. (1990). Polarized sorting in epithelia. Cell 62, 207–210.

Simons, K. & Zerial, M. (1993). Rab proteins and the road maps for intracellular transport. Neuron 11, 789–799.

Sinensky, M. & Lutz, R.J. (1992). The prenylation of proteins. BioEssays 14, 25–31.

Skene, H.P. & Virág, I. (1989). Posttranslational membrane attachment and dynamic fatty association of a neuronal growth cone protein, GAP-43. J. Cell Biol. 108, 613–624.

van der Sluijs, P., Hull, M., Huber, L.A., Mâle, P., Goud, B., & Mellman, I. (1992). Reversible phosphorylation-dephosphorylation determines the localization of rab4 during the cell cycle. EMBO J. 11, 4379–4389.

Spiegel, A.M., Backlund, P.S., Butrynski, J.E., Jones, T.L.Z., & Simonds, W.F. (1991). The G protein connection: Molecular basis of membrane association. TIBS 16, 338–341.

Staufenbiel, M. (1987). Ankyrin-bound fatty acid turns over rapidly at the erythrocyte plasma membrane. Mol. Cell. Biol. 7, 2981–2984.

Steele-Mortimer, O., Gruenberg, J., & Clague, M.J. (1993). Phosphorylation of GDI and membrane cycling of rab proteins. FEBS Lett. 329, 313–318.

Štefanová, I., Horejší, V., Ansotegui, I.J., Knapp, W., & Stockinger, H. (1991). GPI-anchored cell-surface molecules complexed to protein tyrosine kinases. Science 254, 1016–1019.

Stenmark, H., Valencia, A., Martinez, O., Ullrich, O., Goud, B., & Zerial, M. (1994). Distinct structural elements of rab5 define its functional specificity. EMBO J. 13, 575–583.

Sternweis, P.C. (1986). The purified α subunits of G_o and G_i from bovine brain require $\beta\gamma$ for association with phospholipid vesicles. J. Biol. Chem. 261, 631–637.

Stevenson, F.T., Burstein, S.L., Fanton, C., Locksley, R.N., & Lovett, D.H. (1993). The 31-kDa precursor of interleukin 1α is myristoylated on specific lysines within the 16-kDa N-terminal propiece. Proc. Natl. Acad. Sci. USA 90, 7245–7249.

Stick, R., Angres, B., Lehner, C.F., & Nigg, E.A. (1988). The fates of chicken nuclear lamin proteins during mitosis: evidence for a reversible redistribution of lamin B_2 between inner nuclear membrane and elements of the endoplasmic reticulum. J. Cell Biol. 107, 397–406.

Sudo, Y., Valenzuela, D., Beck-Sickinger, AG., Fishman, M.C., & Strittmatter, S.M. (1992). Palmitoylation alters protein activity: Blockade of G_o stimulation by GAP-43. EMBO J. 11, 2095–2102.

Swiezewska, E., Thelin, A., Dallner, G., Andersson, B., & Ernster, L. (1993). Occurrence of prenylated proteins in plant cells. Biochem. Biophys. Res. Comm. 192, 161–166.

Takai, Y., Kaibuchi, K., Kikuchi, A., & Kawata, M. (1992). Small GTP-binding proteins. Intl. Rev. Cytol. 133, 187–230.

Taniguchi, H. & Manenti, S. (1993). Interaction of myristoylated alanine-rich protein kinase C substrate (MARCKS) with membrane phospholipids. J. Biol. Chem. 268, 9960–9963.

Tartakoff, A.M. & Singh, N. (1992). How to make a glycoinositol phospholipid anchor. TIBS 17, 470–473.

Taylor, T.C., Kahn, R.A., & Melançon, P. (1992). Two distinct members of the ADP-ribosylation factor family of GTP-binding proteins regulate cell-free intra-Golgi transport. Cell 70, 69–79.

Thelen, M., Rosen, A., Nairn, A.C., & Aderem, A. (1991). Regulation by phosphorylation of reversible association of a myristoylated protein kinase C substrate with the plasma membrane. Nature 351, 320–322.

Towler, D.A., Gordon, J.I., Adams, S.P., & Glaser, L. (1988). The biology and enzymology of eukaryotic protein acylation. Ann. Rev. Biochem. 57, 69–99.

Tsuchiya, M., Price, S.R., Tsai, S.-C., Moss, J., & Vaughan, M. (1991). Molecular identification of ADP-ribosylation factor mRNAs and their expression in mammalian cells. J. Biol. Chem. 266, 2772–2777.

Turner, J.M., Brodsky, M.H., Irving, B.A., Levin, S.D., Perlmutter, R.M., & Littman, D.R. (1990). Interaction of the unique N-terminal region of tyrosine kinase $p56^{lck}$ with cytoplasmic domains of CD4 and CD8 is mediated by cysteine motifs. Cell 60, 755–765.

Ullrich, O., Horiuchi, H., Bucci, C., & Zerial, M. (1994). Membrane association of Rab5 mediated by GDP-dissociation inhibitor and accompanied by GDP/GTP exchange. Nature, 368, 157–160.

Ullrich, O., Stenmark, H., Alexandrov, K., Huber, L.A., Kaibuchi, K., Sasaki, T., Takai, Y., & Zerial, M. (1993). Rab GDP dissociation inhibitor as a general regulator for the membrane association of Rab proteins. J. Biol. Chem. 268, 18143–18150.

Verner, K. & Schatz, G. (1988). Protein translocation across membranes. Science 241, 1307–1313.

Vorburger, K., Kitten, G.T., & Nigg, E.A. (1989). Modification of nuclear lamin proteins by a mevalonic acid derivative occurs in reticulocyte lysates and requires the cysteine residue of the C-terminal CXXM motif. EMBO J. 8, 4007–4013.

Warren, G.B., Houslay, M.D., Metcalfe, J.C., & Birdsall, N.J.M. (1975). Cholesterol is excluded from the phospholipid annulus surrounding an active calcium transport protein. Nature 255, 684–687.

Weber, K., Plessmann, U., & Traub, P. (1989). Maturation of nuclear lamin A involves a specific carboxy-terminal trimming, which removes the polyisoprenylation site from the precursor; implications for the structure of the nuclear lainina. FEBS Letts. 257, 411–414.

Wedegaertner, P.B., Chu, D.H., Wilson, P.T., Levis, M.J., & Bourne, H.R. (1993). Palmitoylation is required for signaling functions and membrane attachment of $G_q\alpha$ and $G_s\alpha$. J. Biol. Chem. 268, 25001–25008.

Wei, C., Lutz, R., Sinensky, M., & Macara, I.G. (1992). $p23^{rab2}$, a ras-like GTPase with a -GGGCC C-terminus, is isoprenylated but not detectably carboxymethylated in NIH3T3 cells. Oncogene 7, 467–473.

Wickner, W.T. & Lodish, H.F. (1985). Multiple mechanisms of protein insertion into and across membranes. Science 230, 400–407.

Wilson, J.M., Fasel, N., & Kraehenbuhl, J.-P. (1990). Polarity of endogenous and exogenous glycosyl-phosphatidylinositol-anchored membrane proteins in Madin-Darby canine kidney cells. J. Cell Sci. 96, 143–149.

Wilson, A.L. & Maltese, W.A. (1993). Isoprenylation of Rab1B is impaired by mutations in its effector domain. J. Biol. Chem. 268, 14561–14564.

Yang, C., Mollat, P., Chaffotte, A., McCaffrey, M., Cabanié, L., & Goud, B. (1993). Comparison of the biochemical properties of unprocessed and processed forms of the small GTP-binding protein, rab6p. Eur. J. Biochem. 217, 1027–1037.

Yokode, M., Pathak, R.K., Hammer, R.E., Brown, M.S., Goldstein, J.L., & Anderson, R.G.W. (1992). Cytoplasmic sequence required for basolateral targeting of LDL receptor in livers of transgenic mice. J. Cell. Biol. 117, 39–46.

Zhang, F., Lee, G.M., & Jacobson, K. (1993). Protein lateral mobility as a reflection of membrane microstructure. BioEssays 15, 579–588.

Zhu, J.-K., Bressan, R.A., & Hasegawa, P.M. (1993). Isoprenylation of the plant molecular chaperone ANJ1 facilities membrane association and function at high temperatures. Proc. Natl. Acad. Sci. USA 90, 8557–8561.

Zuber, M.X., Strittmatter, S.M., & Fishman, M.C. (1989). A membrane-targeting signal in the amino terminus of the neuronal protein GAP-43. Nature 341, 345–348.

Zurzolo, C., Lisanti, M.P., Caras, I.W., Nitsch, L., & Rodriguez-Boulan, E. (1993). Glycosylphosphatidylinositol-anchored proteins are preferentially targeted to the basolateral surface in Fischer rat thyroid epithelial cells. J. Biol. Chem. 6, 1031–1039.

Zurzolo, C., van't Hof, W., van Meer, G., & Rodriguez-Boulan, E. (1994). VIP21/caveolin, glycosphingolipid clusters and the sorting of glycosyl phosphatidylinositol-anchored proteins in epithelial cells. EMBO J. 13, 42–53.

RECOMMENDED READINGS

Lipid Modification of Proteins: A Practical Approach (1991). (Hooper, N.M. and Turner, A.J., Eds.) IRL Press, Oxford.

Lipid Modifications of Proteins (1992). (Schlesinger, M.J., ed.) CRC Press, Boca Raton, Florida.

Chapter 5

Influence of Diet Fat on Membranes

MICHAEL THOMAS CLANDININ

Principles of Medical Biology, Volume 7A
Membranes and Cell Signaling, pages 93–119.

ISBN: 1-55938-812-9

INTRODUCTION

Interest in membrane models has considerably advanced understanding of membrane structure and its relationship to membrane function as originally proposed (Singer et al., 1972) and reviewed (Stubbs et al., 1984). It has become generally accepted that biological membranes are not of constant composition but are changing, dynamic and responsive structures in terms of membrane constituents (McMurchie et al., 1988; Clandinin et al., 1991). This extensive diversity results in differences among cell types and among similar membrane types of different cells. The complexity characteristic of the biological membrane, complicated by our still incomplete window of knowledge, makes it difficult to definitively describe how specific changes in membrane structure or within individual structural constituents result in step-by-step mechanisms altering membrane-dependent functions. For example, it is hard to envisage within the heterogenous mixture of species of an individual membrane phospholipid how change in the fatty acid species or positional specificity of fatty acids in the phospholipid molecule alters function. As most sites of metabolic regulation are to some degree membrane-dependent or at least membrane-associated, the perplexing nature of interpreting the metabolic implications of a change in membrane composition is apparent. Thus, it is not yet possible to define membrane composition that is most desirable in terms of membrane functions.

In setting the focus for this chapter, topics detailing the relationship of membrane physical chemistry to function will be omitted. This chapter will focus on the role of dietary fat in the nutritionally adequate diet to demonstrate, through the use of several physiological examples, that dietary fat is a normal determinant of membrane structure and thus is a constant modulator of the biological activity of subcellular membranes and processes that may be regulated through membranes. In so doing, emphasis will be placed on the *in vivo* approach. Few *in vivo* studies of the impact of diet fat on membrane structure and function exist for man. However, from studies in animal models, where diets that reflect the normal variation in human diet fat intake have been assessed, it is possible to draw analogous physiological comparisons.

THE MEMBRANE MODEL

The initial fluid mosaic model proposed for the cell membrane bilayer (Singer et al., 1972) advanced the concept of biological membranes resulting in recognition of the dynamic nature of the membrane and that specific organizational heterogeneity serves functional and structural purposes as well as conductive roles. Initial notions of random organization of membrane protein and lipid have been extensively revised by recognition that domains exist in membranes where lipid-protein and lipid-lipid interactions are highly specific and organized to provide the structural-functional characteristics of a particular membrane. It is now generally accepted that the structure of the lipid bilayer is imparted by the hydrophobic effect of the amphipathic structure of polar membrane lipids. The lipid bilayer is anisotropic with both well-ordered regions and lipid-crystal-like regions (Yeagle, 1989).

Membrane-Lipid

The composition of lipids in individual cell membranes is characteristic of the membrane type. Although membrane polar lipid composition and membrane cholesterol content is carefully regulated by the cell, the content of these constituents also varies with the cell cycle or age and in response to a variety of stimuli or changes in environment and physiological state (for example, with diet and in disease states) (Clandinin et al., 1991). Clearly, as membrane phospholipid and cholesterol content may regulate activity of individual membrane proteins (Yeagle, 1989; Clandinin et al., 1991), this type of transition in membrane composition will have functional consequences.

Most biological membranes contain a variety of polar lipids (phosphatidylcholine, phosphatidylethanolamine, phosphatidylserine, phosphatidylinositol, sphingomyelin, and cardiolipin in mitochondria), some of which may be in alkylacyl or glycosylated forms. In many membranes individual phospholipid types are present in a wide variety of species, some of which predominate. As many as 40 different fatty acids may be incorporated into the sn-1 or sn-2 position of the phospholipid molecule (Kuksis, 1978). This vast array of diversity in the structural constituents present causes a variety of analytical complexities that normally results in assessment of only a few, often simple, structural characteristics when change in membrane structure is compared with change in membrane function. Measurement of properties and components of the bulk membrane lipid isolated will need to become much more specific to reveal any kind of integrated perspective of how the heterogenous mixture of phospholipid species interacts with membrane protein to alter biological functions.

Phosphoinositides and sphingomyelins have specific roles as messenger molecules. For example, sphingomyelin is widely distributed in mammalian cells. Fatty acid composition of sphingomyelin is relatively resistant to dietary fat alterations but membrane content of sphingolipids may not be. Recent studies have shown that sphingomyelin and other sphingolipids play a role in cellular functions (Hannun et al., 1987). Sphingosine, a breakdown product of sphingomyelin, ceramides, and other glycolipids, is a potent inhibitor of protein kinase C and cellular events dependent on this enzyme. The relatively higher concentration of sphingolipids in plasma membrane is also consistent with the concept that the breakdown product of sphingolipids (sphingosine) may be involved in signal transduction (Merrill and Stevens, 1989) in a manner that responds to dietary control. This area for potential effect of diet on subcellular regulation has not been extensively explored.

Understanding subcellular sites for synthesis of membrane lipids and targeting of these constituents to membrane sites could form the basis for another review.

Although phospholipid synthesis occurs for membrane assembly, it is also synthesized and modified extensively *in situ* in some cases to provide a role in second-messenger pathways in response to specific stimuli e.g., regulation of protein kinase by diacylglycerol (Kikkawa et al., 1989) and sphingomyelin (Hannun and Bell, 1989) and perhaps to provide molecular species for special purposes (Hargreaves and Clandinin, 1989). Thus research on membrane lipids will need to differentiate between assessment of new membrane synthesis (which may appear quantitatively most important) and the dynamic processes involved in routine maintenance of a wide variety of subcellular membrane lipid pools undergoing change in relation to regulation of subcellular events (which may be most significant in terms of understanding relationships to function).

Membrane-Protein

Membrane proteins may be structural elements or other proteins bound to the charged surface of the bilayer (peripheral) or inserted to varying degrees in the bilayer (integral). Lateral mobility is believed to be dependent on the fluid properties of the membrane lipid and the size of the protein or protein aggregate. Ability of integral proteins to move laterally produces topographic rearrangement of the cell surface as a mechanism for regulation of specific cell-surface properties (e.g., uptake of low-density lipoprotein receptor complex). Transmembrane transport represents another aspect of molecular motion applying to a variety of molecules in addition to lipids. The rate at which lipid molecules shift between bilayer surfaces, i.e., flip-flop, is apparently slower than lateral diffusion of lipid or protein and is a mechanism for maintenance of sidedness or asymmetry as well as membrane stability. Molecular motion in membranes also includes movement of the fatty acyl chains of membrane phospholipids. This type of movement may involve deacylation-reacylation of phospholipid molecules *in situ* and is a mechanism for membrane fluidity (Kimelberg, 1977).

Cytoskeletal protein elements differ between cell types (Nelson and Lazarides, 1984; Bennett, 1985). The role of cytoskeletal proteins in maintenance or ordering of membrane structure and the dynamic state is not well understood nor has the involvement of cytoskeletal elements in maintenance of subcellular organelle structure and transport of organelle components been well defined. This area of cell biology is currently evolving rapidly and has direct implications for many disorders of subcellular metabolism.

It has been estimated that cell membrane protein content is extensive with only a few layers of lipid providing the separation between membrane proteins for nearest neighbors of nonaggregated proteins (Yeagle, 1989). It is likely that this spatial separation and interstitial packing with membrane lipid may also be very dynamic, responding to the physiological factors that control or result in insertion and removal of proteins from the membrane. By separating integral proteins from the native membrane environment, it has been shown that some membrane proteins have an absolute requirement for membrane lipid (e.g., β-hydroxybutyrate dehydrogenase, Na^+-K^+-ATPase, cytochrome oxidase, and insulin receptor, Yeagle, 1989). How this requirement is met or variably expressed in the apparently heterogeneous membrane lipid mixture in the native biological membrane surrounding the protein is far from clear.

In human epithelial cell membranes, impermeability to Cl^- alters incorporation of fatty acids (e.g., C18:2ω6) into membrane constituents (Kang et al., 1992). Thus it is also clear that anion transport plays a role in fatty acid incorporation into cell membrane phospholipids, suggestive of a relationship between the function of membrane anion transport channels and membrane lipid composition.

Maintenance of Membrane Order

Several forms of noncovalent and covalent forces are involved in the interaction of membrane protein with membrane lipid. The specificity of these interactions in terms of a functional protein requiring a specific lipid (summarized by Clandinin et al., 1991, and Yeagle, 1989) has led to the concept of an annulus of lipid providing a specific microenvironment around hydrophobic regions of functional membrane proteins. Lipid polar head group specificity for membrane-bound enzymes supports this concept of specific lipid affinities and sequestration of lipid. This micropolymorphism of membrane structure would enable the lipid to independently influence membrane proteins, permitting precise control of membrane functions and enabling the protein to retain integrity of specific functions while moving laterally through a heterogeneous lipid environment. Tight or close association of a proportion of membrane lipids with functional proteins would provide for sequences of interacting enzymes of multienzyme arrays. Although this concept explains why some lipid dependent functions may not respond to change in bulk lipid properties such as fluidity, it does not provide the overall mechanism for membrane order and the fundamental answer to why cell membranes and organelle membranes have a characteristic shape and domain with typical phospholipid composition, site-specific placement (e.g., sidedness) and a distinctive structural morphology. The answer to these questions will require determination of subcellular traffic in phospholipid and specific peptides to reveal how both membrane elements are placed in specific regions in the cell and the manner by which the cell determines the overall heterogeneity and polarity of this traffic.

Spatial control of many membrane proteins is essential for normal cellular functions. For example, the specificity of G protein-mediated signal transduction may depend on colocalization of membrane receptor, G protein, and/or effector (Neer and Clapham, 1988). In this regard, eukaryotic cells contain cytoskeletal elements that are continuous with a distinct subcortical membrane skeleton, which apparently mediates attachment of the cytoskeleton to integral proteins in the plasma membrane. The accepted model for this interaction involves binding of the cytoplasmic domains of integral membrane proteins to specific components of the cytoskeletal network (summarized by Bennett, 1990). A conceptually different model that provides a tentative framework for understanding protein-phospholipid interactions requires direct associations between the membrane skeleton and a relatively small number of integral membrane proteins. Through the mechanisms outlined above, these membrane proteins could preferentially associate with an annulus of specific phospholipids resulting in spatially restricted membrane subdomains. Conformational changes in other membrane proteins (perhaps due to ligand binding) could then result in preferential association with a different phospholipid subdomain. This might enable spatially restricted changes in protein activity that can be further regulated by altering the relative sizes or region of specific subdomains (as might be induced by change in dietary fat intake). Thus, one could hypothesize that microtubular or cytoskeletal elements may also interact with membrane lipid components, providing form or order to the phospholipid array in the membrane thus altering metabolism. With regard to these models, assembly of membrane components and domains through cytoskeletal elements has been demonstrated (Mahaffey et al., 1990; Pumplin and Bloch, 1990). Analysis of membrane skeletal protein subtypes in different cell types and regions within the cell during cell differentiation may provide the key to how membrane order through membrane cytoskeletal domains may be determined.

DIET AS A DETERMINANT OF SUBCELLULAR STRUCTURE AND FUNCTION IN DIFFERENT ORGANS

Microbial auxotrophs alter membrane lipid composition in response to exogenous factors, resulting in a change in membrane-dependent enzyme activities and processes basic to existence of the cell. For animals capable of adaptive hypothermia (hibernators), reduction in body temperature to facilitate survival is associated with change in membrane composition and associated functions, reducing these functions to a new set point (Aloia et al., 1989). By analogy, for homeotherms normally having a constant high body temperature, extrinsic or dietary influence on membrane lipid fatty acyl tail and polar head group composition has only relatively recently been generally recognized as a consequential physiological mechanism for alteration of membrane structural lipid and thus membrane-dependent functions. This relationship with diet occurs as a consequence of the dietary essential nature of linoleic acid (C18:2ω6) and linolenic acid (C18:3ω3), and the fact that through *de novo* membrane phospholipid synthesis and acyl group turnover in membrane phospholipids new fatty acids of dietary origin can be incorporated into membrane lipids (Clandinin et al., 1991). Within the range of adequate nutritional status, change in the balance of fatty acids forming the dietary fat consumed results in change in membrane structural lipid constituents *in vivo* and in the activity of a wide variety of membrane factors and even some parameters of whole body energetics (Renner et al., 1979). For the illustrations that follow, the diet fats tested have compositional features in common with human fat consumption and are adequate in terms of known requirements for essential fatty acids. The ubiquitous nature and thus potential significance of these observations will be discussed.

Mechanisms by which diet fat has the potential to affect biological processes are many and varied. From the physiological perspective of the homeotherm, initial response to change in dietary fat intake occurs at the level for the enterocyte in the intestinal brush border. Adaptation to diet by the intestinal tract may buffer or modify composition of fatty acids delivered to other tissues and cell type for structural lipid synthesis or metabolism. Response to diet by other tissues may occur at several levels. The first and most obvious function to be identified has involved synthesis of structural lipids of altered fatty acid composition resulting in alteration in membrane functions by specific integral catalytic proteins controlling functions that may regulate biosynthesis of membrane components. Another level of biological function potentially altered by change in dietary fat composition involves change in hormone binding or responsiveness, thus altering either the magnitude of metabolism or synthesis of the cell's metabolic products that may affect activity of other cell types. Finally, as cellular activity is compartmented, it is logical to suppose that a variation in dietary fat intake that induces a change in membrane may also alter expression of nuclear function either by direct interaction with the promoter regions of specific genes or by changing receptor-mediated stimulation of gene transcription or by altering the rate at which gene products are transported out of the nucleus. These levels of interaction of a change in dietary fat intake with function will be illustrated.

Gastrointestinal Tract

Enterocyte

Dietary lipid composition alters intestinal lipid uptake, the lipid composition and transport and digestive function of the brush border membrane and basolateral membrane, and the activity of intracellular lipid synthetic enzymes (Garg et al., 1988a; Flores et al., 1990). Similarly, eicosanoid metabolism and lipid peroxidation may be affected by alterations in the lipid content of the diet. This has importance in gastrointestinal disease mechanisms such as inflammation and free radical induced tissue damage (Turini et al., 1991). The lymphatic export of lipoproteins and their subsequent lipolysis by LPL and LCAT can also be modulated by changes in dietary lipid composition (Grundy, 1983). These factors affect the availability of individual fatty acids to the rest of the body.

The relative availability of saturated and unsaturated fatty acids for phospholipid synthesis is determined by diet and activity of fatty acid elongating and desaturating enzymes (Figure 1). Δ^6-Desaturase converts C18:2ω6 to C18:3ω6 and is a rate-limiting step in synthesis of C20:4ω6 from C18:2ω6. C20:4ω6 is generally present in the *sn*-2 position of membrane phospholipid, plays an important role for membrane physicochemical properties, and is a common precursor for synthesis of pro-

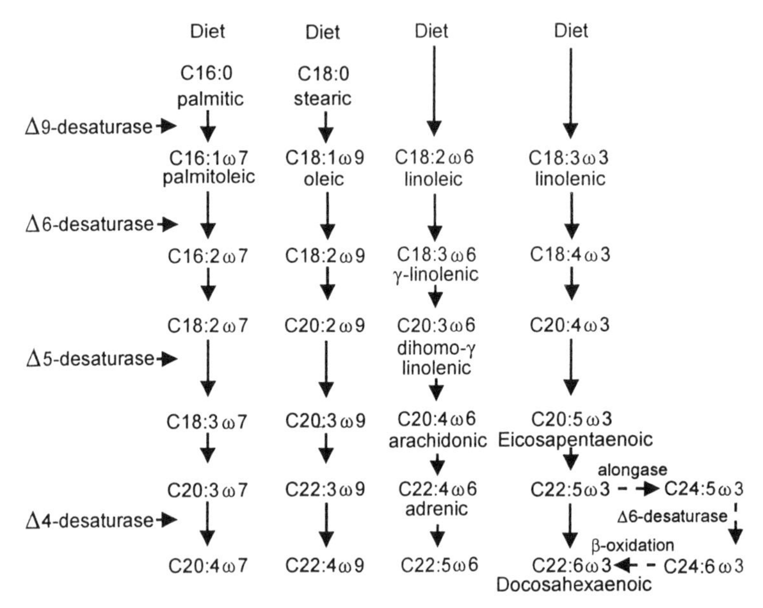

Figure 1. The formation of 22:5ω6 and 22:6ω3 is independent of Δ4-desaturase, alternatively through a retroconversion pathway. The Δ6-desaturase is considered to be a rate-limiting step in the biosynthetic pathway leading to the formation of 20:4ω6.

staglandin, prostacyclin, and thromboxanes. Desaturase enzymes have been documented in liver microsomes and have been demonstrated in the enterocyte (Garg et al., 1988b). By assessing the response of the enterocyte to overnight fasting and short-term transitions in dietary fatty acid intake, it can be also demonstrated that intestinal desaturase activity responds rapidly to normal physiological changes in food intake or the composition of dietary fats fed, thus affecting the fatty acid composition incorporated into enterocyte membrane and products of fat absorption depending on the region of the intestine examined.

Pathological processes are associated with changes in the composition of intestinal brush-border membranes (Thomson et al., 1987). The presence of desaturase enzymes in the enterocyte suggests that specific pathologically induced alteration in brush-border membrane fatty acid composition may be mediated by changes in desaturase activities, thereby altering the type of fatty acids available for brush-border membrane synthesis. The resulting alteration in brush-border membrane lipid composition plays an integral role in the functional changes occurring in the intestine by determining membrane physicochemical and transport properties (McMurchie, 1988). It is clear that adaptation to diet fat intake in the enterocyte modifies transport and metabolic activities of the cell, thus modifying products of absorption passed into the plasma and on to other tissues.

Colonocyte

To examine whether dietary fat alters membrane lipid composition and peroxidation of polyunsaturated fatty acids in "non-proliferative" and "proliferative" cells in the large intestine, Sprague-Dawley rats were fed diets providing a polyunsaturated-to-saturated fatty acid ratio of 1.2 or 0.3 at a high or low level of fat intake for a 25-day period (Turini et al., 1991). Cell populations were isolated and the effect of dietary fat on membrane polyunsaturated fatty acid content and peroxide levels was determined. Neither fat level nor fatty acid composition of diet-influenced total cholesterol, total phospholipids, and percentage of phospholipid classes in membrane phospholipids. Feeding the high-fat and/or high-polyunsaturated-to-saturated fatty acid ratio diet increased polyunsaturated fatty acid content of mucosal cell phospholipids. Membrane content of total saturated fatty acids was not significantly affected by diet. Variation in phospholipid fatty acid composition between "non-proliferative" and "proliferative" cells was observed. Animals fed high-fat diets, compared to groups fed low-fat diets, exhibited higher membrane peroxide levels when results are expressed as nmol/mg protein. Higher peroxide levels were observed in mucosal cells for rats fed high polyunsaturated-to-saturated fatty acid ratio diets when results were expressed per nmol of phospholipid. From these experiments, it is concluded that changes in fat level and fatty acid composition of the diet alters the mucosal cell membrane lipid composition in the rat large intestine and influences susceptibility of mucosal cell lipid to peroxidation. Further research is required to delineate which dietary factors—fat level, polyunsaturated-to-saturated fatty acid ratio, or both—have a primary influence on the degree of lipid peroxidation occurring in this cell type and their relationship to colonocyte inflammation and cancer.

Liver

Can Diet Alter the Structure and Function of Plasma Membranes?

Earlier studies examined the biosynthesis, composition, fatty acyl tail content, turnover or exchange, and change in polar lipids from liver plasma membranes with aging in rats fed diets of undefined fatty acid composition (reviewed by Clandinin et al., 1991). These observations were among the first to suggest that plasma membrane composition may vary with age, cell type and physiological state. The plasma membrane is a primary interface in the homeostatic balance between exogenous influences (e.g., diet) and endogenous control over the biosynthesis or utilization of varied substrates. It is of interest therefore to consider whether dietary fatty acid balance modulates the lipid composition of the plasma membrane, thereby having potential to modify control functions situated in this interface between circulating hormones and hormone-activated functions. Alteration of physical properties of this membrane may influence hormone receptor-mediated functions in a wide variety of cell types.

Glucagon-Stimulated Adenylyl Cyclase

In general, stimulation of adenylyl cyclase requires guanine nucleotides as well as ATP and magnesium. Adenylyl cyclase has a regulatory site, acting as a receptor on the exterior of the cell membrane, a catalytic site, which binds ATP and magnesium and generates cAMP on the interior of the cell membrane; and a coupler which connects the regulatory site to the catalytic site and may be the site of phospholipid action. Earlier studies of liver indicate that dietary fat level affects plasma membrane phospholipid fatty acyl tail composition, with a high-fat diet resulting in decreased glucagon-stimulated adenylyl cyclase activity (Neelands and Clandinin, 1983). By comparing nutritionally adequate diets in which oleate was substituted for linoleate, a dose-response relationship was shown to exist between the fatty acid content of the diet and fatty acyl tail composition of membrane phospholipids. This relationship between diet fat composition and membrane structure was paralleled by simultaneous alterations in glucagon-stimulated adenylyl cyclase activity. As the level of dietary oleic acid increased, the level of C18:1 and total monounsaturated fatty acids in phosphatidylcholine and phosphatidylethanolamine increased and was associated with a corresponding increase in glucagon-stimulated adenylyl cyclase activity (Neelands and Clandinin, 1983). Total ω6 unsaturated fatty acids in phosphatidylcholine correlated inversely with glucagon-stimulated adenylyl cyclase activity. Dietary linolenic acid level can also determine the degree of change in membrane fatty acyl tail composition and glucagon-stimulated adenylyl cyclase activity (Morson and Clandinin, 1986). As the dietary ratio of ω6/ω3 unsaturated fatty acids increased by increasing the C18:2ω6 level, the ratio of ω6/ω3 unsaturated fatty acids in the membrane decreased, altering adenylyl cyclase activity. It is apparent that alterations of the structural lipid constituents of the plasma membrane modify adenylyl cyclase activity, perhaps through conformational interaction with the catalytic unit due to change in the membrane microenvironment, analogous to that proposed for other integral membrane proteins (for example, the mitochondrial, ATPase Innis and Clandinin, 1981).

Clinically, currently available lipid emulsions for total parenteral nutrition may result in cholestasis after prolonged use. Recent research in animal models indicates that addition of 20:5ω3 to the lipid emulsion may prevent or reduce cholestasis (Van Aerde et al., 1993). This observation is accompanied by and may be attributed to change induced in plasma membrane phospholipids within the liver.

Nuclear Envelope

The nuclear envelope is a double membrane structure with both membranes interconnected and it controls nucleocytoplasmic exchange of macromolecules in eukaryotic cells (Herlan et al., 1979). Nucleotide triphosphatase, localized near the nuclear pore complex (Agutter, 1985), provides energy for nucleocytoplasmic translocation of ribonucleic acid. Nucleocytoplasmic transport of ribonucleic acid proceeds exclusively through nuclear pore complexes with lipid-clustering modulating opening and closing of the nuclear pore complex, thus regulating nucleocytoplasmic exchange of ribonucleic acid. Evidence indicates that significant regulation of gene expression occurs at the posttranscriptional nuclear level and most posttranscriptional controls are operated within the nucleus or at the nuclear envelope to regulate qualitative and quantitative flow of ribonucleoproteins to the cytoplasm. Phosphorylation of nuclear envelope components is also involved in regulation of nucleocytoplasmic transport of macromolecules and cell-cycle events. The nuclear envelope actively metabolizes polyphosphoinositides and this metabolism may be important in regulation of nuclear envelope function, NTPase activity and thereby NTPase-modulated RNA efflux (Smith and Wells, 1983).

In liver, the nuclear envelope also possesses binding sites for steroid and thyroid hormones on the nucleus, thereby mediating hormone action (Lefebvre et al., 1984; Venkatraman and Lefebvre, 1985). Thyroid hormones generally act by selective stimulation of gene transcription or pretranslational events. Binding of L-T_3 to the nuclear envelope is dependent on the lipid environment in terms of number of binding sites (Venkatraman et al., 1986). Insulin also binds to the nuclear envelope

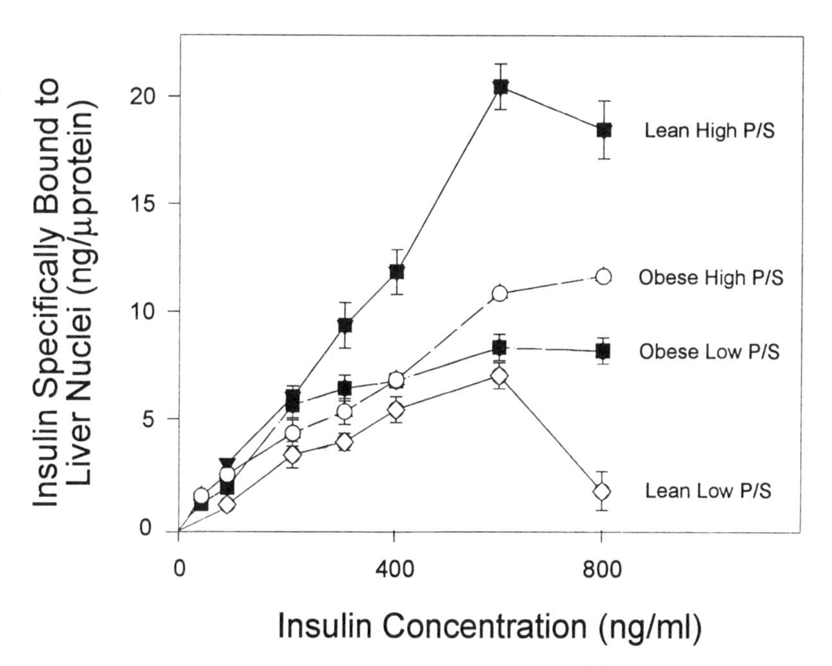

Figure 2. Insulin binding to the nuclear envelope of the ob/ob mouse. After consuming a diet high or low in polyunsaturated to saturated fatty acids for 4 weeks, liver nuclei were isolated and the specific binding of insulin was measured (Lefebvre et al., 1984).

(Cheema et al., 1991) in a manner that is markedly affected by diet fat composition and genetically determined physiologic state in the obese mouse (Figure 2) (Cheema et al., 1991). However, its function in this regard is unknown. Thus, it is likely that the nuclear envelope structural composition is directly associated with the control and expression of nuclear activity integral to cell function.

Phospholipid fatty acid profile of the nuclear envelope phospholipids can be altered in liver by dietary lipid composition, resulting in altered membrane function (Venkatraman and Clandinin, 1988). A specific gene difference in MRL/mp mice also alters nuclear envelope phospholipid fatty acid profile, nuclear envelope nucleoside triphosphatase activity, RNA efflux from isolated nuclei and binding of L-triiodothyronine (Cheema et al., 1991) to the liver nuclear envelope. Diet- and strain induced change in phospholipid fatty acid composition was reflected in change in nuclear envelope function. Lpr/lpr mice exhibit significantly higher NTPase activity and RNA efflux from isolated nuclei when compared with +/+ strain. Animals fed a diet high in linoleic acid exhibit higher NTPase activity and higher RNA efflux compared with animals fed a diet low in linoleic acid (Figure

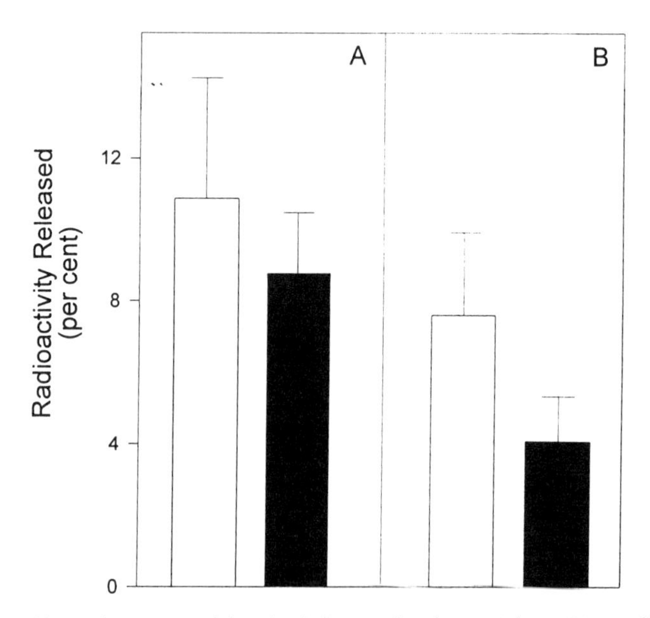

Figure 3. Effect of strain and level of dietary linoleic acid on RNA efflux from prelabeled nuclei. Prelabeled nuclear protein 500μg was incubated with 7 mg cytosolic protein at 30°C for 30 min. Values are mean ± SD for 5 determinations. RNA efflux from unlabeled nuclei was determined by measuring radioactivity in the RNA effluxed and then expressed as a percent of nuclear RNA effluxed. A) Lpr/lpr mice; B) +/+ mice; □, high linoleic acid diet; ■, low linoleic acid; statistical significance: diet: high vs. low linoleic acid = $P < 0.05$; strain: lpr/lpr vs. +/+ = $P < 0.001$.

3). Thus, genetically determined differences in the composition and function of the nuclear envelope are reflected in change in the expression of nuclear activity in liver.

Change observed in NTPase activity and RNA efflux from isolated nuclei could be explained in terms of alteration in physical characteristics of the nuclear envelope due to differences in phospholipid composition. Phospholipid fatty acid composition has also been reported to be altered in nuclei prepared from hepatoma cells (Lemenovskaya et al., 1976) which suggests that lipid dedifferentiation found in membranes of tumor cells also extends to the nuclear envelope.

It is relevant that the lpr/lpr autoimmune mouse develops immunologic abnormalities such as autoantibodies, immune complex, lymphoproliferative disorders, and accelerated aging. Our earlier work with lpr/lpr mice suggested that genetic background and dietary linoleic acid could also alter membrane fatty acid composition of immunologic cells, thereby changing the immunologic functions of these cells (Cinader et al., 1983; Tiwari et al., 1986). Whether these genetically determined physiological changes in expression of cellular activities are the consequence of altered nuclear envelope function and expression of nuclear activity is not known. It is also not clear which specific gene products may be stimulated by hormonal interactions at the level of the nuclear envelope and which gene products may be altered by diet. This type of interaction is likely to provide an important means by which dietary fat alters gene expression. It can be concluded that both genetic background and diet fat composition are important factors influencing phospholipid fatty acid composition of the nuclear envelope, thereby altering the characteristics and function of this membrane in the liver. If this mechanism operates in other tissues, it clearly has the potential to explain the heterogeneity in response to diet observed to date.

Lymphocyte Composition and Immune Suppression

Decrease in immunoresponsiveness and increase in the degree of incidence of autoimmunity are a concomitant of aging in man and mouse (Cinader, 1981). Resistance against experimentally acquired immunological tolerance is an experimental equivalent of autoimmunity and is found in most strains of inbred mice although at vastly varying rates of age-dependent progression (Cinader et al., 1981a,b). The changes of immune responsiveness and tolerance-inducibility are attributable to changes in the balance between various types of T-cells, B-cells and factors. Most mice which develop very early resistance against tolerance-induction later develop glomerulonephritis. Several genes are involved in a process which ultimately results in disease resembling lupus erythematosus. In SJL/J mice, resistance against tolerance is largely (Hosokawa and Cinader, 1983) due to a change in T-suppressor capacity. We have characterized various aspects of this change and have explored the possibility that a gene is being expressed, in this animal, which affects nucleic acid metabolism (Cinader, 1983) and has led to attempts to affect declining suppressor capacity by various nucleic acid analogues. By diet induced alterations in membrane composition, on the assumption that some of the observed functional changes in aging may be the consequence of age-dependent membrane changes, we have demonstrated that low P/S diets appear to lead to an increased immune response against heterologous gamma-globulin. The opposite effect is produced in animals which have been given the heterologous gamma-globulin in a tolerogenic form prior to immunization: the antibody response is lower in animals fed low P/S diets. Clearly, there is a profound effect of diet on the balance which controls immune responsiveness. It seems that specific suppressor capacity is affected by the membrane composition with a change in balance of lymphoid cell population produced by the high versus low P/S diet (Tiwari et al., 1988).

Adipocyte

Plasma Membrane

Little was known until recently (Field et al., 1988; Field et al., 1990) about the role of the adipocyte plasma membrane phospholipid and the influence of fat intake or disease state on the structure and function of this membrane. The insulin receptor is embedded in the plasma membrane bilayer and appears sensitive to the surrounding lipid environment. Altered microsomal rates of fatty acid Δ^6- and Δ^9- desaturation occur in experimental models of diabetes mellitus. Insulin therapy corrects this defect and partially normalizes altered membrane fatty acid composition (Faas and Carter, 1980).

On the premise that composition of adipocyte plasma membrane is an integral determinant of insulin-stimulated functions in the adipocyte, we designed *in vivo* feeding experiments to examine the effect of alterations in dietary fatty acid composition on the fatty acid composition of adipocyte membrane phospholipids, on insulin binding to its receptor in the plasma membrane, and on insulin-stimulated transport and utilization of glucose. Diets representing the physiological range of dietary fatty acid composition consumed by humans were fed. Control and diabetic rats were compared to assess whether increasing consumption of polyunsaturated fatty acids would normalize the decreased amounts of fatty acid and desaturase products expected in membranes in the diabetic state and improve insulin binding and insulin responsiveness. Diets differed only in the proportion of safflower oil and hydrogenated beef tallow to provide polyunsaturated/saturated fatty acid ratios (P/S) of from 0.20 (low P/S) up to 2.0 (high P/S). Linseed oil was added to obtain a total n-3 fatty acid content of 1% (w/w) of the diet fat. After feeding the diet for 21 days, diabetes was induced in animals in each diet by intravenous injection of streptozocin. Diabetic animals were continued on their respective diet treatment for an additional 14 days. Adipocyte plasma membrane was isolated, membrane phospholipids were separated, and fatty acid composition was analyzed (Field et al., 1990). Insulin binding, glucose transport, and glucose utilization were also assessed.

Diet significantly altered the fatty acid composition of the major adipocyte plasma membrane phospholipids in both control and diabetic animals (Field et al., 1990). Feeding a high polyunsaturated fat diet increased the total polyunsaturated fatty acid content and P/S ratio in all five membrane phospholipids analyzed from control animals. Consumption of the high P/S diet by control animals increased the content of C20:4ω6 in phosphatidylethanolamine, phosphatidylinositol, phosphatidylserine, and sphingomyelin, the content of C18:2ω6 in phosphatidylcholine and phosphatidylserine. In diabetic animals the tendency to higher content of polyunsaturated fatty acids in phospholipid for animals fed the high P/S diet was also evident.

Insulin Binding

For all insulin concentrations used to test insulin binding cells from control animals fed the high P/S diet (P/S = 1.0) bound significantly more insulin than cells from control animals fed the low P/S diet (P/S = 0.25). From Scatchard analysis of binding in control animals, increased insulin binding was associated with an increased number of available high-affinity, low-capacity insulin binding sites (Field et al., 1990). At low insulin concentrations (0.1–10 ng/ml), adipocytes from diabetic animals bound significantly more insulin than adipocytes obtained from control animals fed the low P/S diet (Field et al., 1990).

To test that a direct relationship exists between the P/S ratio of dietary fat consumed, the composition of the major adipocyte plasma membrane phospholipids, and insulin binding, animals were fed 1 of 10 diets varying in P/S ratio between 0.14 and 1.8. The response surfaces of the total specific bound insulin was calculated. A significant positive relationship was found between the P/S ratio of the diet and the amount of insulin bound at every insulin concentration (Figure 4). However, when the lower insulin levels within a smaller range of insulin values were examined, the amount of insulin bound increased to attain a plateau with increasing diet P/S ratio (Figure 4).

Relationship between Diet, Membrane Composition and Insulin Binding

Diet P/S ratio influenced content (% w/w) of major membrane fatty acids in adipocyte plasma membrane phospholipids. The content of several fatty acid constituents in phosphatidylcholine and most major fatty acids in phosphatidylethanolamine were related to both change in the diet P/S ratio and the amount of insulin bound. A three-dimensional relationship was obtained to illustrate the relationship between change in dietary P/S ratio, levels of fatty acid constituents in plasma membrane phospholipids, and insulin binding at a low concentration of insulin (1.2 ng/ml) (Figure 5). The amount of insulin bound increased as the content of polyunsaturated fatty acids in phosphatidylethanolamine increased. Insulin

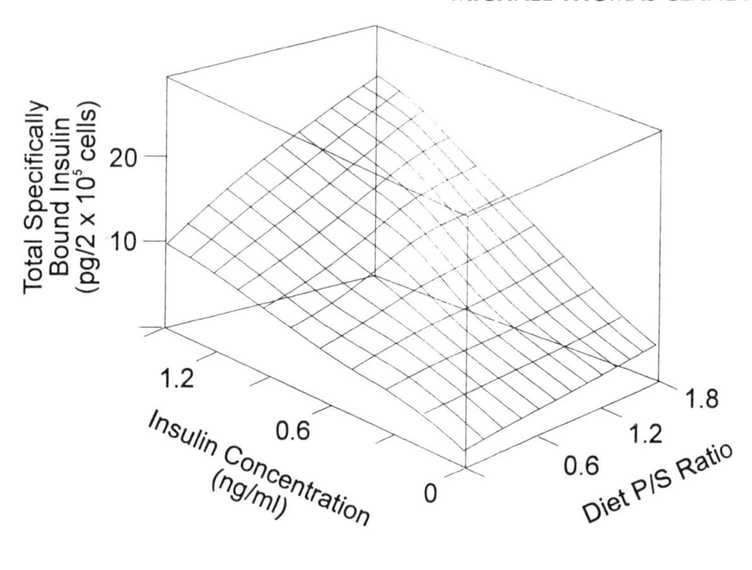

Figure 4. Relationship between the dietary P/S ratio and insulin binding to adipocytes. The response surfaces for the total specific bound insulin for animals fed the 10 experimental diets were calculated over the physiological range (0–1.5 ng/ml) of insulin concentration. Equations to describe the response surfaces for the insulin bound as a function of dietary P/S ratio and insulin concentration were developed (Field et al., 1990).

binding decreased with increasing content of saturated and monounsaturated fatty acids in phosphatidylethanolamine and C16:0 in phosphatidylcholine. Increase in dietary P/S ratio between the level normally consumed by the North American population (P/S = 0.35) and that recommended (P/S = 1.0) results in a remarkable increase in the membrane polyunsaturated fatty acid content (primarily ω6 fatty acids of phosphatidylethanolamine and this increase in polyunsaturated fatty acids is associated with increased binding of insulin to adipocytes (Figure 5).

Insulin Action

Diabetic cells transported less glucose than cells from either control group. At all insulin concentrations tested, adipocytes from control animals fed the high P/S diet transported significantly more glucose than control animals fed the low P/S diet (Field et al., 1990). The amount of glucose transported per insulin bound was reduced in diabetic animals. Diet influenced this relationship. For both control and diabetic animals fed the low P/S diet, glucose transport reached a maximum, whereas in animals fed the high P/S diet, glucose transport continued to increase as more insulin was bound.

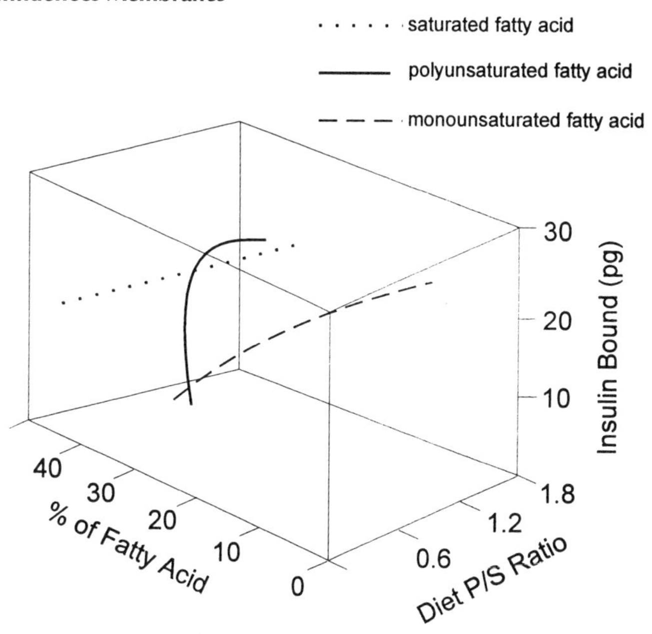

Figure 5. Relationship between dietary P/S ratio, fatty acid composition of phosphatidylethanolamine and insulin binding. Sprague-Dawley rats were fed 1 of 10 purified high-fat nutritionally adequate diets. The diets differed in the proportion of polyunsaturated to saturated fatty acids (P/S ratio) ranging from 0.14 to 1.8. These P/S ratios are typical of the range in dietary P/S ratio consumed by the North American population (abstracted from Field et al., 1990).

These physiological examples clearly demonstrate that feeding animals diets providing a fat content and fatty acid composition analogous to that consumed by segments of the North American population significantly alters the fatty acyl composition of the major phospholipids of the adipocyte plasma membrane in both control and diabetic animals. Characteristically, individual phospholipid classes respond to dietary fatty acid manipulation to different degrees. Although the molecular control mechanism through which dietary fat composition influences the fatty acyl composition of plasma membranes is unknown, differences in rate of synthesis of phospholipids *de novo* (Hargreaves and Clandinin, 1987a), redistribution of fatty acyl chains via phospholipase (Van den Bosch, 1980) or acyltransferases and methyltransferase (Hargreaves and Clandinin, 1987b), and direct desaturation of membrane phospholipid-linked fatty acids (Van den Bosch, 1980) have been demonstrated to be altered by change in the dietary fat consumed. Availability of fatty acids at specific sites of synthesis may also be influenced by dietary fat.

In adipocytes, specific diet-induced transitions in membrane changes insulin binding at what can be considered both physiological and supraphysiological insulin concentrations. The degree of change occurring in membrane polyunsaturated fatty acid content of phosphatidylethanolamine appears to be greater than that observed for other essential fatty acids, especially when related to the concomitant changes in insulin binding. Attaining a plateau for membrane polyunsaturated fatty acid content and the levelling off of monounsaturated fatty acid content of phosphatidylethanolamine for animals fed the higher P/S diet follows a similar trend to insulin binding, increasing the amount of glucose transported per insulin bound. This suggests that diet through change in membrane composition influenced coupling between the insulin receptor and the glucose transporter. Feeding a high P/S diet to diabetic animals tended to normalize this function. Molecular events subsequent to this hormone receptor interaction causing the metabolic action of insulin on glucose transport are unknown. However, due to contact with membrane lipids, several membrane-mediated events might be influenced by diet-induced alterations in the lipid bilayer. Recent research indicates that insulin-stimulated phosphorylation of the tyrosine residues on the β-subunit of the receptor and the activation of endogenous insulin receptor kinase are critical steps in insulin action that are altered by transitions in membrane composition, thus resulting in altered glucose utilization.

Muscle

In muscle feeding, high levels of dietary ω3 fatty acids in the form of fish oil alter muscle membrane composition, cause a marked inhibition of PGE_2 and $PGF_{2\alpha}$ synthesis, suppress rates of protein synthesis and degradation, and increase responsiveness of muscle glucose transport to insulin. Fish-oil feeding may thus have implications for many aspects of muscle metabolism. Diverse processes in skeletal-muscle metabolism are thought to be PG-dependent. These include increased blood flow in active and reactive hyperemia (Hill et al., 1990), inflammation subsequent to traumatic injury (Fisher et al., 1991), fusion and differentiation of myoblasts (Hausman and Berrgrun, 1987) and regulation of protein synthesis and degradation (Palmer, 1990). Rates of protein synthesis and degradation dictate whether skeletal muscle is in protein balance, or shows net protein deposition or protein loss. Both of these processes are highly regulated, and vary in a number of physiological and pathological states (Kettelhut ct al., 1988). The apparent ability of fish-oil feeding to suppress rates of muscle protein synthesis and degradation may have an impact on muscle growth and atrophy and remains to be tested.

Increased glucose transport by muscles of fish-oil fed rats may have practical implications for management of diabetes and other pathological states manifested by insulin resistance at the tissue level. Our results in this regard agree with those of Storlein et al. (1987, 1991), who also fed high levels of fish oil. The minimum level of dietary fish oil, as well as the necessary duration of feeding to observe this effect, remain to be identified. Feeding rats long chain fatty acids significantly increased the content of C22:5 ω3 and C22:6 ω3 in skeletal-muscle membrane phospholipids (Liu et al., 1994). The fatty acyl compositions of individual phospholipids were affected differently. Feeding rats high or low P/S diets did not significantly affect membrane phospholipid fatty acyl composition (Liu et al., 1994). Both dietary long-chain ω3 and polyunsaturated fatty acids can increase insulin binding to skeletal muscle sarcolemma, as a result of the increase in the $B_{max.}$ of the low-affinity high-capacity binding site. The effect of the long-chain ω3 fatty acids is more potent than that of ω6 polyunsaturated fatty acids.

It is clear that normal variation in dietary fat intake alters the composition of structural lipids found in subcellular membranes. These diet-induced alterations in membrane composition appear to influence insulin binding and action in adipocyte membranes, glucose transport and protein turnover in muscle, glucagon-stimulated adenylyl cyclase activity in liver, as well as hormone action in other cell types. Moreover, it is clear that feeding the animal a variation in dietary fat intake that enables normal growth (i.e., not deficient in essential fatty acids) alters a wide range of functions *in vivo* from many cell types examined to date (Clandinin et al., 1991). Studies of the plasma membrane from other cell types in brain, testes (reviewed in Garg et al., 1988c; Clandinin et al., 1991) and muscle, small bowel and colon (data not illustrated) clearly indicate that change in dietary fat composition in a nutritionally adequate diet alters plasma lipid membrane composition and a wide variety of metabolic functions central to the activity of these cell types. Because of the potential for modulation by diet of various analogous control mechanisms for metabolic regulation functioning at the plasma-membrane level throughout the body, it is clear that this relationship between diet fat and receptor-mediated function has significant implications in whole-body metabolism, and for maintenance of homeostatic mechanisms regulated by hormone receptors. The challenge for the current nutrition-oriented cell biologist will be to define how to beneficially manipulate cell functions by integrating dietary change with the mechanisms that control or predetermine chronic disease conditions that are apparently (at least in part) directly modulated through plasma membrane structure and function.

Brain

The brain is clearly sensitive to alteration of dietary lipid intake, even in a nutritionally complete diet, and dietary lipids fed during the early postnatal period are important determinants for structural and functional parameters of developing brain tissues. During this time, diet may also produce changes important for the response of neuronal tissue to challenges encountered later in the life cycle (i.e., aging, disease). Developmental processes involving membrane phospholipid biosynthesis and turnover, as well as the interaction of membrane lipids with lipid-dependent enzymes controlling metabolic events in the brain, may be influenced.

Critical Periods for Development and Composition

Accretion of polyunsaturated long chain metabolites of C18:2(9,12) and C18:3(9,12,15) in human brain has been documented (Friedman, 1980; Clandinin et al., 1980a,b). Biosynthesis of these fatty acids during brain development is not well established and it is not known whether fetal accretion of long chain metabolites of the essential fatty acids occurs as a result of fetal synthesis or of mechanisms involving maternal-placental synthesis and transfer. There is evidence for a maternal source (Naughton, 1981).

The source of these fatty acids for synthesis of neural tissues is critical to feeding infants as current infant formulas do not contain the long chain metabolites of essential fatty acids, whereas human milk provides these components. Some developmental evidence exists in animals (Jumpsen et al., 1994) for an impact of low levels of 20:4ω6 and 22:6ω3 in the diet fat on the composition of brain structural lipid in neuronal and glial cells. The effect of diet is affected by developmental age and is variable in different brain regions. In human infants, an effect of feeding small amounts of 22:6ω3 on functions of the retina have also been shown to be important (Carlson et al., 1993). Thus, it may be concluded that the composition of fats fed in early life are of consequence to the lipid composition and functions of neural tissues.

Brain Phospholipid Biosynthesis

Biosynthesis of phosphatidylcholine *de novo* involves two pathways, the phosphatidylethanolamine methyltransferase pathway and the CDP-choline pathway. It is possible that altering dietary fat could affect the synthesis of phosphatidylcholine through changes in cellular pools of metabolites that activate or inhibit the controlling enzyme, cytidylyltransferase (Vance and Choy, 1979). Dietary fat could also modulate phosphatidylcholine biosynthesis through the methylation pathway (Bremer and Greenberg, 1961). Low levels of methyl-transferase activity in brain microsomes have been demonstrated (Bremer and Greenberg, 1961). More recently, this methylation pathway in synaptosomal preparations appears to be mediated through levels of substrate or products of the reactions (Hoffman and Cornatzer, 1981). It has also been suggested that methyltransferase shows a preference for unsaturated fatty acyl species (Le Kim et al., 1973), with the major products of this pathway being highly unsaturated phosphatidylcholines (Trewhella and Collins, 1973; Salerno and Beeler, 1973). Thus, a mechanism is suggested whereby dietary fat could alter the fatty acid of phosphatidylethanolamine, which in turn could determine the extent of methylation of phosphatidylethanolamine to phosphatidylcholine.

Brain, like other excitable tissue, has characteristically high levels of long-chain polyunsaturated fatty acids, particularly C22:6ω3. The importance of polyunsaturates in brain membranes is emphasized by the fact that changes occurring in essential fatty acid deficiency maintain the overall membrane content of polyunsaturates, despite reductions in the proportion of fatty acids of the linoleic and linolenic acid series (Alling et al., 1972). Feeding a diet containing soybean oil providing ω6 and ω3 fatty acids versus a sunflower oil diet rich in ω6 fatty acids and providing 0.2% ω3 fatty acids produces higher levels of ω3 fatty acids in phosphatidylcholine of microsomal and synaptic plasma membranes and a lower ω6 to ω3 ratio in constituent fatty acids. Phosphatidylcholine ω3 fatty acid levels increased in microsomal membranes of animals fed diets containing 20% (w/w) fat for 4 weeks, compared with weanling animals. Levels of ω3 fatty acids in synaptic plasma membrane were lower for these animals when compared with weanling animals. These observations demonstrate a membrane specific response to nutritional factors during development. It is evident that there is considerably greater turnover than previously conceived, even in so-called stable myelin fractions (Gould and Dawson, 1976). Estimated half-lives are less than 20 days for phosphatidylethanolamine, ethanolamine plasmalogen and phosphatidylcholine (Jungalwala and Dawson, 1971). Diet has also been shown to alter phospholipid fatty acid composition in brain membranes within 24 days (Foot et al., 1982). Studies of brain *in vivo* involving labeled precursors indicate rapid labeling of subcellular membrane lipid fractions in the order of microsomes > mitochondria > myelin. The extent of transfer of lipid from the endoplasmic reticulum to other subcellular fractions, transport of phospholipids down the axon to the nerve terminal, and/or intracellular exchange of lipid is not known. The control mechanisms regulating dynamic changes in levels and composition of phospholipids during brain development are also not understood, but presumably relate to precursor availability and levels of modulation of enzyme activity.

The CDP-choline pathway for phosphatidylcholine biosynthesis is the quantitatively major route for phosphatidylcholine synthesis in brain, with the phosphatidylethanolamine methyltransferase pathway contributing to the total amount of phosphatidylcholine synthesized *de novo*. Phosphocholinetransferase is a lipid-dependent membrane-bound enzyme catalyzing synthesis of phosphatidylcholine from CDP-choline and a diglyceride. Activity of phosphocholinetransferase in the microsomal membrane of weanling rat brain approaches that of adult rat liver, demonstrating the great importance of neuronal phospholipid biosynthesis in the weanling animal (Hargreaves and Clandinin, 1987a). A similar observation is made for phosphatidylethanolamine methyltransferase activity in the synaptic plasma membrane of weanling animals (Hargreaves and Clandinin, 1987b). Higher levels of phosphatidylethanolamine methyltransferase activity observed in synaptic plasma membranes versus microsomal membranes suggests a specific function for phosphatidylcholine produced via this pathway at this subcellular site. The pool of phosphatidylcholine synthesized via the phosphatidylethanolamine methyltransferase pathway also has a fatty acid profile that is distinct from the bulk membrane phosphatidylcholine pool (Strittmatter et al., 1979) and has a rapid metabolic rate of turnover (Mogelson and Sobel, 1981). The phosphatidylethanolamine methyltransferase pathway has been implicated in specific regulatory processes and in processes of neurotransmitter synthesis and release (Mozzi et al., 1982; Blustajn and Wurtman, 1984). Substrate preference by methyltransferases of the phosphatidylethanolamine methyltransferase pathway for polyunsaturated species of phosphatidylethanolamine may contribute to the polyunsaturated pool of phosphatidylcholine produced via this pathway (Hargreaves and Clandinin, 1987a,b). Thus, it is conceivable that nutritional regulation of distinct pools of phosphatidylcholine within the membrane may have significant implications for regulation of varied enzyme-linked cellular events.

Synthesis of Neurotransmitters and Behavior

The interaction between diet, membrane polar head group content, cholesterol concentration, phospholipid fatty acid composition, and function of integral membrane proteins is likely to be extremely complex. In brain, as many enzymes of neurotransmitter metabolism are lipid dependent, it is logical to postulate a possible interaction between dietary lipid and brain neurotransmitter metabolism. These changes in brain membrane composition have been shown to alter the thermotropic behavior of acetylcholinesterase activity (Foot et al., 1983), but have not yet been associated with brain function.

In conclusion, these observations illustrate that in neural tissues dietary fat affects a complex metabolic pathway for phosphatidylcholine biosynthesis in a coordinated fashion. This may be important to aspects of development concerning choline metabolism, or regulatory processes dependent on signals from a changing milieu in the microenvironment of the plasma membrane. In this regard, the stimulatory effect of longer chain polyunsaturated essential fatty acid homologues in membrane phosphatidylethanolamine increases phosphatidylethanolamine methyltransferase and decreases CDP-choline activity. It is also concluded that dietary fat influences phosphatidylethanolamine composition in microsomal and synaptic plasma membrane of the brain. Increasing dietary fat ω6/ω3 ratio increases microsomal and synaptic plasma membrane ω6/ω3 fatty acid ratio in phosphatidylethanolamine fractions. Membrane phosphatidylethanolamine species containing one, four or six double bonds and the ratio of 22:5ω6/22:6ω3 in phosphatidylethanolamine containing four to six double bonds are also affected by the fatty acid composition of the diet. Increase in membrane ω6 fatty acid content is associated with increased phosphatidylethanolamine methyltransferase activity and decreased phosphocholine transferase activity, thus indicating a mechanism by which change in an exogenous factor (i.e., dietary fat intake) may alter brain phospholipid biosynthesis. It is provocative to ponder whether or not diet could be used to induce formation of membrane structures more resistant to specific insults that cause degeneration of brain structural material and/or degeneration of cholinergic neurons or to reverse degenerative changes occurring in neural membrane structure and function.

SUMMARY

The experiments reviewed herein establish that the phospholipid components of cell membranes are not static and can be compositionally altered by changes in the fatty acid composition of the nutritionally adequate diet. This realization has important implications for studies investigating mechanisms for control of membrane-related functions in the body and clearly indicates a basic conceptual pitfall in extrapolating from well-defined homogeneous model membrane systems to the membrane of the homeotherm. It is also apparent that manipulation of diet may serve as a tool to examine the physiological forces involved in forming and maintaining the polymorphic structure of the membrane and the relationship of structure to biological functions of membranes. On an applied level of nutrition it is apparent that manipulations of dietary fatty acid composition, within the range of fat intake consumed by humans, produces compositional differences in membrane structural elements in animal and human subjects and in membrane function in animal models. It is likely that relationships between diet fat and membrane structure-function will evolve as a fundamental basis for diet intervention in human disease processes.

REFERENCES

Agutter, P. (1985). In: Nuclear Envelope Structure and RNA Maturation (Smuckler, E.A. & Clawson, G.A., eds.), pp. 561–578, Alan R. Liss, New York.

Alling, C.B.A., Karlsson, I., & Svennerholm, L. (1972). The effect of dietary levels of essential fatty acids on growth of the rat. Nutr. Metab. 16, 38–50.

Aloia, R.C. & Raison, J.K. (1989). Membrane function in mammalian hibernation. Biochim. Biophys. Acta 988, 123–146.

Bennett, V. (1985). The membrane skeleton of human erythrocytes and its implications for more complex cells. Ann. Rev. Biochem. 54, 273–304.

Bennett, V. (1990). Spectrin: A structural mediator between diverse plama membrane proteins and the cytoplasm, Curr. Opin. Cell Biol. 2, 51–56.

Blustajn, J.K. & Wurtman, R.J. (1984). Alzheimer's disease: Advances in basic research and therapies (Wurtman, R.J., Corkin, S.H., & Growdon, J.H., eds.), pp. 183–198. Center for Brain Science and Metabolism Charitable Trust.

Bremer, J. & Greenberg, D.M. (1961). Methyl transferring enzyme system of microsomes in the biosynthesis of lecithin (phosphatidylcholine). Biochim. Biophys. Acta 46, 205.

Carlson, S.E., Werkman, S.H., Peeples, J.M., Cooke, R.J., & Tolley, E.A. (1993). Arachidonic acid status correlates with first year growth in preterm infants. Proc. Natl. Acad. Sci. USA 90, 1073–1077.

Cheema, S.K., Venkatraman, J.T., & Clandinin, M.T. (1991). Diet fat influences insulin binding to liver nuclei from lean and obese mice. FASEB J. 5,A1305 (abstr.)

Cinader, B. (1981). Antigens, vol. VI. (M. Sela., Ed.), Academic Press, New York.

Cinader, B., Amagai, T., Matsuzawa, T., & Nakano, K. (1981a). In: Recherches Biomedicales sur le Vieillissement, Cahiers de l'ACFAS 7, 5–33.

Cinader, B., Amagai, T., Matsuzawa, T., & Nakano, K. (1981b). In: Cellular and Molecular Mechanisms of Immunologic Tolerance (Hraba, T. & Hasek, M., eds), pp. 187–199, Marcel Dekker, New York.

Cinader, B. (1983). Aging and the immune system. Clin. Biochem. 16,113–124.

Cinader, B., Clandinin, M.T., Hosokawa, T., & Robblee, N.M. (1983). Dietary fat alters the fatty acid composition of lymphocyte membranes and the rate at which suppressor capacity is lost. Immunol. Lett. 6,331–337.

Clandinin, M.T., Chappell, J.E., Leong, S., Heim, T., Swyer, P.R., & Chance, G.W. (1980a). Intrauterine fatty acid accretion rates in human brain: implications for fatty acid requirements. Early Hum. Dev. 4/2, 121–129.

Clandinin, M.T., Chappell, J.E., Leong, S., Heim, T., Swyer, P.R., & Chance, G.W. (1980b). Extrauterine fatty acid accretion in infant brain: implications for fatty acid requirements. Early Hum. Dev. 4/2, 131–138.

Clandinin, M.T., Cheema, S., Field, C.J., Garg, M.L., Venkatraman, J., & Clandinin, T.R. (1991). Dietary fat: Exogenous determination of membrane structure and cell function. Invited review for FASEB J. 5, 2761–2769.

Faas, F.H. & Carter, W.J. (1980). Altered fatty acid desaturation and microsomal fatty acid composition in the streptozotocin diabetic rat. Lipids 15, 953–961.

Field, C.J., Ryan, E.A., Thomson, A.B.R., & Clandinin, M.T. (1988). Dietary fat and the diabetic state alter insulin binding and the fatty acid composition of adipocyte plasma membrane. Biochem. J. 253, 417–424.

Field, C.J., Ryan, E.A., Thomson, A.B.R., & Clandinin, M.T. (1990). Diet fat composition alters membrane phospholipid composition, insulin binding, and glucose metabolism in adipocytes from control and diabetic animals. J. Biol. Chem. 265, 11143–11150.

Fisher, B.D., Reid, D.C., & Baracos, V.E. (1991). Effect of systemic inhibition of prostaglandin synthesis on muscle protein balance after trauma in the rat. Can. J. Physiol. Pharmacol. 69, 831–836.

Flores, C.A., Brannon, P.M., Wells, M.A., Morrill, M., & Koldovsky, O. (1990). Effect of diet on triolein absorption in weanling rats. Am. J. Physiol. 258, G38–G44.

Foot, M., Cruz, T., & Clandinin, M.T. (1982). Influence of dietary fat on the lipid composition of rat brain synaptosomal and microsomal membranes. Biochem. J. 208, 631.

Foot, M., Cruz, T., & Clandinin, M.T. (1983). Effect of dietary lipids on synaptosomal acetylcholinesterase activity. Biochem. J. 211, 507.

Friedman, Z. (1980). Essential fatty acids revisited. Am. J. Dis. Child. 134, 397–408.

Garg, M.L., Wierzbicki, A.A., Thomson, A.B.R., & Clandinin, M.T. (1988a). Fish oil reduces cholesterol and arachidonic acid content more efficiently in rats fed diets containing low linoleic acid to saturated fatty acid ratio. Biochim. Biophys. Acta 962, 337–344.

Garg, M.L., Keelan, M., Thomson, A.B.R., & Clandinin, M.T. (1988b). Fatty acid desaturation in the intestinal mucosa. Biochim. Biophys. Acta 958, 139–141.

Garg, M.L., Sebokova, E., Thomson, A.B.R., & Clandinin, M.T. (1988c). Delta-6 desaturase activity in liver microsomes of rats fed diets enriched with cholesterol and/or omega-3 fatty acids. Biochem. J. 249, 351–356.

Gould, R.M. & Dawson, R.M.C. (1976). Incorporation of newly formed lecithin into peripheral nerve myelin. J. Cell Biol. 68, 480.

Grundy, S.M. (1983). Absorption and metabolism of dietary cholesterol. Ann. Rev. Nutr. 3, 71–96.

Hannun, Y.A. & Bell, R.M. (1987). Lysosphingolipids inhibit protein kinase C: Implications for the sphingolipidoses. Science 235, 670–674.

Hannun, Y.A. & Bell, R.M. (1989). Functions of sphingolipids and sphingolipid breakdown products in cellular regulation. Science 243, 500–507.

Hargreaves, K. & Clandinin, M.T. (1987a). Phosphocholinetransferase activity in plasma membrane: Effect of diet. Biochem. Biophys. Res. Commun. 145, 309–315.

Hargreaves, K.M. & Clandinin, M.T. (1987b). Phosphatidylethanolamine methyltransferase: Evidence for influence of diet fat on selectivity of substrate for methylation in rat brain synaptic plasma membranes. Biochim. Biophys. Acta 918, 97–105.

Hargreaves, K. & Clandinin, M.T. (1989). Coordinate control of CDP-choline and phosphatidylethanolamine methyltransferase pathways for phosphatidylcholine biosynthesis occurs in response to change in diet fat. Biochim. Biophys. Acta 1001, 262–267.

Hausman, R.E. & Berrgrun, D.A. (1987). Prostaglandin binding does not require direct cell-cell contact during chick myogenesis *in vitro*. Exp. Cell Res. 168, 457–462.

Herlan, G., Giese, G., & Wunderlich, F. (1979). Influence of nuclear membrane lipid fluidity on nuclear RNA release. Exp. Cell. Res. 118, 305–309.

Hill, M.A., Davis, M.J., & Meininger, G.A. (1990). Cyclooxygenase inhibition potentiates myogenic activity in skeletal muscle arterioles. Am. J. Physiol. 258, H127–H133.

Hoffman, D.R. & Cornatzer, W.E. (1981). Microsomal phosphatidylethanolamine methyltransferase: Some physical and kinetic properties. Lipids 16, 533–540.

Hosokawa, T. & Cinader, B. (1983). Polymorphism of T- and B- cell sensitization by aggregate-freed heterologous gamma-globulin. Immunol. Lett. 6, 129–136.

Innis, S.M. & Clandinin, M.T. (1981). Mitochondrial membrane polar head group composition is influenced by diet fat. Biochem. J. 198, 231–234.

Jumpsen, J., Clandinin, M.T., & Lien, E. (1994). Change in dietary fatty acid composition alters 20:4(6) (AA) and 22:6(3) (DHA) content of phosphatidylcholine (PC) in neurons and glia during rat brain development. Canadian Federation of Biological Societies Annual Meeting, Montreal, P.Q., June 16–18, 1994.

Jungalwala, F.B. & Dawson, R.M.C. (1971). The turnover of myelin phospholipids in the adult and developing rat brain. Biochem. J. 123, 683.

Kang, J.X., Man, S.F.P., Brown, N.E., Labrecque, P.A., & Clandinin, M.T. (1992). The chloride channel blocker anthracene 9-carboxylate inhibits fatty acid incorporation into phospholipid in cultured human airway epithelial cells. Biochem. J. 285, 725–729.

Kettelhut, I.C., Wing, S.S., & Goldberg, A.L. (1988). Endocrine regulation of protein breakdown in skeletal muscle. Diabetes/Metab. Rev. 4, 751–772.

Kikkawa, U., Kishimoto, A., & Nishizuka, Y. (1989). The protein kinase C family: Heterogeneity and its implications. Ann. Rev. Biochem. 58, 31–44.

Kimelberg, H.K. (1977). The influence of membrane fluidity on the activity of membrane bound enzyme in dynamic aspects of cell surface organization. In: Cell Surface Reviews (Poste, G. & Nicolson, G.L., eds.), vol. 3, pp. 205–293. Elsevier/North Holland Biomedical, New York.

Kuksis, A. (ed.) (1978). Handbook of Lipid Research, vol. 1. Fatty Acids and Glycerides, Plenum, New York.

Le Kim, D., Betzing, H., & Stoffel, W. (1973). Studies *in vitro* and *in vivo* on methylation of phosphatidyl-N-N-dimethylethanolamine to phosphatidylcholine in rat liver. Hoppe Seylers Z. Physiol. Chem. 354, 437–444.

Lefebvre, T.A., Howell, G.M., & Goldsteyn, E.J. (1984). In: Regulation of Androgen Action (Bruchovsky, N., Chapdelaine, A., & Neumann, F., eds.), The Proceedings of an International Symposium, pp. 155–158. Congressdruck R. Bruckner, F.R.G., Montreal, Canada.

Lemenovskaya, A.F., Koen, Y.M., Perevoshchikova, K.A., Zbarski, I.B., Dyatlovitshay, E.V., & Ber-gel'son, L.D. (1976). Phospholipid composition of rat-liver and hepatoma 27 nuclear-membrane and nuclei. Biokhimiya 41, 818–821.

Liu, S., Baracos, V.E., Quinney, H.A., & Clandinin, M.T. (1994). Dietary ω-3 and polyunsaturated fatty acids modify fatty acyl composition and insulin binding in skeletal-muscle sarcolemma. Biochem. J. 299, 831–837.

Mahaffey, D.T., Peeler, J.S., Brodsky, F.M., & Anderson, R.G.W. (1990). Clathrin-coated pits contain an integral membrane protein that binds the AP-2 subunit with high affinity. J. Biol. Chem. 265, 16512–16520.

McMurchie, E.J. (1988). Physiological regulation of membrane fluidity. In: Advances in Membrane Fluidity (Aloia, R.C., Curtain, C.C., & Gordon, M.L., eds.), vol. 3, pp. 189–237. Alan R. Liss, New York.

Merrill, A.H., Jr. & Stevens, V.L. (1989). Modulation of protein kinase C and diverse cell functions by sphingosine—A pharmologically interesting compound linking sphingolipids and signal transduction. Biochim. Biophys. Acta 1010, 131–139.

Mogelson, S. & Sobel, B.E. (1981). Ethanolamine plasmalogen methylation by rabbit myocardial membranes. Biochim. Biophys. Acta 666, 205–211.

Morson, L.A. & Clandinin, M.T. (1986). Diets varying in linoleic and linolenic acid content alter liver plasma membrane lipid composition and glucagon-stimulated adenylate cyclase activity. J. Nutr. 116, 2355–2362.

Mozzi, R., Siepi, D., Adreoli, V., & Porcellati, G. (1982). Biochemistry of SAM and related compounds (Usdin, E., Borchardt, R.T., Creveling, C.R., eds.), pp. 129–138. MacMillan Press, New York.

Naughton, J.M. (1981). Supply of polyenoic fatty acids to the mammalian brain: The ease of conversion of the short-chain essential fatty acids to their longer chain polyunsaturated metabolites in liver, brain, placenta and blood. Int. J. Biochem. 13, 21–32.

Neelands, P.N. & Clandinin, M.T. (1983). Diet fat influences liver plasma-membrane composition and glucagon-stimulated adenylate activity. Biochem. J. 212, 573–583.

Neer, E.J. & Clapham, D.E. (1988). Roles of G protein subunits in transmembrane signalling. Nature 333, 129–134.

Nelson, W.J. & Lazarides, E. (1984). Assembly and establishment of membrane cytoskeleton domains during differentiation: Spectrin as a model system. In: Cell Membranes: Methods and Reviews (Elson, E., Frazier, W., & Glazer, L., eds.), vol. 2, pp. 219–246. Plenum, New York.

Palmer, R.M. (1990). Prostaglandins and the control of muscle protein synthesis and degradation. In: Prostaglandins, Leukotrienes, Essential Fatty Acids 39, 95–104.

Pumplin, D.W. & Bloch, R.J. (1990). Clathrin-coated membrane: A distinct membrane domain in acetylcholine receptor clusters of rat myotubes. Cell Motil. Cytoskeleton 15, 121–134.

Renner, R., Innis, S.M., & Clandinin, M.T. (1979). Effects of high and low erucic acid rapeseed oils on energy metabolism and mitochondrial function of the chick. J. Nutr. 109, 378–387.

Salerno, D.M. & Beeler, D.A. (1973). The biosynthesis of phospholipids and their precursors in rat liver involving de novo methylation and base-exchange pathways, *in vivo*. Biochim. Biophys. Acta 326, 325.

Singer, S.J. & Nicolson, G.L. (1972). The fluid mosaic model of the structure of cell membranes. Science. 175, 720–731.

Smith, D.C. & Wells, W.W. (1983). Phosphorylation of rat liver nuclear envelopes. II. Characterization of *in vitro* lipid phosphorylation. J. Biol. Chem. 258, 9368–9373.

Storlein, L.H., Kraegen, E.W., Chisholm, D.J., Ford, G.L., Bruce, D.G., & Pascoe, W.S. (1987). Fish oil prevents insulin resistance induced by high-fat feeding in rats. Science 237, 885–888.

Storlein, L.H., Jenkins, A.B., Chisholm, D.J., Pascoe, W.S., Khouri, S., & Kraegen, E.W. (1991). Influence of dietary fat composition on development of insulin resistance in rats. Relationship to muscle triglyceride and omega-3 fatty acids in phospholipid. Diabetes 40, 280–289.

Strittmatter, W.J., Hirata, F., & Axelrod, J. (1979). Phospholipid methylation unmasks cryptic beta-adrenergic receptors in rat reticulocytes. Science 204, 1205.

Stubbs, C.D. & Smith, A.D. (1984). The modification of mammalian membrane fluidity and function. Biochim. Biophys. Acta 779, 89–137.

Thomson, A.B.R., Keelan, M., & Clandinin, M.T. (1987). Onset and persistence of changes in intestinal transport following dietary fat manipulation. Lipids 22, 22–27.

Tiwari, R.K., Venkatraman, J.T., Cinader, B., Flory, J., Wierzbicki, A., Goh, Y.K., & Clandinin, M.T. (1988). Influence of genotype on the phospholipid fatty acid composition of splenic T and B lymphocytes in MRL/MpJ-lpr/lpr mice. Immunol. Lett. 17, 151–158.

Tiwari, R.K., Clandinin, M.T., & Cinader, B. (1986). Effect of high/low dietary linoleic acid feeding on mouse splenocytes: modulation by age and influence of genetic variability. Nutr. Res. 6, 1379–1387.

Trewhella, M.A. & Collins, F.D. (1973). Pathways of phosphatidylcholine biosynthesis in rat liver. Biochim. Biophys. Acta 296, 51–61.
Turini, M.E., Thomson, A.B.R., & Clandinin, M.T. (1991). Lipid composition and peroxide levels of mucosal cells the rat large intestine in relation to dietary fat. Lipids 26, 431–440.
Van den Bosch, H. (1980). Intracellular phospholipase A. Biochim. Biophys. Acta 604, 191–246.
Van Aerde, J., Duerkson, D., Chan, G., Thomson, A.B.R., & Clandinin, M.T. (1993). Physiologically modifying the fatty acid (FA) composition of intravenous (IV) lipid emulsions to simulate milk FA reduces cholestasis by total parental nutrition (TPN). Canadian Pediatric Society Annual Meeting, Vancouver, B.C., June 1993.
Vance, D.E. & Choy, P.C. (1979). How is phosphatidylcholine biosynthesis regulated? Trends Biochem. Sci. 4, 145.
Venkatraman, J.T. & Lefebvre, Y.A. (1985). Multiple thyroid hormone binding sites on rat liver nuclear envelopes. Biochem. Biophys. Res. Comm. 132, 35–41.
Venkatraman, J.T., Lefebvre, Y.A., & Clandinin, M.T. (1986). Diet fat alters the structure and function of the nuclear envelope: modulation of membrane fatty acid composition, NTPase activity and binding of triiodothyronine. Biochem. Biophys. Res. Commun. 135, 655–661.
Venkatraman, J.T. & Clandinin, M.T. (1988). Ribonucleic acid efflux from isolated mouse liver nuclei is altered by diet and genotypically determined change in nuclear envelope composition. Biochim. Biophys. Acta 940, 33–42.
Yeagle, P.L. (1989). Lipid regulation of cell membrane structure and function. FASEB. J. 3, 1833–1842.

RECOMMENDED READINGS

Cinader, B., Clandinin, M.T., Hosokawa, T., & Robblee, N.M. (1983). Dietary fat alters the fatty acid composition of lymphocyte membranes and the rate at which suppressor capacity is lost. Immunol. Lett. 6, 331–337.
Clandinin, M.T., Cheema, S., Field, C.J., Garg, M.L., Venkatraman, J., & Clandinin, T.R. (1991). Dietary fat: Exogenous determination of membrane structure and cell function. FASEB J. 5, 2761–2769.
Field, C.J., Ryan, E.A., Thomson, A.B.R., & Clandinin, M.T. (1990). Diet fat composition alters membrane phospholipid composition, insulin binding, and glucose metabolism in adipocytes from control and diabetic animals. J. Biol. Chem. 265, 11143–11150.
McMurchie, E.J. (1988). Physiological regulation of membrane fluidity. In: Advances in Membrane Fluidity (Aloia, R.C., Curtain, C.C., & Gordon, M.L., eds.), vol. 3, pp. 189–237. Alan R. Liss, New York.

Chapter 6

Membrane Fusion and Exocytosis

CARL E. CREUTZ

Principles of Medical Biology, Volume 7A
Membranes and Cell Signaling, pages 121–141.

ISBN: 1-55938-812-9

INTRODUCTION

Cells are separated from their environment and internally compartmentalized by membranes. These membranes have a lipid matrix that prohibits the free passage of molecules that are in aqueous solution either inside or outside of the cell. However, communication with the environment and between internal compartments is essential for many cell functions. Therefore, complex mechanisms have evolved to permit signal transduction across membranes and to catalyze the movement of solutes from compartment to compartment. In the case of small molecules such as inorganic ions, carbohydrates, amino acids, and so forth, protein molecules embedded in the membrane serve as molecular transport devices. However, even though some of these transport molecules are fairly complex, multisubunit proteins, they generally are not capable of promoting the transbilayer movement of folded macromolecules. In eukaryotic cells more complex protein based devices permit the transfer of mRNA through specialized pores in the nuclear membrane, and unique proteins are involved in catalyzing the transfer of newly synthesized, unfolded proteins into the lumina of the endoplasmic reticulum, mitochondria or chloroplasts. However, an important general mechanism that eukaryotic cells rely on for transfer of material through compartments, or, conversely, the sequestration of macromolecules into compartments, is the process of *membrane fusion*. This chapter will present an overview of membrane fusion events in cells and discuss general aspects of the mechanics of this phenomenon. Then current understanding of the molecular biology of protein secretion by exocytosis will be described in more detail. The membrane fusion events in the exocytotic pathway are just beginning to be revealed at the molecular level. Current research hypotheses will be presented. Although some of these hypotheses will be disproved in the near future, ultimately others will prove to be correct and will likely become the basis of novel pharmacological approaches to the treatment of disorders in intercellular communication.

MEMBRANE FUSION EVENTS BETWEEN AND WITHIN CELLS

Biological membrane fusion events can be grouped into two classes: those involving the fusion of the *extracytoplasmic* faces of membranes and those involving the *cytoplasmic* faces of membranes (see Figures 1 and 2). The internal faces of the membranes limiting intracellular organelles, such as Golgi cisternae or secretory vesicles, are topologically equivalent in this context to extracytoplasmic, or "extracellular" faces. There are no obvious examples of fusion between a cytoplasmic face of one membrane with an extracytoplasmic face of another membrane. It is reasonable to consider these two classes of fusion separately because it is likely that they are mediated by very distinct mechanisms. It has been demonstrated that both the plasma membrane and the membranes of intracellular organelles are asymmet-

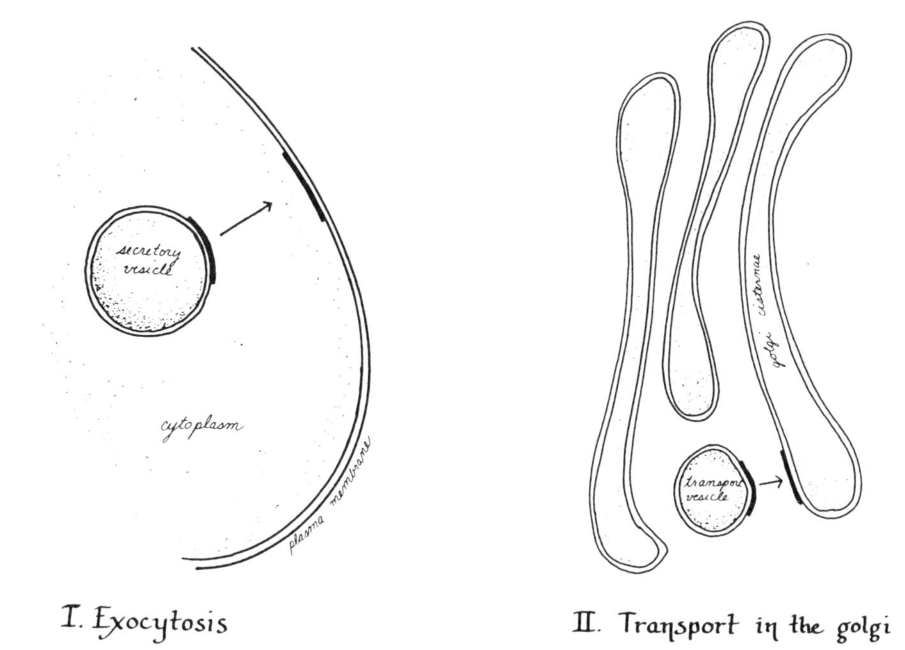

Figure 1. Schematic illustration of two examples of membrane fusion events involving the *cytoplasmic* membrane faces. Drawing by Sandra L. Snyder.

ric structures: The lipid compositions of each monolayer making up the bilayer structure are distinct, with phosphatidylcholine and glycolipids found in greater concentration in the extracytoplasmic face and acidic lipids such as phosphatidylserine and phosphatidylinositol in the cytoplasmic face. In addition, membrane proteins have unique orientations with regard to the cytoplasmic and extracytoplasmic faces. The environments of each of the different faces of a membrane are also dramatically different. The cytoplasmic face is bathed in a medium that has a very low concentration of Ca^{2+}, an ion which may serve an important signaling role, and rich in soluble or cytoskeletal proteins that may be part of complex fusion control mechanisms. The extracytoplasmic face, on the other hand, is exposed to high calcium concentrations in a less controlled environment, probably devoid of regulatory proteins. Fusion events between extracytoplasmic faces may therefore be completely dependent upon macromolecules that are integral parts of the membranes themselves.

To emphasize the breadth of membrane fusion phenomena it is useful to compile a list of some of the major occurrences of this process. Fusion between extracellular faces of membranes is less common since this event violates the integrity of the single cell. Such fusion events are nonetheless essential for the development of tissues that consist of syncytia. For example, during the development of muscle

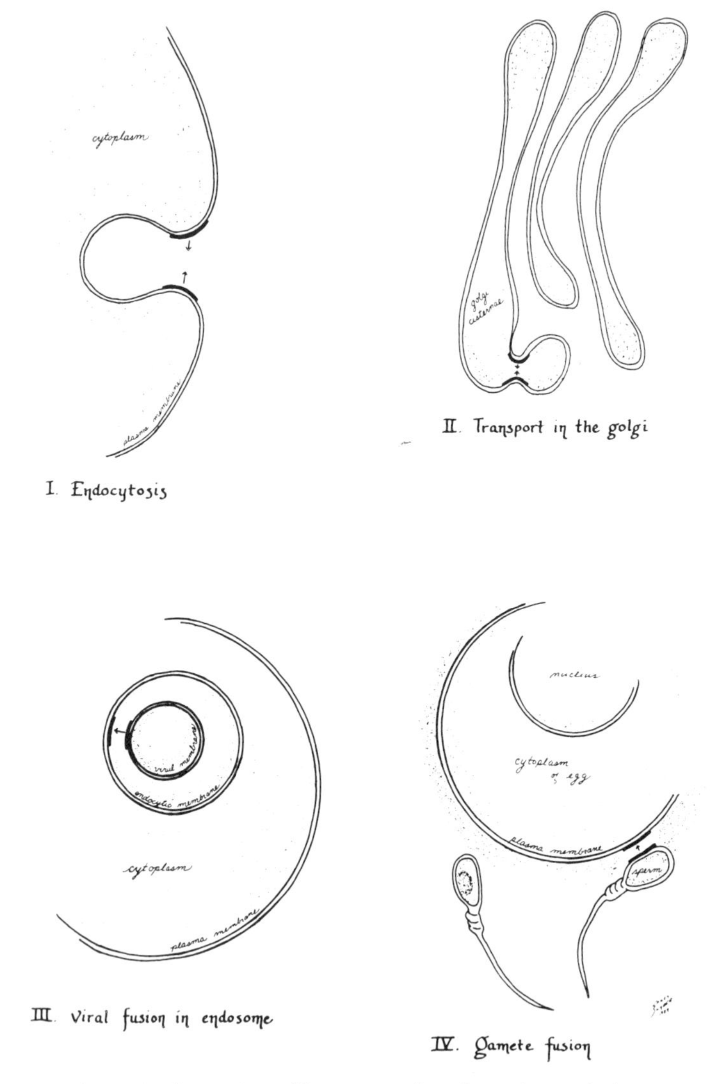

Figure 2. Schematic illustration of four examples of membrane fusion events involving the *extracytoplasmic* membrane faces. Drawing by Sandra L. Snyder.

fibers the membranes of individual myoblasts fuse to form elongated, multinuclear, single muscle cells. The fusion of the membranes of gametes, such as sperm and egg, removes the critical barrier to the mixing of genetic information that is the basis of sexual reproduction. Extracellular membrane fusion is of broad medical importance as it provides the route of entry of enveloped viruses into target cells. The fusion may occur at the plasma membrane of the cell, although the more frequent pathway is fusion with the inner surface of an endocytic vesicle that has encapsulated the virus particle. The fusion event in the case of some viruses is understood at the molecular level. As anticipated above for extracytoplasmic fusion, the viral fusion event does indeed depend on specific fusion proteins that are components of the viral membrane and specific target molecules that are part of the cell membrane.

Recently a protein has been identified on the surface membrane of mammalian sperm that is essential for membrane fusion in fertilization (Blobel, Wolfsberg, Turck et al., 1992). This protein has a region that is similar to a region in the viral fusion proteins that may form a hydrophobic spike. Such a spike might spear the egg membrane to promote fertilization. Therefore, as predicted, there seem to be significant similarities in the mechanisms underlying these two rather different fusion events both of which involve the fusion of extracytoplasmic membrane faces.

As much of the organization of the eukaryotic cell is based on compartmentalization by membranes, it is not surprising that internal membrane fusion events are abundant and frequent between intracellular membranes. Secretory proteins and surface membrane proteins are transported from their site of origin in the rough endoplasmic reticulum to the Golgi by transport vesicles. These vesicles pinch off from the ER membrane by fusion of their inner (extracytoplasmic) membrane surfaces and then fuse with the cytoplasmic surfaces of proximal Golgi cisternae. Transport through the varied manufacturing compartments of the Golgi is similarly accomplished by pinching off and fusing of transport vesicles. Finally, at the distal face of the Golgi additional vesicles are formed by pinching off that go to the cell surface immediately, or fuse with lysosomes, or become storage vesicles destined for fusion with the plasma membrane at a later time as part of the regulated secretory pathway. Endocytic vesicles pinch off from the plasma membrane to bring solutes or membrane components into the cell. These vesicles fuse with one another to form acidic endosomes which may in turn then fuse with lysosome or Golgi elements. In light of the abundance and complexity of this intracellular traffic, it is clear that regulatory mechanisms and "chaperons" must have evolved to assure that only the appropriate partners undergo fusion. These regulatory mechanisms may obscure the features that are common to all fusion events. Because of the fundamental similarity in structure of biological membranes, it is likely that all fusion events in one class (i.e., involving intracellular or extracellular membrane surfaces) may have mediating proteins in common.

ELEMENTS OF MEMBRANE FUSION

The fusion of two biological membranes, in general terms, can be broken down into a series of events. These events are *translocation*, *recognition*, *attachment*, *disruption*, and *resolution*. Classifying the steps in such terms helps focus on events that are mechanistically distinct, that may be subserved by different proteins, and probably are subject to different regulatory mechanisms.

Translocation

Membranes must approach one another to fuse. In some cases this translocation occurs as a simple result of diffusion, as when a virus particle approaches a cell. However, intracellular movement of membrane-bound organelles may be mediated by specific interactions with cytoskeletal components. Both actin filaments and microtubules can interact with some intracellular vesicles and these interactions can result in immobilization through gelation, or in guided movements. Secretory vesicles are frequently sequestered to a pole of a cell where their release will occur. After such localization simple diffusion across a small space may provide efficient translocation.

Recognition

A recognition step may be necessary to prevent non-discriminant membrane fusion. Virus particles have receptor molecules that recognize specific glycoproteins on the surfaces of cells. No such specific targeting molecules have yet been identified that mediate intracellular fusion events, although such proteins have been widely hypothesized to exist. Some proteins that promote membrane attachment (the annexins, see below) interact with specific lipids that are found on intracellular membrane surfaces. The sensitivity of these proteins to activation by calcium appears to depend upon the actual relative concentrations of specific lipids in the membrane. Thus, some degree of targeting may result from the ability of these proteins to attach to membranes of specific lipid composition at a given intracellular calcium concentration. Rigid compartmentalization of organelles into relative positions, such as the stacking of the cisternae of the Golgi by cytoskeletal components, might in some circumstances alleviate the need for specific targeting molecules, since transport vesicles would have little opportunity to go astray before fusing.

Attachment

When the appropriate fusion partners achieve recognition, true fusion begins with an intimate attachment of the membranes. The binding of targeting proteins to their targets may provide this attachment. Viral spike proteins may further secure attachment by inserting hydrophobic peptide extensions into the target cell membrane. The postulated intracellular attachment proteins, the annexins (see below), have multiple recognition sites for membrane lipids. Possibly by attaching different sites to lipids on different membranes they create an effective bridge or attachment between membranes. Alternatively, annexin molecules bound to each membrane of a fusion pair might self-associate, thus resulting in membrane-membrane attachment.

Disruption

The internal structures of the attached membranes must be disrupted in order for fusion to occur. This is a consequence of the topological fact that an aqueous passageway must form through the region that once contained the hydrocarbon-like interior of the lipid bilayer. Again, the viral spike protein is thought to be instrumental in this step in virus-cell fusion. Membranes attached by annexins seem to require the perturbing effects of free, *cis*-unsaturated fatty acids for disruption and complete fusion to occur (Creutz, 1981; Drust and Creutz, 1988; see discussion below). These detergent-like agents may enhance the probability of the formation of "inverted micelles" in the bilayer. These hypothetical structures have the polar lipid head groups facing one another and the hydrophobic tails extended outward. They have been postulated to be intermediates in the fusion process. The need for free fatty acids can be met through the action of specific phospholipases that are activated during exocytosis. Such enzymes may be a critical control point in the fusion process, subject to regulation by calcium or G-proteins.

Resolution

After disruption of the apposed bilayers the membranes must follow a pathway of remodeling and reorganization that results in the appropriate final morphology. For the fusion of a secretory vesicle with the plasma membrane, this pathway leads to the incorporation of the vesicle membrane into the plasma membrane. Fusion starts at a single point of contact between the two membranes and results in the formation of a single continuous (although microscopically heterogeneous) bilayer. Topologically, this is very different from the pathway that the plasma membrane and endosome membranes follow after endocytosis. In this case, the extracellular face of a single membrane, the plasma membrane, invaginates and is drawn upon itself in an annulus: the membrane contact region results from the contraction of a circle to zero radius. After fusion two independent bilayers are formed by remodeling.

A classification of the steps in membrane fusion, as given here, serves to emphasize our ignorance of the molecular basis of the process: Most of these steps remain "black boxes." However, this logical scheme is helpful in organizing information about subcellular components that have been examined in various *in vitro* systems. The classification may also help to define future research goals in this field.

In order to discuss further molecular details of membrane fusion as we are beginning to understand them, it is helpful now to restrict discussion to a specific system that has been particularly fruitful and is of broad medical importance. This is the secretory pathway and exocytosis.

THE SECRETORY PATHWAY AND EXOCYTOSIS

The membrane fusion events in the secretory pathway have been examined in several very informative model systems that will be discussed here. These systems range from simple constitutive secretion in the common baker's yeast, *Saccharomyces cerevisiae*, to the complex regulated secretion of an endocrine cell or a vertebrate synapse. These are obviously quite different systems, involving the biology of organisms separated by hundreds of millions of years of evolution. However, common elements are coming to light in these systems, which may reflect some of the basic features of membrane fusion as described above. In addition, these systems are excellent examples of how research can be thrust forward by careful selection of models that possess unique features amenable to analysis using current technology.

The Yeast *sec* Mutants

Yeast maintain an active secretory pathway for the release of cell wall components and hydrolytic enzymes during growth as well as for the release of polypeptide mating pheromones. A series of temperature sensitive mutants in the secretory pathway have been isolated based upon the ingenious concept that cells that cannot secrete proteins as they are synthesized will increase in density as their cytoplasm becomes increasingly clogged with dense protein molecules (Novick, Field, and Schekman, 1980). After chemical mutagenesis of a yeast culture, secretion, or *sec*, mutants can be isolated on density gradients. Loss of the secretory pathway is lethal to yeast, therefore, it has been necessary to isolate temperature sensitive mutants in which secretion is blocked only at a certain, usually higher, temperature. Twenty-three noncomplementing *sec* mutants were originally isolated, but additional essential complementation groups are continuing to be discovered. The morphology of these cells when shifted to the nonpermissive temperature reveals that the *sec* mutations control various different steps in the overall secretory pathway, including movement from the ER to the Golgi (these mutants accumulate extensive ER), within the Golgi (these mutants elaborate extensive Golgi cisternae), and transport to and fusion with the plasma membrane (these mutants accumulate large numbers of secretory vesicles).

The *SEC4* gene product is an excellent example of the type of protein component of the secretion machinery this genetic approach has brought to light (Salminen and Novick, 1987). This protein is a peripheral component of the cytoplasmic face of both the secretory vesicle membrane and the plasma membrane. It is a small, 20kD protein that shows sequence similarity to the mammalian *ras* gene product and to other GTP-binding proteins such as the coupling, or G, protein of the adenylyl cyclase system. The amino acid residues involved in binding GTP are particularly conserved. The protein binds and hydrolyzes GTP. When one of the residues involved in hydrolysis is changed, the yeast cell cannot carry out secretion and fills up with secretory vesicles.

What could be the function of such a GTP-binding protein in secretion? By analogy with the role of the G protein in the adenylyl cyclase system, the sec4 protein may couple a fusogenic protein or complex to a sensor that recognizes when appropriate membrane-membrane contact has occurred. Alternatively, by analogy with the role of elongation factor Tu in protein synthesis, the sec4 protein could act as a gate that must complete a cycle of hydrolysis of one molecule of GTP for every fusion event. Insight into which model may be correct comes from the experiments with the mutant sec4 protein that cannot hydrolyze GTP. In the cyclase system a G protein with nonhydrolyzable GTP bound is permanently turned "on" and irreversibly activates the cyclase. In protein synthesis, if elongation factor Tu cannot hydrolyze bound GTP, the regulating cycle is broken and protein synthesis cannot continue. Since mutation of the *SEC4* gene so that the sec4 protein cannot hydrolyze GTP results in the accumulation of secretory vesicles, it appears that the sec4 protein acts as an essential one-way ratchet in a more complex process. Interestingly, small GTP-binding proteins of mass 20kD have now been identified in mammalian secretory tissues and even localized to the chromaffin granule membrane (Burgoyne and Morgan, 1985) and the membrane of synaptic vesicles (Sudhof and Jahn, 1991). It seems likely these proteins will be found to perform similar roles in the mammalian secretory pathway to that of sec4 in yeast secretion.

The yeast *SEC18* gene product is apparently a component of a fairly general intracellular fusion machine. This protein was independently investigated as an essential protein for the reconstitution of transport between components of the mammalian Golgi *in vitro* (Wilson, Wilcox, Flynn et al., 1989). Its activity is sensitive to the sulfhydryl reagent N-ethylmaleimide (NEM) so it has been called NSF for NEM-Sensitive Factor. The native protein is a tetramer of 70kD subunits. It binds to receptors on membranes encoded by the *SEC17* gene (also referred to as SNAPS, Soluble NSF Attachment Proteins). NSF has been found to be essential for the *in vitro* reconstitution of ER to Golgi transport and for the fusion of endosomes with one another as well as intra-Golgi fusion events. Yeast cells with temperature-sensitive mutations in the *SEC18* gene exhibit greatly expanded ER in their terminal morphology at the non-permissive temperature. Although the sec18 protein may be essential at several steps in the secretory pathway, the terminal morphology may naturally reflect only the first step in the sequential pathway in which the gene product is essential. The actual function of the sec18 protein is completely obscure, although it is now known to form complexes with several other proteins localized on the secretory vesicle membrane or the plasma membrane (see below). The structure of the sec18 protein appears hydrophilic so it is unlikely to insert into membranes, and thus act to disrupt membrane structure. Each subunit has a consensus sequence common to several ATP-binding proteins, and indeed *in vitro* reconstitution systems dependent on this protein also require ATP. The protein does have some sequence similarity to an ATP-dependent bacterial protease, and therefore it may be involved in activating another component of a complete "fusion machine" by proteolysis.

Regulated Secretion in the Adrenal Chromaffin Cell

The yeast cell does not have a large storage pool of secretory vesicles. Secretion occurs in a constitutive manner as secretory products are synthesized. However, endocrine and exocrine cells typically store large amounts of secretory product to be released upon specific stimulation of the cell by a hormone or neurotransmitter. The adrenal chromaffin cell has been a popular model for studies of the biochemistry of such "regulated" exocytosis, because the secretory vesicles, called chromaffin granules, can readily be isolated in milligram quantities from bovine tissue. These are complex and interesting organelles (see Figure 3) which have on the order of 60 polypeptides in their membranes and another 20 (the *chromobindins*) which can reversibly associate with them in the presence of calcium (Creutz, Dowling, Sando et al., 1983). In addition, short-term primary cultures of chromaffin cells can be maintained and will respond to secretogogues with the release of catecholamines and secretory proteins. The process of exocytosis was in fact originally defined in this cell by morphological criteria, i.e., the fusion of secretory vesicle membrane with the plasma membrane (De Robertis and Vaz Ferreira, 1957), and biochemical criteria, i.e., the release of vesicle contents but not membrane components (Kirshner, Sage, Smith, and Kirshner, 1966). The plasma membrane of the chromaffin cell can be permeabilized with detergents e.g., digitonin (Dunn and Holz, 1983; Wilson and Kirshner, 1983) or saponin (Brooks and Treml, 1983), or by electrical discharge (Knight and Baker, 1982) yielding a model cell that still carries

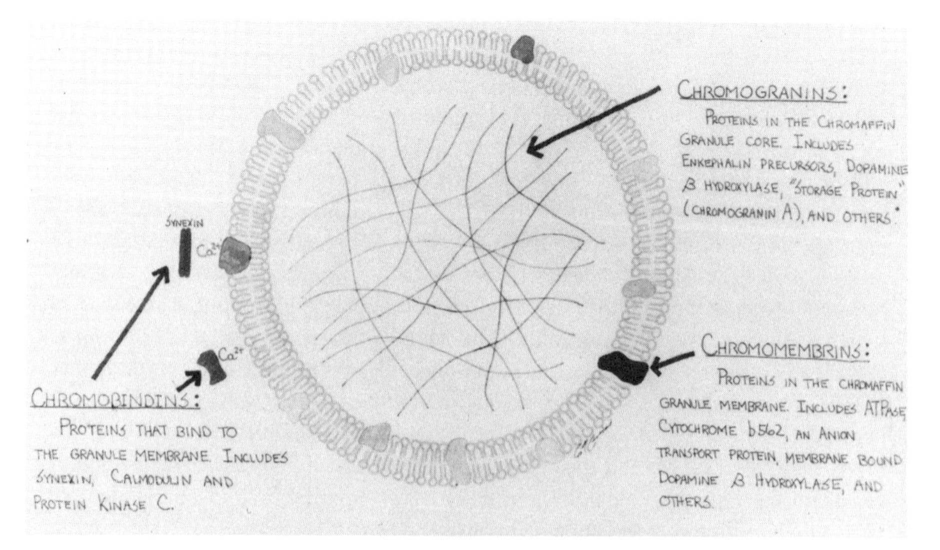

Figure 3. Schematic illustration of the chromaffin granule, the secretory vesicle of the adrenal medulla. The terms *chromogranins, chromomembrins,* and *chromobindins* are defined. Drawing by Carol Ceriani. (From Creutz, Drust, Hamman et al., 1990).

out exocytosis but in which the cytoplasmic milieu can be altered by changing the external medium. In this way the level of calcium necessary to activate the machinery of exocytosis has been determined to be on the order of 1 uM.

To enable a biochemical analysis of exocytosis one would ideally like to be able to reconstruct *in vitro* a model system involving minimal components in which secretory vesicles fuse with plasma membranes. Unfortunately a reliable model of this type has not yet been developed. Two major problems seem to be: (1) obtaining properly oriented plasma membrane sheets or vesicles and (2) the development of a simple biochemical assay for events that are essentially morphological in nature. However, as detailed below, considerable progress has been made in developing models for the fusion of secretory vesicles with one another. These models may be physiologically relevant since in many activated secretory cells the vesicles cannot approach the plasma membrane because of the shear number of vesicles in the cytoplasm and therefore they fuse with one another. In this way they create channels through vesicles that have already fused with the plasma membrane and through which secretory products can escape.

Annexin-Mediated Chromaffin Granule Fusion

Isolated chromaffin granules have a negative surface charge and do not interact with one another unless exposed to unphysiologically high levels of divalent cations. Therefore, it is apparent that during their isolation some factor is lost that permits their interaction with one another in the cell. A number of homologous soluble proteins, called annexins, have been isolated that are able to overcome this repulsive barrier and attach chromaffin granules to one another in the presence of calcium (Creutz, 1992). The first member of this family to be identified was a 47kD protein, synexin, that was isolated from adrenal medullary cytosol using turbidity increases in chromaffin granule suspensions as an assay reflective of membrane aggregation (Creutz, Pazoles, and Pollard, 1978). The name synexin embodies the concept of membrane contact as it is based on the Greek word *synexis* which means "a meeting." The other members of this protein family have been isolated in a variety of contexts. For example, lipocortin (now also called annexin I) was isolated as a steroid-inducible inhibitor of phospholipase A2 (Wallner, Mattaliano, Hession et al., 1986). Calpactin (annexin II) was isolated as a component of the submembranous cytoskeleton in the intestinal epithelial cell brush border (Gerke and Weber, 1984). Both annexins I and II were identified as major substrates for tyrosine-specific protein kinases. The transforming *src* kinase in the case of annexin II (Glenney and Tack, 1985), and the epidermal growth factor receptor associated kinase in the case of annexin I (Schlaepfer and Haigler, 1987, 1988). Thus, both proteins have been associated with membrane signaling events related to growth control.

Endonexin I (annexin IV) and 68kD calelectrin (annexin VI) were identified as mammalian homologs of an electric ray protein, called calelectrin, that binds to membranes in the presence of calcium and might underlie the release of acetylcholine by exocytosis (Walker, 1982; Sudhof, Ebbecke, Walker et al., 1984). Endonexin II (annexin V) was identified as a placental inhibitor of blood coagulation and another substrate for the epidermal growth factor receptor kinase (Kaplan, Jaye, Burgess et al., 1988). It exhibits anticoagulant activity because of its ability to cover membrane surfaces essential for activation of clotting factors. The common element among this confusing array of activities is apparently the ability of all the annexins to bind to membranes in the presence of calcium. This property is shared by some unrelated proteins such as protein kinase C. However, the annexins are unusual in that their action is in a sense "bivalent": They can bind to two membranes simultaneously and bring them together.

It is evident that the different annexins are truly related from their amino acid sequences which show 40 to 60% identity. The proteins have a common structural theme: They have a core domain consisting of four homologous repeats approximately 70 amino acids long and each has a unique N-terminal region 20 to 50 amino

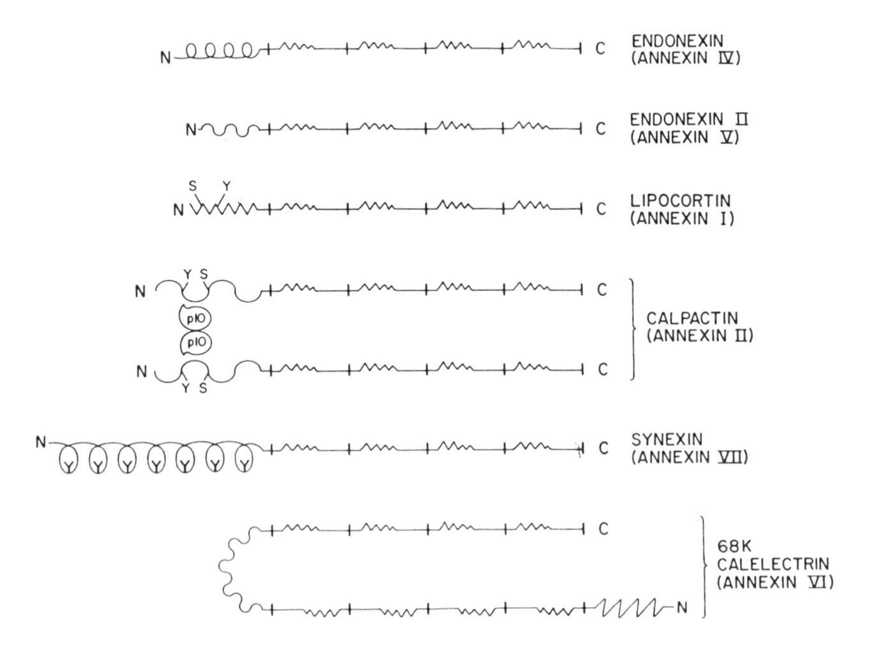

Figure 4. Schematic illustration of the structures of the annexins. The four homologous domains each contain the 17 amino acid endonexin fold sequence represented by a sawtooth line. The unique N-terminal structures are on the left (or right in the case of annexin VI). Y and S represent phosphorylation sites in the tails of annexins I and II. The annexin II tetramer is drawn showing the association of the N-termini of the heavy chains with the light chain dimer. The Y's inside the loops in the tail of synexin represent characteristic repeating tyrosines. (From Creutz, 1992).

acids long. However, there are some interesting variations on this theme. Synexin (annexin VII) has an unusually long amino terminal extension that is highly hydrophobic and repetitive in sequence (Creutz, Snyder, Husted et al., 1988; Burns, Magenzo, Srivistana et al., 1989). Annexin VI was apparently formed by duplication of the gene for one of the other annexins as it has eight repeating domains rather than four. Annexin II associates with a small, 11kD subunit that is itself a dimer homologous in structure to the calcium-binding protein S-100 (Glenney and Tack, 1985). As a consequence, the native state of annexin II is apparently a tetramer composed of two 36kD heavy chains and two 11kD light chains. These structural features of the annexins are summarized schematically in Figure 4.

Although the annexins appear to be able to promote the fusion of pure lipid vesicles with one another (Hong, Duzgunez, Ekerdt, and Papahadjopoulos, 1982), they are much less effective at promoting complete fusion of biological membranes such as that of the chromaffin granule. Thus, as discussed above, they are essentially aggregating agents, capable of mediating the *attachment* step in fusion, but dependent upon free fatty acids as cofactors to promote the *disruption* step (see Figure 5).

Although all of the annexins may be able to promote intermembrane contacts, they show a great variation in the levels of calcium needed to activate this process *in vitro*. Annexin II is the most sensitive to calcium and promotes membrane contacts at levels of calcium (threshold at 0.7 μM) that are comparable to the levels of calcium that activate exocytosis from permeabilized chromaffin cell models (Drust and Creutz, 1988). Synexin and annexin IV require 100 to 200 μM calcium for half-maximal activation which seems extraordinarily high for intracellular activation. Annexins I and V are even less sensitive to calcium which is one reason extracellular roles for these proteins in phospholipase or coagulation inhibition are still entertained. However, the precise sensitivity to calcium of an annexin depends upon the lipid surface with which the protein is interacting. Higher concentrations of phosphatidylserine or free fatty acids in the membrane result in higher affinities for calcium. This indicates there is a tight coupling between the calcium binding site(s) and the lipid binding site(s). One model that could explain this behavior proposes that one of the phosphate groups of the phospholipids in the bilayer provides additional direct coordination of the calcium in the calcium-binding pocket.

Recently, the three dimensional structures of two of the annexins have been determined (Huber, Berendes, Burger et al., 1992; Weng, Luecke, Song et al., 1993). The calcium binding sites are novel and quite distinct from the classical EF-hand structure found in calmodulin. Typically three of the four homologous domains appear capable of binding calcium with high affinity. In each domain the calcium is chelated by two cooperating loops of protein, and it is likely that the phosphate of the lipid headgroup provides an additional coordination site, although this has not yet been directly visualized.

Although similar to one another in their basic structure, the annexins do seem to perform distinct functions in cells. This is suggested by the fact that although

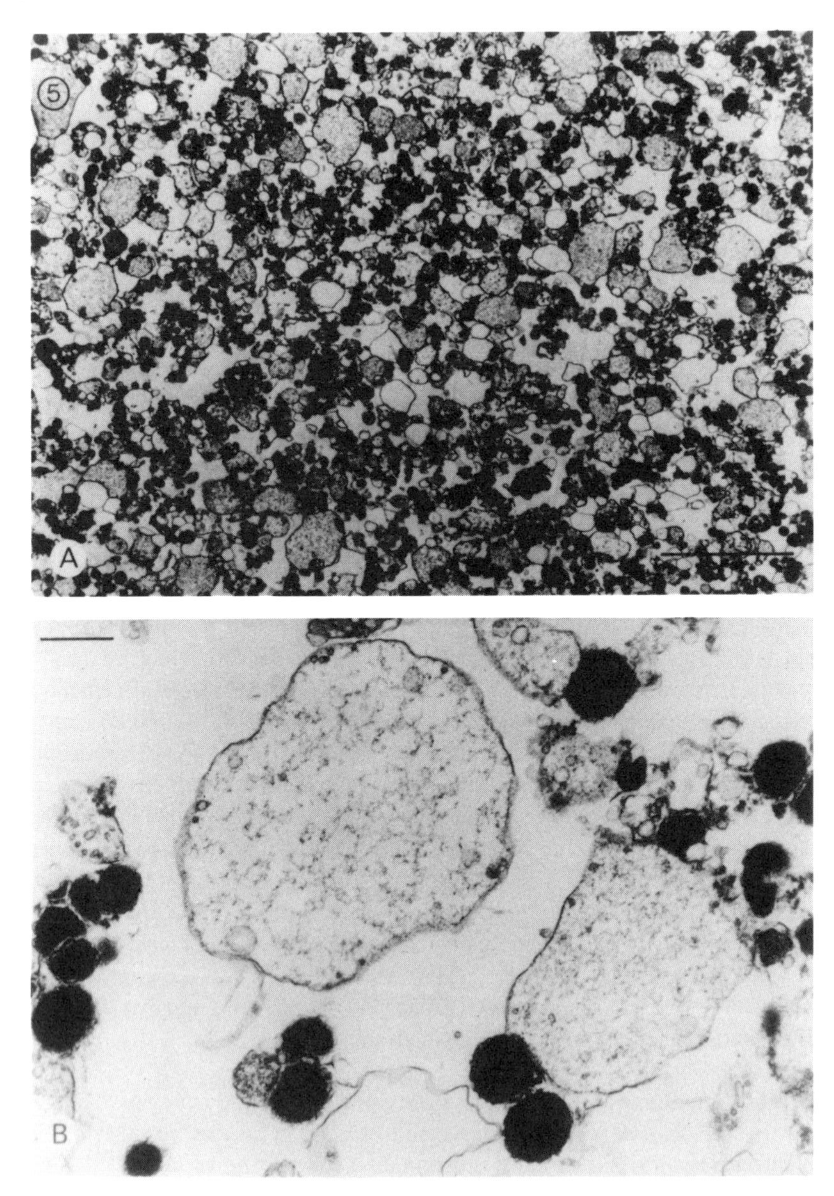

Figure 5. Low and high magnification electron micrographs of chromaffin granules that have undergone synexin-mediated attachment and arachidonic acid-dependent fusion. The fused vesicles swell to encompass the maximum volume. The flocculent material in the large vesicles represents the retained but diluted chromogranins (secretory product). Length of *bar* in A, 5 μm; in B, 0.3 μm. (From Creutz, 1981).

virtually all members of the family seem to be present in all tissues, their relative concentrations vary widely. In addition, within single cells their subcellular distributions are unique. Analysis of the distribution of annexins in chromaffin tissue by sucrose gradient analysis of tissue homogenates has revealed that some members of the family exhibit a significant degree of association with membranes even in the absence of calcium (Drust and Creutz, 1991). While annexins IV and V are largely soluble, annexins II and VI are predominantly membrane bound. Interestingly, significant amounts of annexin II (~20%) are found irreversibly associated with the secretory vesicle membrane. Such calcium-independent annexin-membrane interactions may be due to post-translational modifications such as acylation or may be due to significant conformational changes in the proteins that permit them to enter the hydrophobic portion of the bilayer to some extent.

The differences between the distributions and properties of the various annexins suggest they play non-overlapping roles in the cell. Perhaps all are involved in membrane fusion, but at different locations and involving different membranes. Alternatively, they may perform unrelated functions that have some common element, such as association with a lipid surface. The variations in sequence in the core domains could result in preferential interaction of each annexin with a membrane of different lipid composition. The unique N-terminal extensions could interact with unique proteins, just as the annexin II heavy chain interacts with the light chain. Of the group as a whole, perhaps the best candidate for a direct mediator of membrane fusion in exocytosis is the calpactin tetramer. This molecule has the most appropriate sensitivity to calcium for this role, and has been localized by immunocytochemistry to the region of cytoplasm immediately underlying the plasma membrane of the chromaffin cell (Burgoyne and Cheek, 1987; Nakata, Sobue, and Hirokawa, 1990).

Membrane Protein Mediators of Exocytosis

An important approach to gaining a molecular understanding of the process of exocytosis has been to identify proteins that are components of the membranes that undergo fusion. This is easiest to do with the secretory vesicle membrane since it can be obtained in high purity, thus making it unequivocal whether a given protein is a component of the relevant organelle. Both the chromaffin granule and the synaptic vesicle have been favored organelles for characterization of these proteins. In the synaptic vesicle several proteins have been identified that are uniquely expressed in this organelle or related secretory granules (Sudhof and Jahn, 1991; Damer and Creutz, 1994). Some are involved in the transport of ions or secretory products through the vesicle membrane. For others the functions are less clear. These include *synaptophysin*, which has four transmembrane helices and an extended, proline rich, carboxyl terminus that extends into the cytoplasm. *Synaptotagmin*, which has a single transmembrane domain and a large cytoplasmic domain that contains two repeats of amino acid sequence homologous to the calcium and lipid binding domain of protein kinase C. And *synaptobrevin* (also called *VAMP* for Vesicle Associated Membrane Protein) is a small protein which has a single transmembrane domain and a short cytoplasmic extension.

Recently, genetic evidence has been obtained from the study of lower eukaryotes that synaptotagmin plays a direct role in exocytosis at the synapse. In *Drosophila*, and the nematode, *C. elegans*, mutations in the genes for synaptotagmin lead to severe deficiencies in cholinergic neurotransmission (Littleton, Stern, Schulze et al., 1993; Nonet, Grundahl, Meyer et al., 1993).

Another important route to the identification of protein components of the machinery underlying exocytosis has been the elucidation of the mechanism of action of certain neurotoxins produced by bacteria. In particular, botulinum toxins and tetanus toxin have been determined to act through proteolysis of specific proteins involved in transmitter release after the toxins gain entry into the nerve terminal. Tetanus toxin and Botulinum toxin serotype B both cleave the cytoplasmic portion of synaptobrevin (Schiavo, Benfenati, Poulain et al., 1992). Competitive inhibition of the cleavage by synthetic peptides that have the sequence of the cleavage site are protective against the effects of the toxins. This provides strong evidence that the effects of these toxins on neurotransmission are through cleavage of this protein, and therefore, synaptobrevin must play a critical role in exocytosis.

Strikingly, several other related Clostridial neurotoxins seem to act by cleaving other protein components of the synapse. For example, Botulinum toxin type C1 cleaves a synaptic plasma membrane protein, *syntaxin* (Blasi, Chapman, Yamasaki et al., 1993), and types A and E cleave *SNAP25* (Binz, Blasi, Yamasaki et al., 1994).

The Exocytosis Complex

The different approaches outlined above, including *in vitro* models for membrane fusion, molecular characterization of genes essential for secretion, and elucidation of the targets for neurotoxins that block exocytosis, have led to the identification of multiple components of the complex machinery that regulates and mediates exocytosis in eukaryotic cells. However, a critical remaining challenge is to understand how these pieces fit together. The function of each component must be identified in order to do this. The genetic and toxin-based approaches generally do not give detailed insights along these lines. A fine Swiss watch can be made to fail by removing any number of a vast array of components. Putting a watch together requires greater insight. *In vitro* models, as used extensively to characterize the annexins, will be needed to understand how all these components function cooperatively. Some initial work along these lines has demonstrated specific binding interactions between some of these components *in vitro* (Sollner, Bennett, Whiteheart et al., 1993; Sollner, Whiteheart, Brunner et al., 1993). For example, the synaptobrevin on the vesicle membrane binds to the NSF/SNAP (*sec18*/*sec17*) complex described above, along with syntaxin and SNAP 25 on the plasma membrane. This complex therefore provides a linkage between components of the

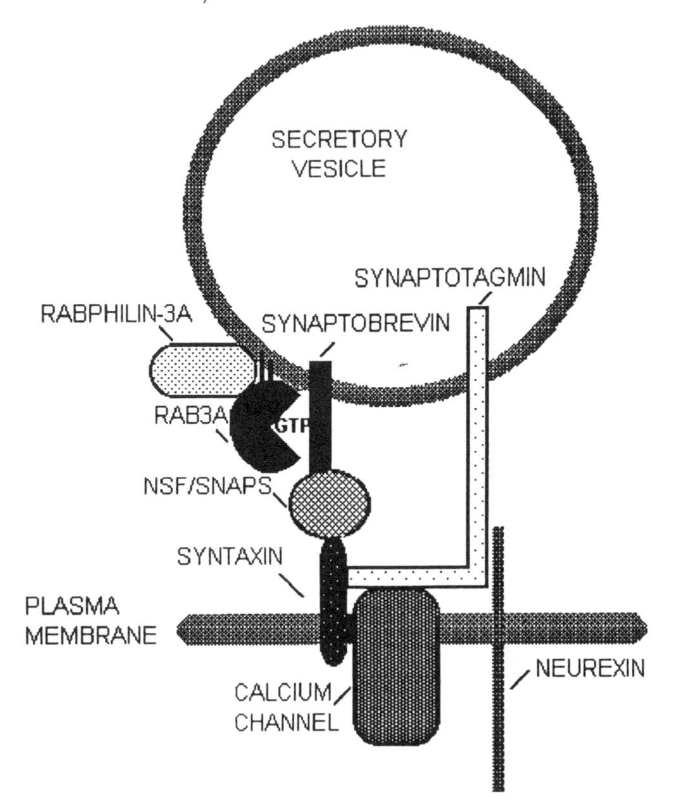

Figure 6. Hypothetical model of protein-protein interactions underlying regulated secretory vesicle exocytosis. NSF and SNAPS interact with synaptobrevin on the vesicle membrane and syntaxin on the plasma membrane forming a fusion complex. Synaptotagmin interacts with neurexin, syntaxin, and the calcium channel, docking the vesicle to the plasma membrane and acting as an inhibitory regulator of vesicle fusion. Synaptobrevin may interact with the small GTP-binding protein, rab3a, which in turn may interact with rabphilin 3A on the vesicle membrane to target vesicles to the plasma membrane. (From Damer and Creutz, 1994).

vesicle and the plasma membrane. Some components of this complex can also interact with synaptotagmin. Perhaps it is through synaptotagmin that the complex may be positively, or negatively, regulated by calcium, although this has not yet been shown *in vitro*. Since all of these components function in an environment rich in annexins, it is possible their functions are regulated by the ability of annexins to sequester lipids and bring membranes into close proximity. Figure 6 provides a schematic map to some of the interactions between these and other components of the exocytosis complex that have been recently discovered (Damer and Creutz,

1994). We can now look forward to rapid progress in the biochemical analysis of the exocytotic machinery as the list of critical components may be nearing completion.

SUMMARY

The lipid bilayer membranes that separate cells and organelles undergo fusion to permit the exchange or compartmentalization of macromolecules. The fusion events involve contact between cytoplasmic or between extracytoplasmic membrane surfaces. These two types of fusion may depend on very different mechanisms because of the different environments of the membrane surfaces. The membrane fusion process can be broken down into the following steps: (1) translocation, the positioning of membranes adjacent to one another, (2) recognition, initial specific interaction between the appropriate fusion partners, (3) attachment, a close physical sealing between the two bilayers, (4) disruption, alteration of bilayer structure to create an aqueous passageway through the membranes, (5) reorganization, the formation of new bilayer(s) from the disrupted membrane components. Any or all of these steps may be mediated by specific proteins and be subject to regulation.

The secretory pathway includes a number of steps that involve membrane fusion: pinching off and fusing of transport vesicles in the endoplasmic reticulum and the Golgi pinching off of secretory vesicles from the Golgi; fusion of secretory vesicles with the plasma membrane during simple exocytosis or with one another during compound exocytosis, and pinching off of endocytic vesicles to recycle secretory vesicle membrane. The molecular basis of these events has been studied using several important model systems that can be subjected to biochemical or genetic analyses. These include the constitutive secretory pathway of yeast, the regulated secretory pathway of the endocrine chromaffin cell, and the action of Clostridial neurotoxins on synaptic transmission. A group of yeast genes (the *SEC* genes) have been identified that mediate several steps in the secretory pathway. These genes have mammalian counterparts that may play similar roles. The annexins are a group of calcium-dependent, lipid-binding proteins that might mediate the binding of the secretory vesicle to the plasma membrane during regulated secretion. Protein components of the secretory vesicle or plasma membrane that may function in exocytosis are targets for the proteolytic activities of neurotoxins that block exocytosis. These membrane proteins include synaptobrevin, syntaxin, and SNAP25. Genetic analysis of lower eukaryotes indicates another secretory vesicle membrane protein, synaptotagmin, may provide regulation of the exocytosis complex of proteins in response to changes in the intracellular calcium concentration. These newly described protein mediators of membrane fusion in exocytosis are promising targets for the development of novel pharmacological agents to manipulate hormone or neurotransmitter release.

REFERENCES

Binz, T., Blasi, J., Yamasaki, S., Baumeister, A., Link, E., Sudhof, T.C., Jahn, R., & Niemann, H. (1994). Proteolysis of SNAP-25 by type-E and type-A botulinal neurotoxins. J. Biol. Chem. 269, 1617–1620.

Blasi, J., Chapman, E.R., Yamasaki, S., Binz, T., Niemann, H., & Jahn, R. (1993). Botulinum neurotoxin-C1 blocks neurotransmitter release by means of cleaving HPC-1/syntaxin. EMBO J. 12, 4821–4828.

Blobel, C.P., Wolfsberg, T.G., Turck, C.W., Myles, D.G., Primakoff, P., & White, J.M. (1992). A potential fusion peptide and an integrin ligand domain in a protein active in sperm-egg fusion. Nature 356, 248–252.

Brooks, J.C. & Treml, S. (1983). Catecholamine secretion by chemically skinned cultured chromaffin cells. J. Neurochem. 40, 468–473.

Burgoyne, R.D. & Cheek, T.R. (1987). Reorganisation of peripheral actin filaments as a prelude to exocytosis. Bioscience Rep. 7, 281–288.

Burgoyne, R.D. & Morgan, A. (1989). Low molecular mass GTP-binding proteins of adrenal chromaffin cells are present on the secretory granule. FEBS Lett. 245, 122–126.

Burns, A.L., Magenzo, K., Srivistana, M., Cheung, B., Seaton-Johnson, D., Shirvan, A., Alijani, R., Rojas, E., & Pollard, H.B. (1989). Purification of human synexin calcium channel protein and structure of the human synexin gene. Proc. Natl. Acad. Sci. USA 86, 3798–3802.

Creutz, C.E. (1981). *Cis*-unsaturated fatty acids induce the fusion of chromaffin granules aggregated by synexin. J. Cell Biol. 91, 247–256.

Creutz, C.E. (1992). The annexins and exocytosis. Science 258, 924–931.

Creutz, C.E., Pazoles, C.J., & Pollard, H.B. (1978). Identification and purification of an adrenal medullary protein (synexin) that causes calcium-dependent aggregation of isolated chromaffin granules. J. Biol. Chem. 253, 2858–2866.

Creutz, C.E., Snyder, S.L., Husted, L.D., Beggerly, L.K., & Fox, J.W. (1988). Pattern of repeating aromatic residues in synexin. Similarity to the cytoplasmic domain of synaptophysin. Biochem. Biophys. Res. Commun. 152, 1298–1303.

Creutz, C.E., Drust, D.S., Hamman, H.C., Junker, M., Kambouris, N.G., Klein, J.R., Nelson, M.R., & Snyder, S.L. (1990). Calcium-dependent membrane-binding proteins as potential mediators of stimulus-secretion coupling. In: Stimulus-Response Coupling: The Role of Intracellular Calcium (Dedman, J. & Smith, V., eds.), pp. 279–310. CRC Press.

Creutz, C.E., Dowling, L.G., Sando, J.J., Villar-Palasi, C., Whipple, J.H., & Zaks, W.J. (1983). Characterization of the chromobindins: Soluble proteins that bind to the chromaffin granule membrane in the presence of calcium. J. Biol. Chem. 258, 14664–14674.

Creutz, C.E., Zaks, W.J., Hamman, H.C., Crane, S., Martin, W.H., Gould, K.L., Oddie, K., & Parsons, S.J. (1987). Identification of chromaffin granule binding proteins: Relationship of the chromobindins to calelectrin, synhibin, and the tyrosine kinase substrates p35 and p36. J. Biol. Chem. 262, 1860–1868.

Damer, C.K. & Creutz, C.E. (1994). Secretory and synaptic vesicle membrane proteins and their possible roles in regulated exocytosis. Progress in Neurobiology 43, 511–536.

De Robertis, E. & Vaz Ferreira, A. (1957). Electron microscopic study of the excretion of catechol-containing droplets in the adrenal medulla. Exp. Cell. Res. 12, 568–574.

Drust, D.S. & Creutz, C.E. (1988). Aggregation of chromaffin granules by calpactin at micromolar levels of calcium. Nature 331, 88–91.

Drust, D.S. & Creutz, C.E. (1991). Differential subcellular localization of calpactin and other annexins in the adrenal medulla. J.. Neurochem. 56, 469–478.

Dunn, L.A. & Holz, R.W. (1983). Catecholamine secretion from digitonin-treated adrenal medullary chromaffin cells. J. Biol. Chem. 258, 4989–4993.

Gerke, V. & Weber, K. (1984). Identity of p36k phosphorylated upon Rous sarcoma virus transformation with a protein purified from brush borders: calcium-dependent binding to non-erythroid spectrin and F-actin. EMBO J. 3, 227–233.

Glenney, J.R. & Tack, B.F. (1985). Amino terminal sequence of p36 and associated p10: Identification of the site of tyrosine phosphorylation and homology with S-100. Proc. Natl. Acad. Sci. USA 82, 7884–7888.

Hong, K., Duzgunes, N., Ekerdt, R., & Papahadjopoulos, D. (1982). Synexin facilitates fusion of specific phospholipid membranes at divalent cation concentrations found intracellularly. Proc. Natl. Acad. Sci. USA 79, 4642–4644.

Huber, R., Berendes, R., Burger, A., Schneider, M., Karshikov, A., & Luecke, H. (1992). Crystal and molecular structure of human annexin V after refinement. J. Mol. Biol. 223, 683–704.

Kaplan, R., Jaye, M., Burgess, W.H., Schlaepfer, D.D., & Haigler, H.T. (1988). Cloning and expression of cDNA for human endonexin II, A calcium and phospholipid binding protein. J. Biol. Chem. 263, 8037–8044.

Kirshner, N., Sage, W.J., Smith, W.J., & Kirshner, A.J. (1966). Release of catecholamines and specific protein from the adrenal gland. Science, 154, 529–531.

Knight, D.E. & Baker, P.F. (1982). Calcium dependence of catecholamine release from bovine adrenal medullary cells after exposure to intense electric fields. J. Membr. Biol. 68, 107–140.

Littleton, J.T., Stern, M., Schulze, K., Perin, M., & Bellen, H.J. (1993). Mutational analysis of Drosophila *synaptotagmin* demonstrates its essential role in Ca^{2+}-activated neurotransmitter release. Cell 74, 1125–1134.

Nakata, T., Sobue, K., & Hirokawa, N. (1990). Conformational change and localization of calpactin I complex involved in exocytosis as revealed by quick freeze, deep etch electron microscopy and immunocytochemistry. J. Cell Biol. 110, 13–25.

Nonet, M.L., Grundahl, K., Meyer, B.J., & Rand, J.B. (1993). Synaptic function is impaired but not eliminated in *C. elegans* mutants lacking synaptotagmin. Cell 73, 1291–1305.

Novick, P., Field, C., & Schekman, R. (1980). Identification of 23 complementation groups required for post-translational events in the yeast secretory pathway. Cell 21, 205–215.

Salminen, A. & Novick, P.J. (1987). A *ras*-like protein is required for a post-Golgi event in yeast secretion. Cell 49, 527–538.

Schiavo, G., Benfenati, F., Poulain, B., Rossetto, O., Polverino de Laureto, P., DasGupta, B.R., & Montecucco, C. (1992). Tetanus and Botulinum B neurotoxins block neurotransmitter release by proteolytic cleavage of synaptobrevin. Nature 359, 832–835.

Schlaepfer, D.D. & Haigler, H.T. (1987). Characterization of calcium-dependent phospholipid binding and phosphorylation of lipocortin I. J. Biol. Chem. 262, 6931–6937.

Sollner, T., Bennett, M.K., Whiteheart, S.W., Scheller, R.H., & Rothman, J.E. (1993). A protein assembly-disassembly pathway *in vitro* that may correspond to sequential steps of synaptic vesicle docking, activation, and fusion. Cell 75, 409–418.

Sollner, T., Whiteheart, S.W., Brunner, M., Erdjument-Bromage, H., Geromanos, S., Tempat, P., & Rothman, J.E. (1993). SNAP receptors implicated in vesicle targeting and fusion. Nature 362, 318–324.

Sudhof, T.C., Ebbecke, M., Walker, J.H., Fritsche, U., & Boustead, C. (1984). Isolation of mammalian calelectrins: A new class of ubiquitous Ca^{2+}-regulated proteins. Biochemistry 23, 1103–1109.

Sudhof, T.C. & Jahn, R. (1991). Proteins of synaptic vesicles involved in exocytosis and membrane recycling. Neuron 6, 665–677.

Walker, J.H. (1982). Isolation from cholinergic synapses of a protein that binds to membranes in a calcium-dependent manner, J. Neurochem. 39, 815–823.

Wallner, B.P., Mattaliano, R.J., Hession, C., Cate, R.L., Tizard, R., Sinclair, L.K., Foeller, C., Chow, E.P., Browning, J.L., Ramachandran, K.L., & Pepinsky, R.B. (1986). Cloning and expression of human lipocortin, a phospholipase A_2 inhibitor with potential anti-inflammatory activity. Nature 320, 77–81.

Weng, X., Luecke, H., Song, I.S., Kang, D.S., Kim, S.-H., & Huber, R. (1993). Crystal structure of human annexin I at 2.5 A resolution. Protein Sci. 2, 448–458.
Wilson, D.W., Wilcox, C.A., Flynn, G.C., Chen, E., Kuang, W.J., Henzel, W.J., Block, M.R., Ullrich, A., & Rothman, J.E. (1989). A fusion protein required for vesicle-mediated transport in both mammalian cells and yeast. Nature 339, 355–359.
Wilson, S.P. & Kirshner, N. (1983). Calcium-evoked secretion from digitonin-permeabilized adrenal medullary chromaffin cells. J. Biol. Chem. 258, 4994–5000.

Chapter 7

Endocytosis

THOMAS WILEMAN

Principles of Medical Biology, Volume 7A
Membranes and Cell Signaling, pages 143–169.

ISBN: 1-55938-812-9

INTRODUCTION

Cells take up proteins from the extracellular fluid by endocytosis. During endocytosis, localized areas of the plasma membrane pinch off and form vesicles that carry small droplets of extracellular fluid into the cell. Endocytic vesicles are relatively small, and are unable to take up significant quantities of proteins present at low concentration within the extracellular media. To get around this problem cells use high affinity receptors to capture proteins from the extracellular fluid. This specialized form of endocytosis is called receptor-mediated endocytosis. The cell surface receptors concentrate in endocytically active areas of the plasma membrane called coated pits and enter the cell via coated vesicles. The receptor-mediated pathway is very efficient, and a single cell can take up millions of protein molecules per hour. In the absence of receptors, proteins are taken into cells passively by fluid phase endocytosis. Because there is no concentration of extracellular protein at the cell surface, fluid phase endocytosis is very inefficient. Fluid phase endocytosis is however useful for the uptake of proteins and solutes that are present at high concentration in the extracellular fluid.

RECEPTOR DISTRIBUTION AND RECEPTOR RECYCLING

Membranes and Receptors Recycle During Endocytosis

Endocytosis results in the internalization of relatively large amounts of plasma membrane. Fibroblasts, for example, ingest up to 50% of their plasma membrane each hour. In most cases the content of endocytic vesicles is delivered to lysosomes for degradation. If the endocytosed membrane were not replaced continuously, then endocytosis would cause cells to get smaller and smaller. Interestingly, a high percentage of the internalized membrane, and associated receptors are not degraded, but are returned to the cell surface by a process called receptor recycling (Steinman et al., 1983). This ability to recycle membrane means that cells can maintain constant dimensions during hours of endocytosis without synthesizing new membrane on a massive scale. Receptor recycling is a good example of intracellular sorting. Membrane, and membrane associated material are sorted from the contents of endocytic vesicles before fusion with lysosomes. Sorting occurs in an intermediate membrane compartment called an endosome. The internal pH of endosomes is acidic and the low pH is used to regulate the sorting of internalized ligands. In several cases the fall in pH in endosomes accelerates the dissociation of ligands from their receptors.

There are Three Main Pathways for Recycling Receptors

It is possible to place endocytic pathways into three general groups. These pathways are illustrated diagrammatically in Figure 1, and physiological examples are described below.

Receptors Recycle, but Ligands are Degraded in Lysosomes. Endocytosis of Low Density Lipoprotein and the Uptake of Cholesterol

Cholesterol is a hydrophobic sterol that is a major component of cell membranes. Cholesterol is highly insoluble in water, and when present at high levels in the blood, cholesterol precipitates to form atherosclerotic plaques. Individuals with high blood cholesterol levels become predisposed to atherosclerosis and heart attacks early in life. In normal individuals the majority of the total blood cholesterol is esterified and stored within a large protein-lipid complex called low density lipoprotein (LDL). LDL prevents the deposition of cholesterol in the circulatory system, and at the same time, targets cholesterol for uptake into cells where it is needed for membrane synthesis.

Endocytosis of LDL is mediated by a specific cell surface receptor called the LDL receptor. LDL receptors are concentrated in coated pits and take up LDL particles by receptor-mediated endocytosis. The LDL receptor follows pathway 1 described in Figure 1. LDL-LDL receptor complexes enter the cell through coated pits and coated vesicles and are delivered to endosomes. The low pH of endosomes causes dissociation of LDL from the receptor. The receptor recycles to the cell surface and is used to internalize more LDL; meanwhile, the LDL ligand is transported to lysosomes and is degraded. Degradation releases cholesterol from the LDL particle, and the sterol passes across the lysosome membrane into the cytosol where it is used in the biosynthesis of new membrane. The whole pathway is remarkably rapid, LDL receptors recycle every 10 minutes or so, and almost complete degradation of LDL particles occurs within 20 minutes of entry into the cell.

Many individuals with familial hypercholesterolemia inherit a mutated gene encoding the LDL receptor. Some of these mutations result in receptors that are unable to cluster in coated pits. Recent advances in recombinant DNA technology have enabled many of these defective genes to be cloned and sequenced. In the J.D. mutation, a change in a single base pair converts a tyrosine residue in the cytoplas-

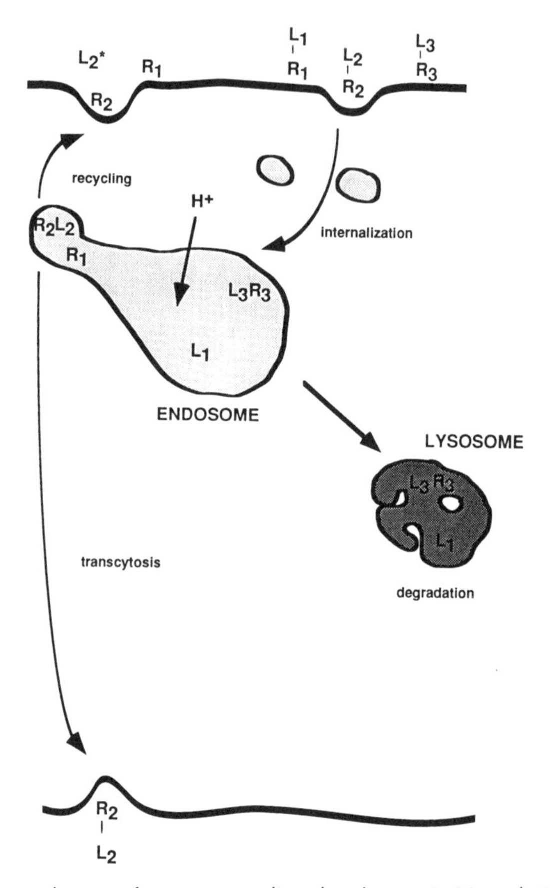

Figure 1. Three pathways of receptor-mediated endocytosis. Ligands (L1, L2, L3) bind receptors (R1, R2, R3) at the plasma membrane of cells and are taken up into endosomes. Receptor-ligand sorting is regulated by pumping protons into the lumen of the endosome. Pathway 1. The low pH of endosomes causes the dissociation of receptor ligand complexes. Dissociated ligands (L1) travel to lysosomes, while receptors (R1) recycle to the cell surface. Pathway 2. Receptor ligand complexes (R2L2) do not dissociate at low pH, but recycle to the cell surface. In polarized cells receptor ligand complexes may be delivered to basolateral (lower) or apical (upper) domains. Modified ligands (L2*) may dissociate at the plasma membrane. Pathway 3. Receptor-ligand complexes (R3L3) do not dissociate at low pH and travel from endosomes to lysosomes and are degraded.

mic tail of the receptor to a cysteine. As will be described below, this mutation occurs in a region of the receptor that is crucial for entry coated pits. The defective receptor binds LDL at the plasma membrane, but is unable to enter the endocytic pathway and cannot, therefore, clear LDL or cholesterol from the plasma.

Receptors and Ligands Recycle through Endosomes

The transferrin cycle and the uptake of iron. The important feature of this pathway is that the receptor-ligand complexes formed at the cell surface are stable in the acidic lumen of endosomes after internalization. Receptors and ligands therefore recycle to the cell surface (pathway 2 of Figs. 1 and 2). Transferrin is a serum glycoprotein that transports iron to all the tissues of the body. The uptake of transferrin by cells starts when two molecules of iron bind to an empty (apo) transferrin molecule in the serum. The binding of iron increases the affinity of

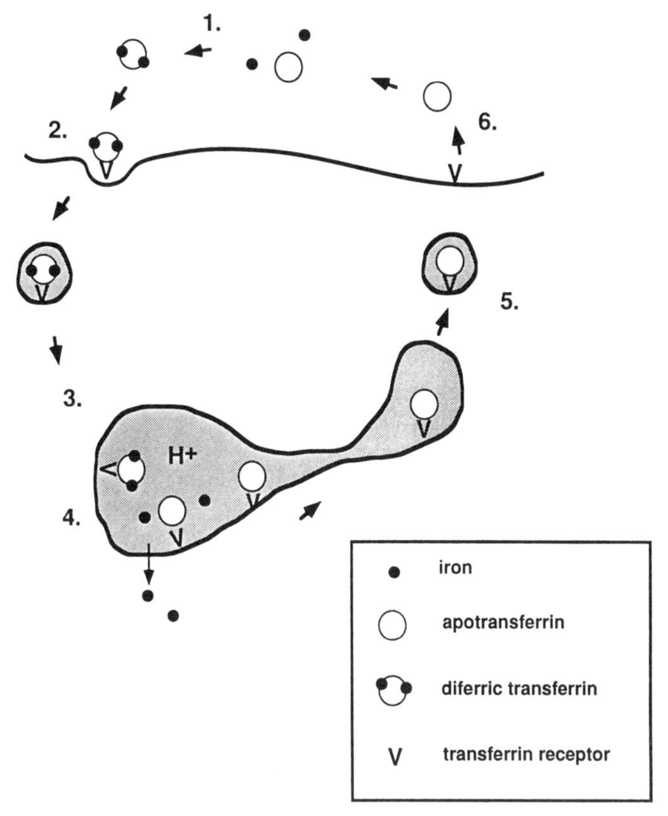

Figure 2. The transferrin cycle and the uptake of iron. 1. Iron binds apotransferrin in the plasma. 2–3. The resulting diferric transferrin has a high affinity for the transferrin receptor, and enters the cell by receptor-mediated endocytosis. 4. The low pH of endosomes causes iron to dissociate from transferrin and pass across the endosomal membrane into the cell. 5. The resulting apotransferrin remains bound to the transferrin receptor at acid pH, and recycles to the cell surface. 6. Apotransferrin dissociates from the transferrin receptor at physiological pH and is free to bind more iron.

transferrin for the cell surface transferrin receptor and the resulting diferric transferrin molecule is taken into cells by receptor mediated endocytosis. On reaching the endosomal compartment, the pH surrounding diferric transferrin falls. This causes iron to dissociate from transferrin and pass across the endosomal membrane into the cytoplasm. Importantly, apotransferrin remains bound to the transferrin receptor at low pH and recycles with the receptor to the plasma membrane. On reaching the cell surface, the apotransferrin-transferrin receptor complex encounters physiological pH (7.4) causing apotransferrin to dissociate. The released apotransferrin is now free to adsorb more iron, bind to the transferrin receptor, and initiate another round of uptake.

Transcytosis in polarized cells. Epithelial cells have specialized upper (apical) and lower (basolateral) surfaces. This polarity is maintained because epithelial cells direct proteins specifically to one surface or the other. The polymeric immunoglobulin (polyIg) receptor is a good example. The poly Ig receptor is synthesized by the endoplasmic reticulum and then sorted via the Golgi apparatus to the basolateral plasma membrane where the receptor binds dimeric IgA (Figure 3). Next, the receptor-IgA complex enters the endocytic pathway and appears at the apical surface. A proteolytic clip releases the ectodomain of the receptor, called secretory component, and the bound IgA, into the apical media. The net result is the transport of IgA across the cell. In tissues this corresponds to the uptake of IgA from the blood and its secretion in milk, saliva, and bile.

Polarized transport also travels in the opposite direction. Epithelial cells in the gut of suckling rats express a receptor that binds maternal IgG secreted into milk. Importantly, the receptor binds IgG at low pH, the pH of the gut, but releases the immunoglobulin molecule at physiological pH. Receptor-IgG complexes are taken from the gut lumen into epithelial cells by receptor mediated endocytosis (Figure 3). Receptor-ligand complexes remain intact in the acidic lumen of the endocytic pathway, and recycle to both apical and basolateral surfaces. Recycling to the basolateral surface results in exposure of the receptor to physiological pH, and the release of the immunoglobulin. In this way a pH gradient across the gut epithelium can direct the one way traffic of a ligand into the blood.

Receptors and Ligands Travel to Lysosomes

Down regulation of peptide growth factor receptors. Peptide hormones and growth factors, such as epidermal growth factor and insulin, bind to cell surface receptors and deliver a mitogenic stimulus to cells. Receptor ligand complexes cluster in coated pits, and enter cells by receptor-mediated endocytosis. In contrast to the two examples described above, the peptide hormone receptors do not recycle, but remain with the endosome and are eventually delivered to lysosomes where they are degraded (pathway 3, Figure 1). This causes the removal of receptors from the surface of cells, and prevents the constitutive stimulation of signal transduction

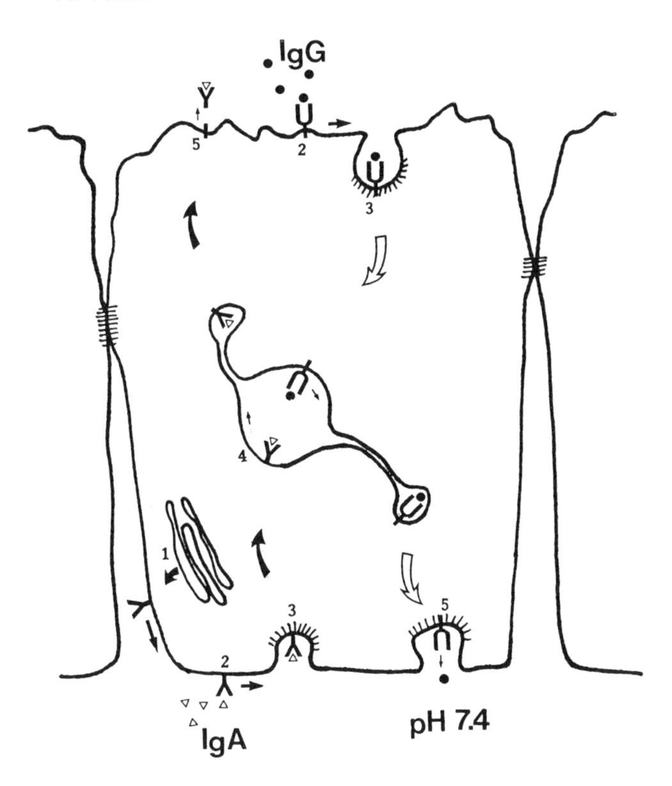

Figure 3. Receptor-mediated endocytosis in polarized cells: The transport of immunoglobulin molecules across gut epithelial cells. Polymeric IgA. (*1*) IgA receptors are synthesized in the endoplasmic reticulum and pass through the Golgi to the basolateral cell surface. (*2–3*) IgA receptors bind polymeric IgA, and receptor-ligand complexes are internalized into endosomes. (*4*) Receptor-ligand complexes are stable at acid pH and recycle to the apical surface. (*5*) A protease at the apical surface cleaves the ectodomain of the receptor. The receptor fragment, now called secretory component, and the ligand are released as a complex into the gut. IgG. (*2–3*) Immunogobulin receptors bind IgG in the gut at low pH and receptor-ligand complexes are taken up into endosomes. (*4*) Receptor-ligand complexes are stable at acid pH and recycle to the basolateral surface. (*5*) Immunoglobulin dissociates at physiological pH, and is released into the blood.

pathways by mitogens. This process is sometimes called receptor "down modulation." The cytoplasmic tails of many peptide hormone receptors have protein tyrosine kinase activities that phosphorylate key proteins in signal transduction pathways. Receptor stimulation also produces a high level of autophosphorylation. Phosphorylation of cytoplamsic tyrosine residues may provide the molecular signal that leads to the transport of this class of receptor from endosomes to lysosomes.

Cells Often Contain Large Intracellular Pools of Endocytosis Receptors

Ligands enter cells continuously at 37°C. When the rate of uptake is calculated, it becomes apparent that many more molecules enter cells than can be accounted for by the binding of ligands to cell surface receptors. These results suggest that receptors are reused after internalization, and have led to the idea of receptor recycling (Goldstein et al., 1979). A detailed analysis of the recycling of the asialoglycoprotein receptor of hepatocytes shows that the receptor recycles with a half time of 7.5 minutes, and does so a total of 250 times in its lifetime. Interestingly, receptors internalize faster ($T_{1/2}$ 1–2 minutes) than they can be recycled to the surface ($T_{1/2}$ 7.5 minutes). This discrepancy predicts that there should be a transient loss of cell surface binding activity immediately following a wave of endocytosis. Quantitative experiments demonstrate, however, that there is very little loss of binding activity during this period. This subtle point suggests that there are internal pools of receptors that move quickly to the cell surface to replace those that have just been internalized. In many cases internal pools of receptors hold the majority of the receptors synthesized by cells. Hepatocytes, for example, have 860,000 asialoglycoprotein receptors, but at any particular time point, only 80,000 are at the cell surface.

COATED PITS AND COATED VESICLES

Coated Pits and Coated Vesicles Form the Entrance to the Endocytic Pathway

Receptors and receptor-ligand complexes are concentrated in specialized domains of the plasma membrane called coated pits. When the plasma membrane is viewed in cross section by the electron microscope, a protein coat can be seen attached to the cytoplasmic face of the pit. The coat, which will be described in detail below, is made from a protein complex called clathrin. The clathrin lattice causes coated pits to invaginate to form coated vesicles. Given that approximately 1500–3000 pits enter the cell each minute, coated pits and coated vesicles provide a very efficient means of delivering receptors and ligands into the endocytic pathway. Coated pits do not show a preference for individual receptors and many different receptors, including those that will ultimately follow different intracellular pathways, enter the cell via the same coated pit.

Coated Pits and Coated Vesicles are Coated with Clathrin

Coated vesicles can be purified from tissue sources such as placenta and brain (Pearse, 1987). Isolated vesicles are covered by a complex latticework of protein arranged into polygons. These same polygons can be seen attached to the cytoplasmic face of coated pits that bud from the plasma membrane of cells (Figure 4 and Heuser and Evans, 1980). The major component of both latticeworks is again clathrin. Clathrin is composed of a heavy chain of 180 kD, and two light chains,

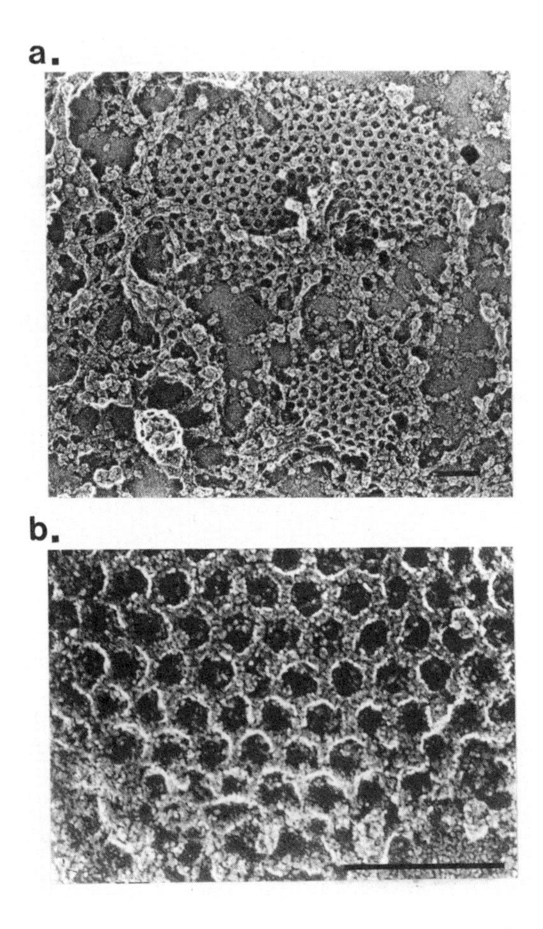

Figure 4. Coated pits revealed by electron microscopy. The plasma membrane of human fibroblasts was analyzed by rapid freeze deep etch microscopy. The images show the cytoplasmic face of the coated pit attached to the plasma membrane. The intricate clathrin coat is clearly visible. Bar, 0.1 μm. b). Higher magnification of the clathrin coat. The coat is seen arranged into regular polygons, each is approximately 30 nm across. (From Heuser and Evans, 1980, with permission).

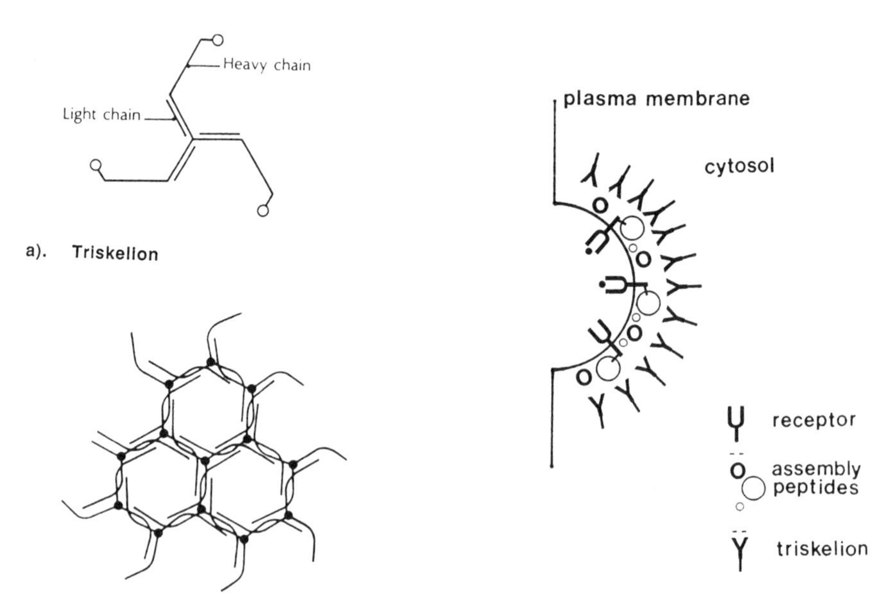

Figure 5. Arrangement of clathrin, adaptins and receptors in coated pits. (*a*) Three heavy and three light chains assemble into each clathrin triskelions. (*b*) An intricate arrangement of triskelions produces the hexagonal clathrin coats seen on the cytoplasmic face of coated pits. (*c*) Adaptins, or assembly peptides, bind to receptor tails exposed on the cytoplasmic face of the plasma membrane. Note that the adaptins lie between the plasma membrane and the clathrin coats. (Adapted from Pearse, 1987).

LC_a and LC_b, of 23 and 27 kD respectively. The proteins of clathrin are arranged into three-legged structures called triskelions (Figure 5a). A triskelion is constructed from three heavy chains with their carboxyl termini joined at a central hub. The light chains interact predominantly with the center of the triskelion, leaving the remainder of the heavy chains to extend to form legs (Pearse, 1987; Brodsky, 1988). Triskelions form the structural unit of the polygonal latticework and purified triskelions will condense, under appropriate *in vitro* conditions, to form polygonal clathrin cages. The results suggest that polygonal lattice formation is driven by the correct packing of heavy chains (Figure 5b). The light chains are not needed for the assembly of triskelions.

Clathrin Coats Contain Adaptin Proteins

A detailed analysis of coated vesicles shows that proteins other than clathrin are associated with triskelions. These proteins are called adaptin or assembly proteins

(Robinson, 1992). Adaptins assemble into heterotetrameric complexes, and two structurally related adaptor complexes, AP1 and AP2, have been described (Figure 6). AP1 contains two different 100 kD proteins called γ and β′ adaptin; these associate with light chains of 47 and 19 kD. AP2 complexes also contain two 100 kD proteins, α and β, and these associate with light chains of 47 and 19 kD. The structure of adaptin complexes has been probed using the electron microscope (Heuser and Keen, 1988). Rotary shadowed images of plasma membrane adaptins show a central protein core with two earlike extensions. The ears are formed from the C-termini of the two 100 kD chains, the N-termini form the protein core, and associate strongly with the light chains. The β chain of AP2 is 85% identical to the β′ chain of AP1. The α chain of AP2, however, shows little (25%) structural relationship with the γ chain of AP1 suggesting the α and γ subunits may have different functions. α and γ adaptins appear to have different subcellular distributions. Antibodies raised against the α adaptin of AP2 complexes bind to coated pits at the plasma membrane, but antibodies specific for the γ subunit of AP1 complexes bind to coated pits formed in the trans Golgi network. The results suggest that the α and γ subunits target the different adaptin-clathrin complexes to specific subcellular locations (Robinson, 1992).

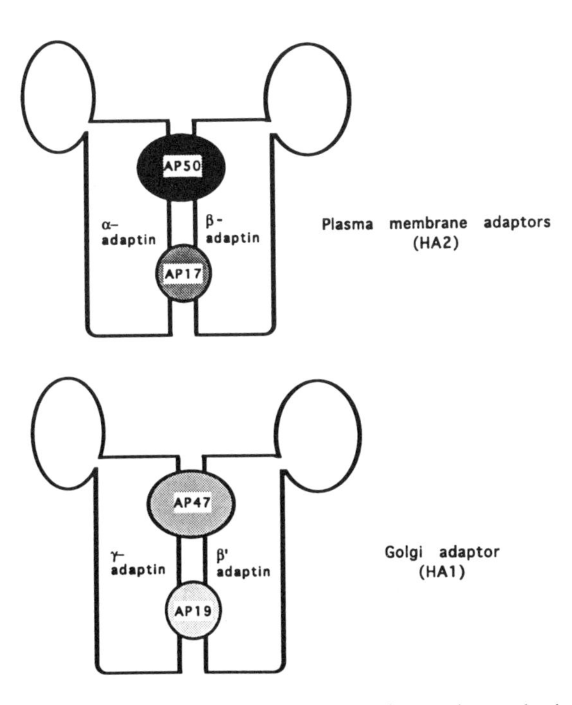

Figure 6. Adaptin complexes. The diagrams show the packing of adaptin proteins into AP1 (also called HA2 or Golgi adaptins) and AP2 (also called HA1 or plasma membrane adaptins) complexes. (Adapted from Robinson, 1992).

It is thought that adaptin proteins promote the assembly of clathrin cages. When mixed with purified triskelions, adaptor proteins stimulate coat formation, and are incorporated into the interior of the polygonal cages. This position allows adaptins to interact directly with the cytoplasmic face of the membrane contained within coated pits or coated vesicles (Figure 5c). This model is supported by *in vitro* experiments demonstrating that adaptin proteins bind directly to the cytoplasmic tails of receptors that concentrate in coated pits (see below).

The Cytoplasmic Tails of Receptors Contain Internalization and Intracellular Sorting Motifs

It is now known that specific amino acid sequences present within the cytoplasmic tails of receptors cause them to cluster in coated pits. The importance of a tyrosine (**Y**) residue was first deduced from the sequence of the naturally occurring (JD) mutant of the LDL receptor. The JD mutant receptor, which is unable to cluster in coated pits, carries a single point mutation that changes a tyrosine residue in the cytoplasmic tail to a cysteine (Brown and Goldstein, 1986). Tyrosine-based internalization signals have now been studied in some detail (Lazarovits and Roth, 1988, reviewed in Vaux 1992; Trowbridge et al., 1993) and have been found in the cytoplasmic tails of several receptors that enter the endocytic pathway. In some instances the tyrosine residue can be replaced by phenylalanine without loss of function, suggesting that aromatic residues are important. The generation of an internalization signal, however, requires that the aromatic residues are arranged within a correct amino acid context. Studies on the LDL receptor have identified a minimal internalization sequence of asparagine-proline-?-tyrosine, called an NPX**Y** motif. Other tyrosine-based internalization motifs have been defined for the mannose-6-phosphate receptor (**Y**K**Y**SKV) and for the transferrin receptor (**Y**XRF). Recently, peptides containing NPX**Y** and related sequences have been analyzed by NMR and have been shown to form tight β turns in solution (Bansal and Gierash, 1991; Eberle et al., 1991). The tight turn is thought to be maintained by a strong interaction between the tyrosine/aromatic residue and the asparagine (**N**). Significantly, the degree of tight turn formed by synthetic peptides *in vitro* correlates well with the ability of the same sequence to induce the internalization of receptors *in vivo*. A second, and structurally unrelated internalization motif has been defined recently. The cytoplasmic tails of several proteins that enter the endocytic pathway contain a leucine-leucine or leucine-isoleucine internalization motif (Letourneur and Klausner, 1992). In some cases the dileucine motifs complement tyrosine-based internalization signals; for other receptors, they function independently.

As described above, tyrosine and dileucine based internalization motifs cause receptors to cluster in coated pits and enter the endocytic pathway. Considerable evidence now suggests that tyrosine and dileucine-based internalization motifs also determine the intracellular fate of proteins once they have entered the endocytic pathway. Internalization motifs have been shown to direct proteins from endosomes to lysosomes, and to be important for sorting membrane proteins to apical or basolateral membranes in polarized cells. How is it then that the internalization motifs of receptors that enter the endocytic pathway, but then return to the cell surface, fail to be recognized by lysosomal or polarized sorting mechanisms? The answer may lie within the amino acids that surround internalization motifs. Different neighboring amino acids may alter the affinity of internalization motifs for intracellular sorting pathways. The cytoplasmic tail of the LDL receptor contains two NPXY motifs with different flanking amino acids. Both motifs act as basolateral sorting signals, but only one can operate independently as an internalization sequence. For the poly Ig receptor there are two tyrosine based internalization/endocytosis signals, and *separate basolateral* sorting signal, that does not contain a crucial tyrosine residue.

Phosphorylation May Alter the Activity of Internalization and Sorting Signals

The cytoplasmic tails of poly IgG and mannose-6-phosphate receptors are phosphorylated on serine residues during endocytosis. Interestingly, these serine residues are found close to known sorting motifs (Figure 7). It is possible that phosphorylation switches sorting signals on and off. Significantly, kinase activities capable of phosphorylating the cytoplasmic tails of receptors have been found in purified coated vesicles. It is thought that, for some receptors, these kinases may regulate the interaction of receptors with adaptin proteins found within coated pits and vesicles.

Adaptins Bind to Internalization Motifs

In vitro binding data show that purified adaptin proteins bind to tyrosine-based internalization motifs. Importantly, the specificity of adaptin binding correlates well with what is known about the intracellular distribution of adaptin complexes.

^{653} R-A-R-H-R-**R-N-V-D**-R-V-**S***-I-G-S-Y-R (*a*)

β-turn

^{2489}H-D-D-**S***-D-E-D-**L-L**-H-V (*b*)

dileucine

Figure 7. Phosphorylation of serine residues close to internalization motifs. (*a*) β-Turn implicated in the sorting of the IgA receptor to the basolateral surface of gut epithelial cells. (**S*** phosphorylated serine) (*b*) Dileucine sequence important for targeting of the man-6 P receptor from endosomes to the lysosomal pathway. (**S*** phosphorylated serine)

As described above, cells contain two adaptin complexes, AP1 and AP2, which bind to the trans Golgi or plasma membrane respectively. Purified plasma membrane adaptins (AP2) bind to the cytoplasmic tails of receptors such as the LDL receptor and mannose 6-P receptor that concentrate in coated pits formed at the plasma membrane. Importantly, purified plasma membrane adaptins do not bind to the cytoplasmic tail of the LDL receptor carrying the J.D. mutation. The J.D. mutation causes the loss of the tyrosine-based internalization motif, and produces a receptor that cannot cluster in coated pits *in vivo*. Equally compelling is the observation that Golgi (AP1) adaptors do not bind to the LDL receptor tail *in vitro*, but do bind to the internalization motif of the mannose-6-P receptor. Again, this specificity correlates well with *in vivo* observations. The mannose-6-P receptor would be expected to bind to both adaptins because it is found in the Golgi apparatus and at the plasma membrane. The LDL receptor is only found at the plasma membrane, and would not be expected, therefore, to bind to Golgi AP1 adaptors.

Adaptins Seed the Assembly of Coated Pits and Coated Vesicles

Several elegant *in vitro* assays have started to define the biochemistry of coated pit and coated vesicle assembly (Moore et al., 1987; Schmid et al., 1992, 1993). The cytoplasmic face of the plasma membrane contains a finite number of assembly sites that seed the formation of coated pits (Figure 8). It is thought that assembly is initiated by adaptins. Adaptin proteins are located on the inside of the clathrin lattice, where they could easily interact with both the plasma membrane and with clathrin. Moreover, purified adaptins bind to plasma membranes in the absence of clathrin, and to purified clathrin cages *in vitro*. Importantly, clathrin does not bind to membranes in the absence of adaptin proteins.

The precise binding site for the adaptin proteins on the plasma membrane is not known. Above, it was described how adaptin proteins bind specifically to tyrosine-based internalization motifs on the cytoplasmic tails of endocytosis receptors. It was once thought that internalization motifs would be ideal candidates for seeding coated pit formation. Interestingly, clathrin coats are not found on endocytic vesicles. Endocytic vesicles, nevertheless, expose receptor tails to the cytosolic pools of adaptins and clathrin. There must, therefore, be some way of preventing the interaction of clathrin with internalization motifs once receptors have entered the endocytic pathway. The binding of adaptins to receptor tails is probably regulated by an intermediate protein. The nature of this protein is unknown, but it is possible that it is an integral membrane protein of the plasma membrane (Figure 8).

The next stage in the endocytic cycle is the invagination of the clathrin coated area of membrane, and the budding of a coated vesicle. Even though the biochemistry of these steps is not well understood some *in vitro* experiments suggest that invagination and budding require Ca^{2+}, ATP, cytosolic proteins, and the hydrolysis of GTP. A candidate protein that may regulate the budding of coated vesicles through the hydrolysis of GTP is cytoplasmic dynamin (reviewed by Trowbridge,

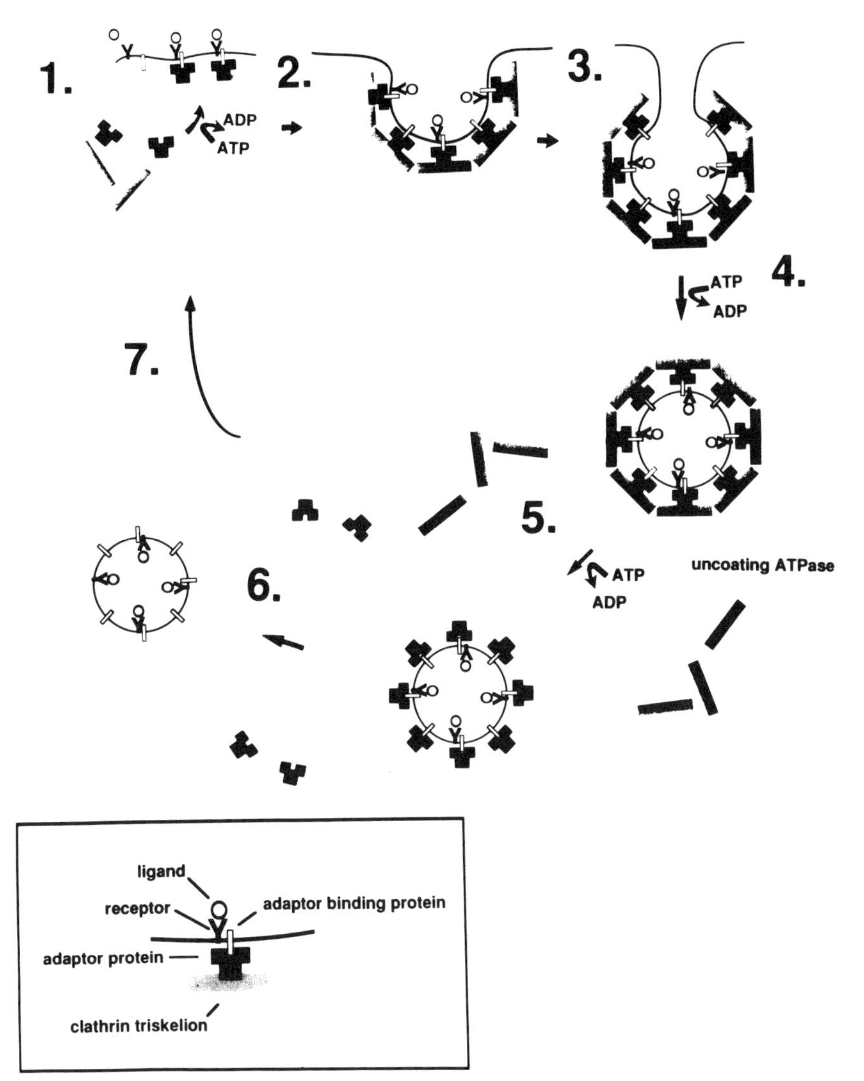

Figure 8. Cycles of coated pit and coated vesicle assembly. (*1*) The cytoplasmic tails of receptors associate with adaptor-binding proteins. Adaptor-binding proteins have not been identified, but it is thought that they are integral membrane proteins of the plasma membrane. This interaction regulates the association of adaptin complexes with the cytoplasmic tails of receptors. The adaptin-tail complex seeds the assembly of clathrin coats onto the cytoplasmic face of the plasma membrane. These steps may require hydrolysis of ATP. (*2*) Rearrangement of clathrin triskelions into regular polygons causes invagination of the membrane and formation of a coated pit. (*3–4*) Coated pits pinch off from the membrane and form coated vesicles. This may require hydrolysis of ATP and GTP and the binding of dynamin and annexin IV. (*5*) Clathrin coats are removed by uncoating ATPase. (*6*) Adaptin complexes may also be lost from coated vesicles; if and how, this occurs is not known. (*7*) Soluble pools of clathrin triskelions and adaptin complexes are available to repeat the cycle. (Adapted from Schmid, 1993.)

1993; Kelly, 1995). Dynamin was first implicated as being important for receptor recycling from studies on the Shibire mutation of Drosophilia. The mutation causes defective recycling of synaptic vesicles at the nerve endings and leads to paralysis of the fly. The *Shibire* gene is homologous to mammalian dynamin. This information led several groups to determine the effects of mutant dynamin molecules on endocytosis in mammalian cells *in vivo*. Significantly, mutant dynamin molecules that cannot bind GTP, inhibit endocytosis. These results provide a direct link between dynamin, the hydrolysis of GTP, and endocytosis. The precise site of action of dynamin is unknown, but it is suspected that dynamin regulates the budding of coated vesicles from the plasma membrane (Takei et al., 1995; Hinshaw and Schmid, 1995).

Annexin IV is a second cytosolic protein that may be involved in the budding of coated vesicles (Lin et al., 1992). Vesicle budding *in vitro*, for example, requires annexin and Ca^{2+}. Annexin IV binds tightly to Ca^{2+}, and also to phophatidylethanolamine, a phospholipid enriched on the cytoplasmic face of the plasma membrane. It is thought that annexin may participate in the severing of the stalk that attaches invaginated coated pits to the plasma membrane (Steps 3 & 4 of Figure 8).

Removal of Clathrin Coats Requires an ATPase

Clathrin coated vesicles are transient structures that lose their coats within minutes of pinching off from the cell membrane. This would appear to be in conflict with the studies described above that demonstrate the spontaneous coating of plasma membrane with clathrin coats *in vitro*. An enzyme capable of removing clathrin coats from vesicles has been purified. The enzyme uses ATP as an energy source, and has been called uncoating ATPase. Uncoating ATPase binds to intact cages, hydrolyzes ATP, and releases itself in a stoichiometric complex with intact triskelions. The way in which the clathrin is processed for another round of coated pit formation is not known. Furthermore, it is not known if, or when, adaptin molecules, or the putative adaptin regulatory protein leave coated vesicles.

ENDOSOMES

Uncoated vesicles fuse with small tubular and vesicular endosomes located near the cell surface. The endosome compartment controls the subsequent destination of internalized receptors and ligands.

Protein Sorting in Endosomes Requires Acidification

It has been known for several years that the uptake of many ligands by receptor-mediated endocytosis is slowed when cells are incubated with drugs such as amines and proton ionophores that collapse intracellular pH gradients. These experiments provided early evidence that intracellular sorting in endosomes required endosomal acidification. Support for this model came from studies on the physical chemistry of receptor-ligand interactions. In several cases, purified receptors were found to have a high affinity for ligand at neutral pH, and a low affinity at acid pH. Putting these two observations together, it seemed likely that a lowering of pH in the lumen of endosomes would accelerate the dissociation receptor-ligand complexes *in vivo* and provide the first step for the physical separation of receptors and ligands into different membrane compartments. Direct evidence for endosomal acidification has been provided by fluorescein-labeled ligands. Fluorescein acts as a sensor of pH because the fluor emits strongly at acidic pH, and only weakly at neutral pH. The fluorescence emitted from labeled ligands rises sharply after they enter cells, suggesting entry of ligands into an acidic compartment. The average pH of endosomes is approximately 5.5. Peripheral endosomes near the cell surface are slightly less acidic (pH 6.0) than late endosomes and lysosomes (pH 5.5).

Endosomes are Acidified by ATP-Dependent Proton Pumps

Fluorescent probes have been used in *in vitro* experiments to determine the mechanism of endosome acidification. When purified endosomes, or endosomes in permeabilized cells, are mixed with physiological buffers they have a neutral pH, but significantly, the internal contents acidify when ATP is added to the incubation mixture. These experiments demonstrate that endosomes lower their internal pH by means of ATP-dependent proton pumps. *In vitro* acidification experiments also demonstrated that coated vesicles contain proton pumps. Because large quantities of coated vesicles can be isolated from animal tissues such as calf brain, it has been possible to purify the proton pump (Arai et al., 1987; 1988). The ATP-dependent proton pump is a large multisubunit complex containing nine polypeptides with molecular weights ranging from 17–100 kD, (Figure 9). The 17 kD subunit is very hydrophobic and present as six copies in each pump. It is likely that these subunits form an ion channel, and regulate the transport of protons across the membrane. The functions of the other subunits are largely unknown, but the 59 and 78 kD subunits are exposed to the cytoplasm and may be ATP-binding proteins.

Endosomes are Dynamic Organelles

Endosomal acidification provides the correct physiological conditions for the dissociation of ligands and receptors. Evidence for the subsequent physical separation of receptors and ligands into different membrane compartments has come from elegant immunogold electron microscopy experiments. By using different sized gold beads attached to antibodies specific for ligands or internalized receptors, Geuze and colleagues (Geuze, 1983) have demonstrated the movement of receptors into tube-like extensions attached to endosomes. Significantly, ligands appear to remain within the lumen of the endosome, suggesting that the tubular extensions separate receptors and ligands and transport recycling receptors back to the plasma membrane.

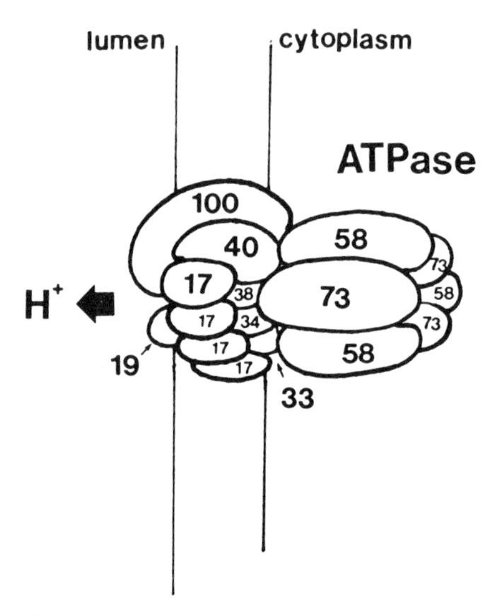

Figure 9. Structure of the vacuolar ATPase. The diagram shows the proposed arrangement of the many subunits of the vacuolar proton translocating ATPase. (From Arie et al., 1988, with permission).

In most cases internalized ligands enter lysosomes within 30–60 minutes of entering the cell and are degraded. It is not known what happens to the endosomal carrier during this period. There are two extreme models for the fate of endosomes (Helenius et al., 1983; Murphy, 1991). In the first, endosomes are considered to move in a conveyerbelt-fashion through the cell and are then destroyed by fusing with lysosomes. In the second, endosomes are considered to be permanent structures that use vesicle shuttles to transfer specific proteins to lysosomes. Current experimental data (Stoorvogel et al., 1991) support an intermediate model for the biogenesis of endosomes (Figure 10). It is thought that early endosomes fuse with large numbers of incoming vesicles, and at the same time recycle the bulk of the internalized receptors to the cell surface. Most of the receptor-ligand sorting occurs in this compartment. Early endosomes are dynamic structures that fuse with one another and there is an extensive mixing of content. There is relatively little direct fusion of early endosomes with late endosomes. Data from *in vitro* experiments suggest that the contents of early endosomes are transported to late endosomes by intermediate carrier vesicles, and that fusion requires intact microtubules (Aniento et al., 1993). Late endosomes also receive small quantities of ligands directly from the coated vesicle pathway, and recycle some proteins to the cell surface. The precise fate of the late endosome is less well understood. They probably mature slowly through the gradual loss of content into carrier vesicles, and ultimately fuse with lysosomes.

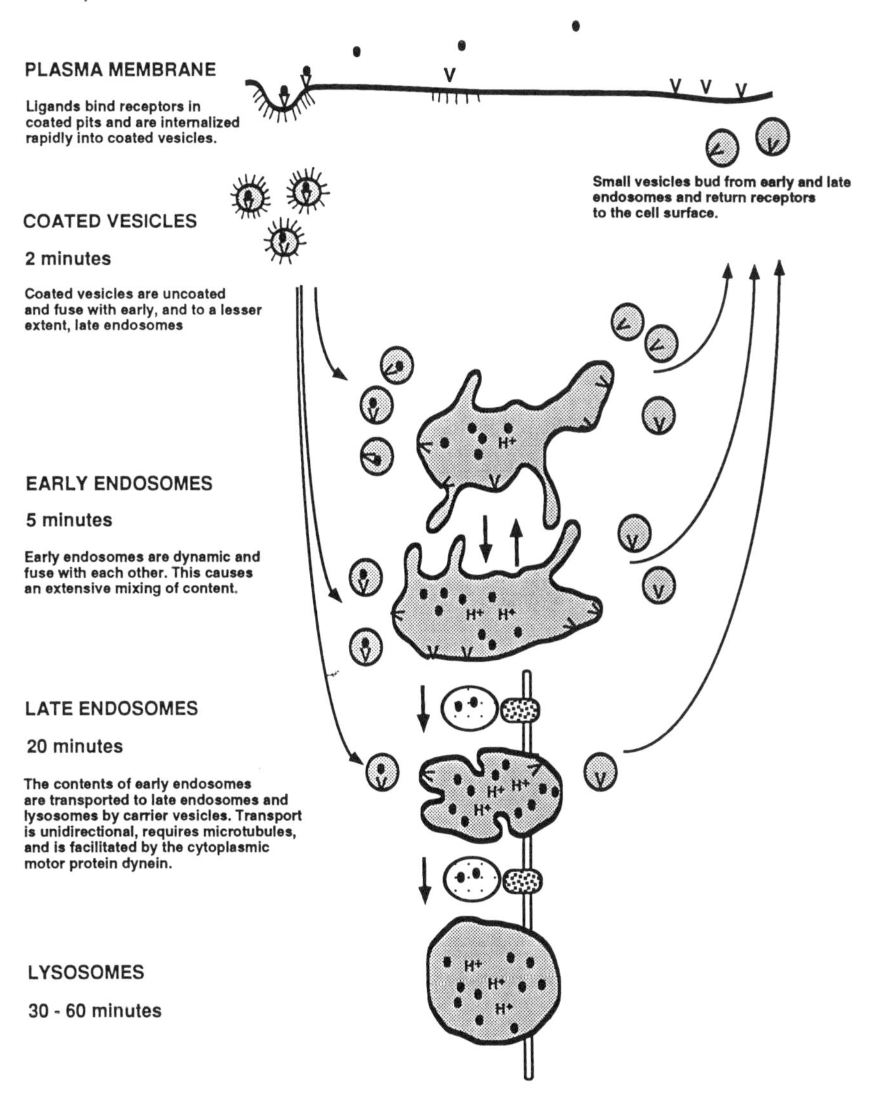

Figure 10. Biogenesis of endosomes. The annotated diagram shows the movement of membrane, receptors and ligands through the endocytic pathway.

Endocytosis is Regulated by Small GTP-binding Proteins Called Rabs

Small GTP binding proteins share homology with the 21 kD GTP-binding protein, *ras*. The conformation of *ras*-like GTP-binding proteins is determined by the extent of phosphorylation of a bound guanine nucleotide. Given that small GTP-binding proteins have GTPase activity, it is possible that they act as molecular switches regulated by cycles of GTP-binding, GTP hydrolysis and GDP release (Bourne, 1988). An extensive family of small GTP-binding proteins, called rab proteins, have been identified that may act as molecular switches within intracellular transport pathways. Three lines of evidence suggest that rab proteins regulate vesicular transport. First, specific rab proteins localize to distinct intracellular membrane compartments. Second, yeasts carrying mutations in rab homologues show specific blocks in secretion. Third, *in vitro* assays that measure specific membrane fusion events of the endocytic and secretory pathway are inhibited by GTPγS: a non-hydrolyzable analog of GTP, rab-specific antibodies and by peptides that mimic rab sequences.

Several experiments implicate rab4 and rab5 as being important in the regulation of the endocytic pathway (Bucci et al., 1992; van der Sluijs et al., 1992). Rab4 and rab5 localize to early endosomes and the plasma membrane, and rab5 can be recovered from isolated coated vesicles. Importantly, rab5 is required for endosome fusion *in vitro* and mutant rab4 or rab5 proteins that are unable to hydrolyze GTP disrupt endocytosis *in vitro*. Current models predict that rab5 regulates the inward arm of the endocytic pathway, and controls the formation of coated vesicles and their fusion with early endosomes. Rab4, on the other hand, may control the recycling pathway back to the plasma membrane. A third rab protein, rab7, concentrates on the membrane of late endosomes, and may be involved in the fusion of late endosomes with lysosomes. Recent experiments identify rab8 on the basolateral membrane of epithelial cells, and in vesicles carrying proteins to the basolateral surface. It is possible that rab8 is important for the transport of proteins to the basolateral surface of cells.

POTOCYTOSIS

Molecules Taken up by Potocytosis Enter Cells Via Caveolae

Potocytosis is a second way for cells to take up molecules from the extracellular media (Anderson, 1992; 1993). The ligands that are taken up by potocytosis are normally of low molecular weight (<1000 daltons) and bind to receptors that are attached to the plasma membrane by glycosylphosphatidylinositol (GPI) anchors (see Chapter 4). During potocytosis, patches of the plasma membrane enriched in GPI anchored proteins invaginate to form vesicles called caveolae. Potocytosis differs from endocytosis because the vesicles formed during potocytosis are thought to remain attached to the plasma membrane. The vesicles (caveolae) nevertheless close, and seal off their content from the extracellular fluid. Several GPI anchored receptors and enzymes are known to concentrate in caveolae. Caveolae can, therefore, accumulate specific extracellular ligands, and convert co-internalized substrates into products. Internalized ligands, and processed substrates, are either stored by caveolae, or delivered into the cytoplasm. Alternatively, they are released from cells when caveolae open.

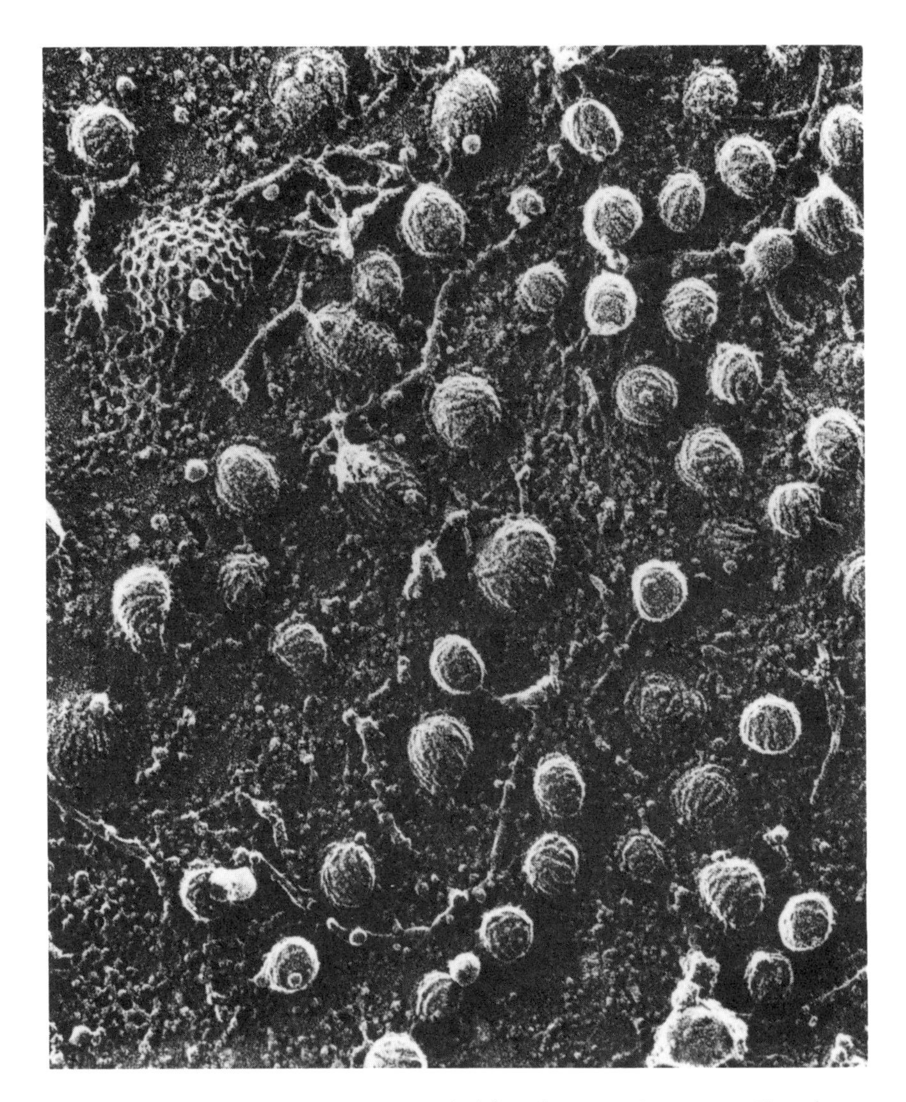

Figure 11. Structure of caveolae revealed by electron microscopy. The plasma membrane of human fibroblasts was analyzed by rapid freeze deep etch microscopy. The images show the striated coat of the caveolae membrane. Each caveolae is between 50 and 100 nm across. A clathrin coated pit can be seen in the top left hand corner. Note that the structure of the caveolae coat is very different from the clathrin coat. (From Rothberg et al., 1992 with permission).

Caveolae are Coated by Caveolin

Caveolae are flask-shaped vesicles of between 50–100 nm diameter attached to the plasma membrane. Caveolae have a distinctive striated protein coat that is easily distinguished from the hexagonal clathrin coats found on endocytic vesicles (compare Figures 4 and 11). The major coat of protein of caveolae is called caveolin (Rothberg et al., 1992). Cell lines that do not express caveolin are unable to form caveolae. Such results suggest that caveolin controls the assembly and invagination of caveolae, and that caveolae are important for the intracellular transport of GPI-anchored proteins. Potocytosis via caveolae is highly dependent on the lipid composition of cells. Potocytosis, for example, requires cholesterol, and isolated caveolae contain high levels of cholesterol and sphingolipids. It is thought that sterols and sphingolipids are important for the clustering of GPI anchored proteins into caveolae. In addition, the assembly of caveolae may require a tripartite interaction between caveolar lipids, the GPI anchors of receptors, and the hydrophobic C-terminus of caveolin.

Although most cell types contain caveolae, the distribution of caveolae within a given cell is not always uniform (Partan, 1996). Caveolae often cluster together into patches containing several hundred vesicles. In fibroblasts, caveolae cluster at the leading edge of the cell. It is thought that the heterogeneous distribution of caveolae within cells is organized through binding to actin and other molecules of the cytoskeleton.

There are Three Types of Potocytosis

These are represented diagrammatically in Figure 12, and physiological examples are described below.

1. Receptor-mediated potocytosis. During receptor-mediated potocytosis, high affinity GPI-anchored receptors concentrate extracellular ligands in patches at the cell surface. Clustered receptor-ligand complexes internalize into caveolae. The caveolae close, and changes within the lumen of the caveolae, such as a decrease in pH, or loss of Ca^{2+}, facilitate the dissociation of ligands from receptors. Ligand molecules reach high concentrations within caveolae and diffuse down concentration gradients into the cytosol. The best characterized example of receptor mediated potocytosis is the uptake of the essential vitamin, 5-methyltetrahydrofolate. Individual caveolae contain 600–700 high affinity folate receptors. When caveolae close, a fall in luminal pH triggers the release of 5-methyltetrahydofolate from the folate receptor. Folate concentrations reach mMolar concentrations, and these high concentrations force folate molecules through anion channels into the cell.

2. Enzyme-mediated potocytosis. Several GPI-linked plasma membrane proteins have enzyme activities. The accumulation of these enzymes in caveolae provides a means of localizing enzyme reactions to specialized sites at the cell

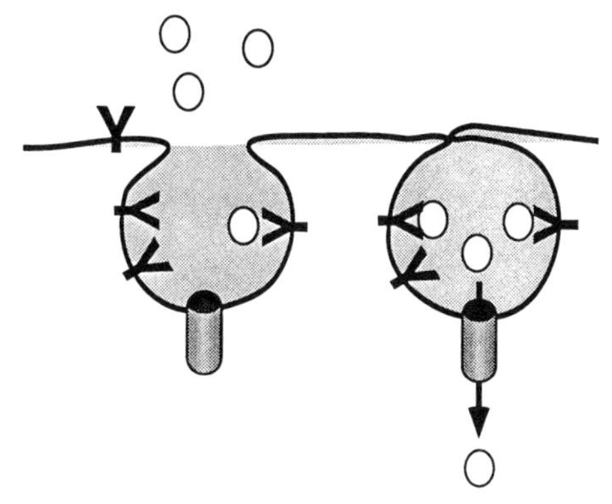

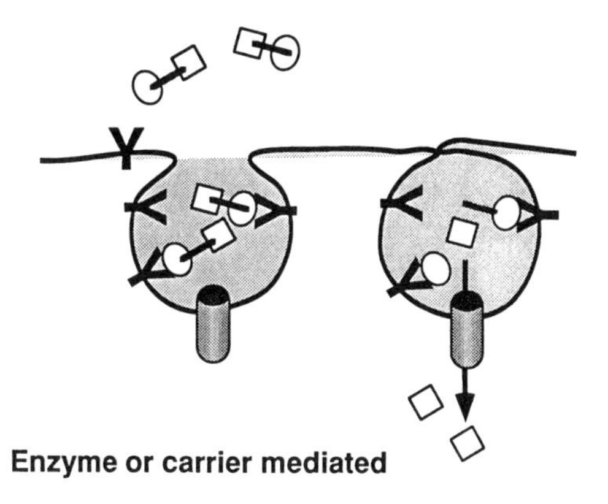

Figure 12. Three pathways of potocytosis. (*1*) Receptor-mediated potocytosis. Ligands (○) bind to high affinity receptors (**Y**) and concentrate in open caveolae at the cell surface. The caveolae close, but do not pinch off from the membrane. The ionic composition in the lumen of the caveolae changes. This could involve activation of proton or Ca^{2+} pumps. Ligands dissociate from receptors and enter the cell by passing down a concentration gradient. This may involve passage through specific transporters. (*2*) Enzyme and carrier mediated potocytosis. These pathways are similar and are represented by a single diagram. During carrier-mediated potocytosis, protein carriers with bound ligand (○-□) bind to cell surface receptors and enter caveolae. The caveolae close and changes in ion composition cause dissociation of the ligand from the carrier, the ligand (□) enters the cell by passing down a concentration gradient. The free carrier (○-) is released when caveolae open. During enzyme-mediated potocytosis, the substrate (○-□) binds to a cell surface enzyme (**Y**) that concentrates in caveolae. When the caveolae close, the enzyme cleaves the substrate. The product (□) either enters the cell as shown, or is released into the intercellular space, when the caveolae open. (Adapted from Anderson, 1992).

surface. If, at the same time, sealed caveolae contain high concentrations of substrate, then specific products can be produced by enzyme-catalyzed reactions within the caveolae. These molecules can be stored, introduced into the cell interior, or released into intercellular spaces.

5′-nucleotidase is a GPI-linked ectoenzyme that catalyzes the production of adenosine from adenosine monophosphate (AMP). Several cells have receptors that bind adenosine with high affinity. The bound adenosine causes changes in intracellular cAMP levels suggesting that adenosine acts as an intercellular messenger. It is possible that the concentration of 5′-nucleotidase within caveolae could provide a mechanism for controlling the localized production and storage of adenosine from extracellular AMP.

3. Carrier mediated potocytosis. Several low molecular weight molecules and ions bind to carrier proteins within the blood. This slows the excretion of low molecular weight compounds and/or prevents the precipitation of hydrophobic molecules. Importantly, low molecular weight compounds cannot be used by cells until they are separated from carrier proteins. In the case of cholesterol and iron this problem is solved by receptor mediated endocytosis. LDL-cholesterol complexes are separated by proteolysis in lysosomes while iron dissociates from transferrin in acidic endosomes (see above). Given that the ionic composition within the luminal space of caveolae can be regulated, it is possible that certain protein-ligand complexes dissociate within caveolae. Thyroxine, for example, circulates within the plasma as a complex with prealbumin. It is known that prealbumin-thyroxine complexes accumulate in caveolae, and it is suspected that thyroxine dissociates from prealbumin in the caveolae lumen. Carrier mediated potocytosis may facilitate the entry of thyroxine into the cell.

Caveolae may be Involved in Intracellular Signal Transduction Pathways

Caveolae sample small aliquots of extracellular fluid during potocytosis. Importantly, inert molecules taken up during potocytosis can be processed by caveolae into agonists that are able to stimulate signal transduction pathways. The production of adenosine from AMP by 5′-nucleotidase is a good example and has been described above. Calcium ions have been implicated as second messengers in numerous signal transduction pathways. Millimolar concentrations of Ca^{2+} are present on the extracellular fluid, and it is likely that the lumen of caveolae take up significant concentrations of Ca^{2+} during potocytosis. Interestingly, it has been shown that caveolae contain an ATP-dependent Ca^{2+} pump, and an inositol trisphosphate-gated Ca^{2+} channel. The asymmetric distribution of caveolae at the plasma membrane could result in the localized introduction of Ca^{2+}, and the generation of Ca^{2+} gradients across the cytoplasm.

It has been known for some time that antibodies against certain GPI anchored proteins can activate cells. GPI-anchored receptors do not span the plasma membrane, and cannot interact directly with cytoplasmic proteins. Because of this it has been unclear how the signal is transduced. Antibodies against GPI-anchored receptors nevertheless stimulate intracellular kinases and activate phosphorylation cascades. Recent studies show that protein tyrosine kinases associate with the cytoplasmic face of caveolae. It is possible that caveolae provide the physical link between GPI-anchored proteins and intracellular phosphorylation pathways. Sargiacomo et al. (1993) have suggested that caveolin acts as the transmembrane adaptor that links external stimuli to cytoplasmic kinases.

SUMMARY

At the start of the 1980s (Goldstein et al., 1979) it was known that receptor mediated endocytosis was a very efficient and specific means of taking proteins into cells. In addition, several high affinity receptors able to concentrate in coated pits, and recycle through the endocytic pathway, had been characterized (Steinman et al., 1983). At that time, the molecules regulating endocytosis, such as clathrin, adaptin and ATP-dependent proton pumps were beginning to be isolated. Since then, a rapid expansion of recombinant DNA technology, and the development of assays that reconstitute endocytosis *in vitro*, have allowed major advances in the understanding of endocytosis to be made.

Recombinant DNA methods have revealed the primary amino acid sequences of many receptors that enter cells by endocytosis. Internalization or, sorting motifs, have been identified in the cytoplasmic tails of receptors. Internalization motifs bind to adaptin complexes, and adaptin seeds the assembly of clathrin coats on the cytoplasmic face of the plasma membrane. Taken together, these data provide a molecular basis for the clustering of receptors in coated pits. Structural rearrangements within the clathrin lattice then cause the invagination of the coated pit, and the formation of the coated vesicle. Clathrin coats are removed by an uncoating ATPase, and uncoated vesicles fuse with endosomes. The lumen of endosomes is acidic. This acid environment, which is maintained by an ATP-dependent proton pump, is essential for the physical dissociation of receptor-ligand complexes. Sorting of receptors and ligands is further enhanced by the movement of receptors into tubular endosomal elements that recycle receptors to the plasma membrane.

Membrane fusion throughout the endocytic pathway requires ATP, GTP, Ca^{2+}, and cytosolic proteins. Many of these cytosolic proteins are required for membrane fusion in general, and are used in vesicular transport steps throughout the cell. Specific cytosolic proteins, called small GTP-binding proteins, or rabs, however, regulate the specificity of membrane fusion. Rabs 4 and 5 bind GTP and act as molecular switches that regulate the formation and recycling of endocytic vesicles. The formation of coated vesicles, and their fusion with early compartments of the endocytic pathway, may also require dynamin. Fusion of endosomes with lysosomes may require the additional help of microtubules.

A specialized form of endocytosis, called potocytosis, has been characterized recently. Ligands taken up by potocytosis bind to GPI-anchored receptors that concentrate in specialized domains of the plasma membrane called caveolae. Caveolae are coated with caveolin and invaginate to form a membrane compartment that is sealed off from the extracellular fluid. Caveolae do not pinch off from the cell surface and concentrate and process molecules within specialized domains of the plasma membrane. Caveolae may be used to take up low molecular weight ligands, such as vitamins, into cells, or for the localized generation of extracellular signaling molecules.

REFERENCES

Aniento F., Emans, N., Griffths, G., & Greuenberg, J. (1993). Cytoplasmic dynein-dependent vesicular transport from early to late endosomes. J. Cell Biol. 123, 1373–1387.

Arai, H., Terres, G., Pink, S., & Forgac, M. (1988). Topology and subunit stoichiometry of the coated vesicle proton pump. J. Biol. Chem. 263, 8796–8802.

Arai, H., Berne, M., Terres, G., Terres, H., Puopolo, K., & Forgac, M. (1988). Subunit composition and ATP site labeling of the coated vesicle proton translocating ATPase. J. Biol. Chem. 263, 8796–8802.

Bansal, A. & Gierash, L.M. (1991). The NPXY internalization signal of the LDL receptor adopts a reverse-turn conformation. Cell 67, 1195–1201.

Bucci, C., Parton, R.G., Mather, I.H., Stunnenberg, H., Simons, K., Hoflack, B., & Zerial, M. (1992). The small GTPase rab5 functions as a regulatory factor in the early endocytic pathway. Cell 70, 715–728.

Eberle, W., Sander, C., Klaus, W., Schmidt, B., Von figura, K., & Peters, C. (1991). The essential tyrosine of the internalization signal in lysosomal acid phosphatase is part of a β turn. Cell 67, 1203–1209.

Ferrow-Novick, S. & Novice, P. (1993). The role of GTP-binding proteins in transport along the exocytic pathway. Ann. Rev. Cell Biol. 9, 575–599.

Geuze, H.J., Slot, J.W., Strous, G.A.M., Lodish, H., & Schwartz, A.L. (1983). Intracellular site for asialoglycoprotein receptor-ligand uncoupling. Cell 32, 277–287.

Helenius, A., Mellman, I., Wall, D., & Hubbard, A. (1983). Endosomes. Trends in Biochem. Sci. 8, 245–250.

Heuser, J. & Evans, L. (1980). Three dimensional visualization of coated pit formation J. Cell Biol. 84, 560–583.

Heuser, J. & Keen, J.H. (1988). Deep-etched visualization of proteins involved in clathrin assembly. J. Cell Biol. 107, 877–886.

Hinshaw, J.E. & Schmid, S.L. (1995). Dynamin self-assembles into rings suggesting a mechanism for coasted vesicle budding. Nature 374, 190–192.

Lazarovits, J. & Roth, M. (1988). A single amino acid change in the cytoplasmic domain allows influenza virus hemagglutinin to be endocytosed through coated pits. Cell 53, 743–752.

Letourneur, F. & Klausner, R.D. (1992). A novel dileucine motif and a tyrosine-based motif independently mediate lysosomal targeting and endocytosis. Cell 69, 1143–1157.

Lin H.C., Sudhof T.C., & Anderson, R.G.W. (1992). Annexin VI is required for budding of clathrin coated pits. Cell 70, 283–291.

Moore, M.S., Mahaffey, D.T., Brodsky, F.M., & Anderson, R.G.W. (1987). Assembly of clathrin-coated pits onto purified plasma membranes. Science 263, 558–563.

Murphy, R.F. (1991). Maturation models for endosome and lysosome biogenesis. Trends Cell Biol. 1, 77–82.

Rothberg, K.G., Heuser, J., Donzell, W.G., Ying, Y-S., Glenny, J.R., & Anderson, R.G.W. (1992). Caveolin, a protein component of caveolae membrane coats. Cell 68, 673–682.

Sargiacomo, M., Sudol, M., Tang, Z-L., & Lisanti, M. (1993). Signal transducing molecules and glycosyl-phophatidylinositol-linked proteins form a caveolin-rich insoluble complex in MDCK cells. J. Cell Biol. 122, 789–807.
Schmid, S.L. (1992). The mechanism of receptor-mediated endocytosis: More questions than answers. BioEssays 14, 589–596.
Schmid, S.L. (1993). Coated vesicle formation *in vitro*: Conflicting results using different assays. Trends Cell Biol. 3, 145–148.
Schmid, S.L. (1993). Biochemical requirements for the formation of clathrin- and COP-coated transport vesicles. Curr. Biol. 5, 621–627.
Steinman, R.M., Mellman, I.S., Muller, W.A., & Cohn, Z. (1983). Endocytosis and recycling of the plasma membrane. J. Cell Biol. 96, 1–27.
Stoorvogel, W., Strous, G.J., Geuze, H.J., Oorschot, V., & Schwartz, A.L. (1991). Late endosomes derive from early endosomes by maturation. Cell 65, 417–427.
Takei, K., McPherson, P.S., Schmid, S.L., & De Camilli, P. (1995). Tubular membrane invaginations coated by dynamin rings are induced by GTPγs in nerve terminals. Nature 374, 186–190.
Trowbridge, I.S. (1993). Dynamin, SH3 domains and endocytosis. Current Biology 3, 773–775.
Van der Sluijs, P., Hull, M., Webster, P., Male, P., Gould, B., & Mellman, I. (1992). The small GTP-binding protein rab5 controls an early sorting event on the endocytic pathway. Cell 70, 729–740.

RECOMMENDED READINGS

Anderson, R.G.W. (1992). Caveolae: Where incoming and outgoing messages meet. Proc. Natl. Acad. Sci. USA 90, 10909–10913.
Anderson, R.G.W. (1993). Potocytosis. Trends in Cell Biol. 3, 69–71.
Bourne, H.R. (1988). Do GTPases direct membrane traffic in secretion? Cell 53, 669–671.
Brodsky, M.F. (1988). Living with clathrin: Its role in intracellular membrane traffic. Science 242, 1396–1402.
Brown M.S. & Goldstein, J.L. (1986). A receptor-mediated pathway for cholesterol homeostasis. Science 232, 34–47.
Goldstein, J.L., Anderson, R.G.W., & Brown, M.S. (1979). Review article: Coated pits, coated vesicles and receptor-mediated endocytosis. Nature 279, 679–684.
Kelly, R.B. (1995). Endocytosis: Ringing necks with dynamin. Nature 374, 116–117.
Mostov, K. & Simister, N.E. (1985). Transcytosis. Cell 43, 389–390.
Parton, R.G. (1996). Caveolae and caveolins. Curr. Opin. Cell Biol. 8, 542–548.
Pearse, B.M.F. (1987). Clathrin and coated vesicles. EMBO J. 6, 2507–2511.
Pearse, B.M.F. & Robinson, M.S. (1990). Clathrin, adaptors and sorting. Ann. Rev. Cell Biol. 6, 151–171.
Robinson, M.S. (1992). Adaptins. Trends Cell Biol. 2, 293–297.
Trowbridge, I.S., Collawan J.F., & Hopkins, C.R. (1993). Signal-dependent membrane trafficking in the endocytic pathway. Ann. Rev. Cell Biol. 9, 129–161.
Vaux, D. (1992). The structure of endocytosis signals. Trends Cell Biol. 2, 189–192.

Chapter 8

Eukaryotic Signal Sequences

MARK O. LIVELY

Principles of Medical Biology, Volume 7A
Membranes and Cell Signaling, pages 171–183.

ISBN: 1-55938-812-9

SIGNAL SEQUENCE STRUCTURE AND FUNCTION

Cell Compartmentation Requires Sorting Information

Nucleated cells are complex, highly organized structures composed of numerous internal compartments known as organelles that are bounded by lipid barriers containing embedded proteins. The first major membrane encountered by many proteins is the endoplasmic reticulum (ER), a highly interconnected network of membranes that enclose the earlier regions of the secretory pathway. Many proteins must be transported completely across these water-impermeable barriers to be exported from the cell or to be localized (compartmentalized) within organelles such as mitochondria, nuclei, lysosomes, peroxisomes, and chloroplasts. Integral membrane proteins are inserted into the membrane lipid bilayer in such a way that their polypeptide chains cross the bilayer one or more times so that the protein becomes integrated into the membrane lipid where the protein remains. Following initiation of protein synthesis in the cytoplasm, cells must immediately "decide" whether to continue synthesis of a given protein on ribosomes that will remain free in the cytoplasm or to redirect the ribosomes to binding sites on the cytoplasmic surface of the ER. Protein synthesis on membrane-bound ribosomes results in movement of the protein into the interior or lumen of the ER. Mitochondrial proteins that are not made in the mitochondrion are encoded in the nucleus of the cell and are completely synthesized on cytoplasmic ribosomes before being transported into the mitochondria (Hartl and Neupert, 1990). Similarly, proteins that reside in the nucleus itself are made in the cytoplasm, then transported into the nucleus (Silver, 1991; Gerace, 1992). Bacterial cells have similar mechanisms to incorporate proteins into their membranes and to secrete others from the cells (Inouye and Halegoua, 1980). In both eukaryotic and prokaryotic organisms, proteins are targeted to the proper cellular compartments by mechanisms that interpret information contained within the amino acid sequences of the proteins to be targeted. These sequences will be referred to here as signal sequences but they are also known in the literature by a number of different names including signal peptides, leader peptides, presequences, and targeting sequences. These amino acid sequences are like postal codes: different cellular destinations require signal sequences with distinct structural characteristics. Components of the protein transport system recognize the information contained in the signal sequences and direct the proteins along the proper pathways to the correct destinations. This chapter will focus specifically on eukaryotic signal sequences that target ribosomes to the ER and on the role of naturally occurring genetic mutations within these signal sequences that are associated with certain diseases in humans.

Protein Targeting to the Endoplasmic Reticulum

Synthesis of all proteins encoded by the nucleus begins on ribosomes that are free in the cytoplasm. Ribosomes synthesizing most secretory proteins and mem-

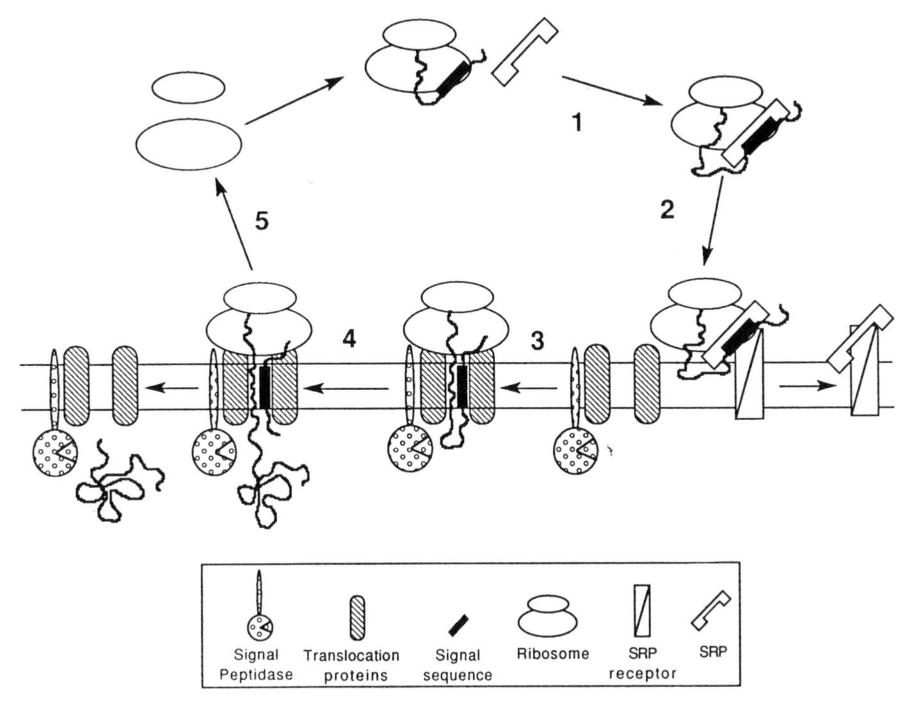

Figure 1. Protein Targeting Model. Protein synthesis is initiated on free ribosomes. Proteins synthesized with signal sequences are bound by SRP (step 1) then directed to the SRP receptor on the cytoplasmic side of the ER membrane (step 2). Once the ribosome is engaged by the protein translocation site, SRP and the SRP receptor dissociate. Continued synthesis results in translocation of the nascent protein into the lumen of the ER (step 3). Before the completion of synthesis, signal peptidase removes the signal sequence (step 4) and the finished protein is discharged into the lumen. The ribosome then dissociates from the ER (step 5) and is free to engage another mRNA. (see Nunnari and Walter, 1992).

brane proteins are subsequently targeted to the ER (Figure 1) where the protein being made is inserted into the lipid bilayer (Rapoport, 1992). After polymerization of fewer than 100 amino acids, the amino terminus of the nascent (growing) protein bearing its signal peptide emerges from the ribosome to be specifically recognized and bound by the signal recognition particle (SRP), which is a complex of six proteins and a small RNA molecule (Nunnari and Walter, 1992). Binding of SRP to the signal sequence temporarily halts polypeptide chain elongation until the entire ribosomal complex becomes bound to the SRP receptor on the cytoplasmic face of the ER. Once bound to the ER, SRP and its receptor dissociate from the ribosomal complex leaving the ribosome and its nascent polypeptide chain associated with the membrane. Continued synthesis results in movement of the protein through the bilayer by a process referred to as translocation. The mechanisms by which proteins are physically moved through the membrane are not well understood but current information indicates that translocation occurs through proteinaceous pores at specialized translocation sites found on the ER (Nunnari and Walter, 1992).

In most cases, the amino terminal signal sequences that target the complex to the ER are temporary features that do not remain a part of the mature protein product. Soon after the ribosome becomes associated with the translocation site, the signal sequence is removed by a membrane-bound proteolytic enzyme known as signal peptidase (see Chapter 9; Müller, 1992). In some cases, signal sequences are not removed from the protein and they become membrane spanning regions that anchor integral membrane proteins in the ER. These uncleaved peptides are known as signal anchor sequences. The major function of signal sequences is to target proteins to the ER, to facilitate insertion into the translocation site, then to provide the information needed to specifically direct removal of the signal from the protein, if it is a cleavable sequence (Randall and Hardy, 1989). Signal sequences do not appear to play a role in the subsequent process of translocation of the remainder of the protein across the membrane.

Overview of Signal Sequences

Cellular components that bind signal sequences must recognize common features of the shape, or secondary structure, of the peptides. It is usually possible to define binding domains of proteins in terms of the amino acid sequence of the protein being bound. In the case of signal sequences, there are now more than 5000 signal sequences described in the combined Genbank (Release # 76.0) and EMBL (Release # 34.0) DNA sequence databases. In spite of this wealth of sequence information, it is still not possible to describe a simple amino acid sequence that precisely defines the minimal information required for function.

At a time when far fewer sequences were known, comparison of 161 different eukaryotic signal sequences revealed a characteristic three-domain structure (von Heijne, 1986). This pattern (Figure 2) appears to be maintained in all signal sequences described to date and can be used to predict the site of cleavage by signal peptidase. In the majority of cases, signal sequences occur at the amino termini of the proteins. Eukaryotic signal sequences vary in length from 15 to 30 amino acid residues. There is usually a short, positively charged region of 1 to 5 amino acids at the amino terminus followed by a longer hydrophobic core region containing 7 to 15 amino acids. The carboxyl terminal region is usually from 3 to 7 amino acids in length and contains amino acids with more polar character. This overall "positive-hydrophobic-polar" (von Heijne, 1990) motif is found in almost all naturally occurring signal sequences. This pattern must exhibit some degree of flexibility since it has been shown that 20% of essentially random peptide sequences introduced in place of the normal signal sequence of yeast invertase can function as

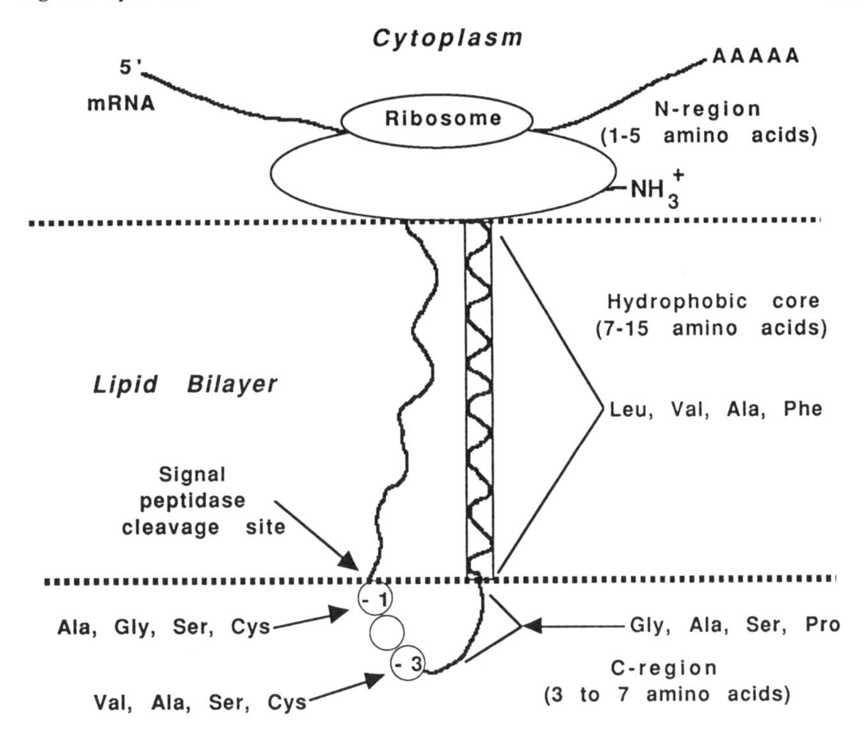

Figure 2. Model Signal Sequence. Signal sequences have three distinct domains each of which contributes to function. The common amino acids found in each region as analyzed by von Heijne (1986) are indicated for each region.

export signals in yeast (Kaiser et al., 1987). These randomly constructed sequences did appear to have a generally hydrophobic character.

Signal sequences are assigned negative numbers beginning at the carboxyl-terminus of the sequence and counting toward the amino-terminus of the precursor protein (Figure 2). Thus the C-terminal amino acid of the signal sequence is the –1 position while +1 denotes the first residue of the mature protein.

The Hydrophobic Core Region

The hydrophobic core is the most easily recognized feature of eukaryotic signal sequences. Three examples of human signal sequences are shown in Figure 3. The hydrophobic core may serve to initiate insertion of the nascent protein into the bilayer and then to position the signal sequence for cleavage by signal peptidase (Figure 1). It is hydrophobic because its amino acid sequence is almost exclusively composed of amino acids with apolar side chains. In von Heijne's 1986 compilation of eukaryotic sequences (von Heijne, 1986), the six most prevalent amino acids in this region (position -13 through -6), in decreasing order of occurrence, were Leu, Val, Ala, Phe, Ser, and Ile (see Figure 2). Cys and Thr also occur often in this region. Because so many different amino acid sequences can form functional hydrophobic cores, recognition of this domain by proteins that bind signal sequences is thought to depend on the net hydrophobic character, the shape (secondary structure) of the region, its overall length, and not on the primary sequence itself.

The hydrophobic core is thought to form an α-helix in the apolar environment of the ER membrane (Briggs and Gierasch, 1984). This conclusion is consistent with the conformations of known membrane spanning domains seen in X-ray crystal structures of integral membrane proteins. Studies of secretion by yeast of human lysozyme containing genetically engineered model signal sequences have shown that functional signal peptides can be constructed with hydrophobic domains composed only of 8 to 10 Leu residues in a row (Yamomoto et al., 1987). Leucine is an amino acid that has a high frequency of occurrence in α-helices (Chou and Fasman, 1978). Amino acid substitutions in the hydrophobic core that introduce charged amino acids or amino acids that disrupt α-helices usually produce non-functional signal sequences.

The Amino-Terminal Region

All eukaryotic signal sequences contain at least one positive charge at the amino terminus contributed by the α-amino group of the polypeptide. While additional positive charges in the region preceding the hydrophobic core are not essential, at least one additional positively charged amino acid seems to be preferred within this region in eukaryotic proteins because Lys or Arg are often found there. The role of the amino-terminal positive charge in signal sequence function has not been determined. When the signal sequence becomes bound to the ER, it is thought to form a hairpin loop such that the amino-terminus is positioned on the cytoplasmic side and the hydrophobic region extends down into the translocation site to signal peptidase on the opposite side (Inouye and Halegoua, 1980; Müller, 1992). In this orientation, a positively charged region near the cytoplasmic face of the membrane could form ion pairs with phospholipid head groups in the membrane, thus properly positioning the signal sequence within the bilayer. Similarly placed positive amino acids play an important role in defining membrane anchor sequences of integral membrane proteins (von Heijne, 1990).

The C-Terminal Region

Once the protein is engaged at the translocation site, the final requirement of the signal sequence is to specify its removal. The site of cleavage is mainly defined by the C-terminal domain of signal sequences although amino acid changes in the amino-terminal domain and hydrophobic core can influence cleavage. Statistical analyses have revealed a deceptively simple pattern near the cleavage site (von

Heijne, 1983; Perlman and Halvorson, 1983; Nielson et al., 1996). Amino acids with small side chains are preferred at positions -1 and -3 of the signal peptide. The cleavage pattern has been called the "(-3,-1) rule" because positions -3 and -1 appear to play a significant role in specifying the site of cleavage by signal peptidase. Amino acids at these positions must have relatively small, neutral side chains. Ala, Gly, Ser, and Cys predominate at position -1 while Ala, Val, Ser, Cys, Thr, and Ile are found most often at position -3 (von Heijne, 1986). Substitutions of amino acids with larger or charged side chains at either of these positions usually block cleavage at the normal site.

The C-terminal region is also the site at which the signal sequence must bend to form the hairpin loop (Figure 2). Gly, Ala, Ser, and Pro, amino acids that occur frequently in β-turns of proteins (Chou and Fasman, 1978), are commonly found in the -4 and -5 positions of signal sequences (von Heijne, 1986) so it is likely that the transmembrane helix is terminated by a turn in the peptide. Since the nascent peptide chain must be attached to the ribosome following insertion of the signal sequence into the lipid bilayer, it is clear that the C-terminus of the peptide extends back through the translocation site through the hydrophobic core of the membrane (Fig. 1). This conformation would permit the cleavage site to be positioned at the catalytic site of signal peptidase which is located in the interior of the ER.

SIGNAL SEQUENCE MUTATIONS ASSOCIATED WITH HUMAN DISEASES

Recent discoveries have shown that mistakes in the processing of signal sequences that arise through genetic errors can cause disease in humans. While signal peptidase itself is probably an essential gene whose loss is incompatible with life, a genetic error that results in a change in the signal sequence of a single protein is likely to affect production of only that protein. The effects of the loss of that protein would determine the severity of the mutation. To date, there have been three reports of human diseases that are directly attributable to genetic errors in signal sequences: a mutation in pre-proparathyroid hormone (pre-proPTH) is associated with a form of familial isolated hypoparathyroidism (Arnold et al., 1990); a signal sequence mutation in blood coagulation factor X causes a severe form of hemophilia in the affected individual (Watzke et al., 1991; Racchi et al., 1993); and a form of familial central diabetes insipidus is associated with a mutation at the cleavage site of pre-provasopressin (Ito et al., 1993). In addition to these, there are three reports of genetic mutations in humans that alter the structure of the signal sequence or alter its site of cleavage without any apparent functional consequences for the affected individuals. The genes affected encode serum albumin (Brennan et al., 1990), antithrombin (Daly et al., 1990), and apolipoprotein B (Visvikis et al., 1990). The mutants that are associated with diseases will be discussed because each illuminates important structural features of signal sequences that are important for function.

Characterization of Signal Sequence Mutants

Mutations of signal sequence function are identified by examining the DNA sequence of the gene encoding the secretory protein whose function is affected. The common observation in the three known examples studied thus far is that the affected patients produced little or none of a particular protein. To determine the reasons for the failure to produce a given gene product, investigators must directly investigate the DNA sequences encoding the protein that is missing in the affected patients. If the complete DNA sequence of the normal protein is known, as it was in each of the cases considered here, the gene sequence can be used to design synthetic DNA probes that will permit clones of the affected gene to be selected. Once a clone of the target protein has been isolated and the DNA sequence determined, comparison of the mutant sequence with the normal DNA sequence should reveal any structural differences. However, simply knowing the predicted amino acid sequence may not necessarily explain why the cell fails to secrete the protein, so additional studies may be needed to further characterize the effect of the mutation. In the cases considered here, DNA sequence analysis of the affected genes revealed mutations that occurred in the signal sequences of each.

Cell-free protein synthesis systems that allow reconstitution *in vitro* of components of the secretory pathway make it possible to directly analyze the early stages of secretion of a given protein (Scheele, 1983). Investigators can engineer plasmid vectors containing the cloned cDNAs encoding both the normal and mutant forms of the secretory protein to be tested. These vectors can be used to produce mRNA for each protein (Melton et al., 1984) and the mRNA can be translated in a cell-free protein synthesis system to produce the proteins for analysis. Microsomes, vesicular fragments derived from rough ER of secretory tissues such as pancreas (Walter and Blobel, 1983), can be included in the cell-free protein synthesis reactions and they will function like ER in intact cells (Scheele, 1983). These systems contain the cellular components that faithfully reproduce the initial stages of protein targeting to the ER. Translation of mRNAs encoding secretory proteins results in targeting and insertion of the proteins into the vesicles where signal peptidase removes the signal sequence (see Figure 1). Assays of this type are known as co-translational protein synthesis and targeting assays and they have been extremely useful for dissection of the processes involved in protein secretion.

Pre-proparathyroid Hormone

A mutation within the hydrophobic core region of human pre-proparathyroid hormone (preproPTH) is associated with a form of familial isolated hypoparathyroidism (Arnold et al., 1990). The DNA of a single affected patient with this condition was examined for mutations that could affect production of PTH. The patient's serum levels of PTH were at the low end of the normal range and she had been treated since infancy with calcium and vitamin D to maintain serum calcium

	N-region	Hydrophobic Core	C-region
Normal preproPTH	MIPAKDMAK	VMIVMLAICFL	TKSDG-
Normal preproPTH	MIPAKDMAK	VMIVMLAIRFL	TKSDG-
Normal factor X	MGRPLH	LVLLSASLAGLLLL	GES-
Mutant factor X	MGRPLH	LVLLSASLAGLLLL	RES-
Normal preproVP	MPDT	MLPACFLGLLAF	SSA-
Mutant preproVP	MPDT	MLPACFLGLLAF	SST-

Figure 3. Human Signal Sequence Mutants. The sequences of the normal and mutant forms of three human signal sequences are aligned by each of the three structural regions. The sites of mutation are shown by larger letters. The sequences are: pre-proparathyroid hormone (pre-proPTH; Arnold et al., 1990); coagulation factor X (Watzke et al., 1991); and pre-provasopressin (pre-proVP; Ito et al., 1993). The standard one-letter amino acid abbreviations are used.

concentrations near normal levels. To examine the nature of this defect, DNA was isolated from peripheral lymphocytes and used to construct a genomic DNA library. The patient's pre-proPTH gene was cloned from this library using synthetic DNA probes based on the known sequence of the *PTH* gene. DNA sequence analysis of the mutant gene revealed a single base substitution of C (cytosine) for T (thymine) in codon 18 of the *pre-proPTH* gene. This change causes an amino acid substitution of Arg for the normal Cys at position -8 of the signal sequence (Figure 3). The -8 position is at, or near, the C-terminal end of the hydrophobic region and the mutation places the positively charged Arg in a region that must be hydrophobic for proper function, suggesting that the effect of the mutation could be on targeting and or cleavage by signal peptidase.

Analysis of the effect of the Cys to Arg mutation using a co-translational assay synthesis system showed that the mutant pre-proPTH molecules were not properly cleaved by signal peptidase (Arnold et al., 1990). Subsequent analysis of this mutation showed that the substitution of Arg for Cys impairs the interaction of the nascent pre-proPTH with SRP as well as the translocation mechanism. In addition, signal peptidase is unable to cleave the mutant protein (Karaplis et al., 1995).

Coagulation Factor X

Coagulation factor X occupies a central point where the two separate arms of the blood clotting pathway converge (Jackson and Nemerson, 1980). Without this protein, blood coagulation is slowed severely and the affected individual experiences a particularly severe form of hemophilia. Human factor $X_{\text{Santo Domingo}}$ is a mutant form of factor X identified in a single patient with a serious bleeding diathesis characterized by less than 1% of normal factor X enzyme activity and less than 5% circulating factor X protein (Watzke et al., 1991). Factor X is a secreted protein that is synthesized as a single polypeptide chain precursor with a 23 amino acid signal sequence (Leytus et al., 1986). To determine the basis for the factor $X_{Santo Domingo}$ mutation, the polymerase chain reaction was used to amplify and clone the coding regions of the mutant factor X gene from DNA obtained from the patient. This approach revealed that a single point mutation in exon 1 (the first coding region of the gene) caused a substitution of Arg for Gly at the critical -3 position of the signal peptide of prepro factor X (Figure 3). Furthermore, these studies revealed that the patient was homozygous for this gene. Examination revealed that the parents were heterozygous for the same allele. Their coagulation function was compromised but was not clinically severe.

Arg is not observed at the -3 position of normal, functional signal sequences (von Heijne, 1986). Analysis of the effect of this mutation in a co-translational assay system *in vitro* revealed that the mutant prepro Factor X is properly targeted to the ER and that it is completely translocated into the lumen (Racchi et al., 1993). Once transferred into the lumen, signal peptidase is unable to cleave the mutant factor X so the uncleaved precursor remains anchored to the membrane via its uncleaved signal sequence. Further studies of the mutant factor X expressed transiently in cells in culture showed that factor $X_{Santo Domingo}$ was not secreted under conditions where normal factor X was properly secreted. Thus, the affected patient is unable to produce factor X because signal peptidase cannot remove the signal peptide. Consequently, the mutant factor X remains associated with membranes inside the cell and is probably degraded without ever leaving the ER. A single amino acid change at a critical position in the signal sequence is the cause of a serious, life threatening condition in this patient.

Pre-provasopressin

Vasopressin is a nine amino acid polypeptide antidiuretic hormone that controls water reabsorption in the kidney. Partial or complete absence of vasopressin in the circulation causes central diabetes insipidus (CDI), a syndrome characterized by excessive thirst and polyuria (Moses, 1985). A Japanese family has been identified in which familial CDI is transmitted in an autosomal dominant mode (Ito et al., 1993). DNA cloning of the vasopressin gene from an affected member of this family revealed a single base substitution that causes the substitution of Thr for Ala at the -1 position of the signal sequence of the vasopressin precursor (Figure 3). This same mutation was also identified in a caucasion kindred with familial CDI (McLeod et al., 1993). Further analysis of the effect of this mutation on secretion of vasopressin using a co- translational assay revealed that targeting and translocation into microsomal vesicles is not measurably affected but cleavage by signal peptidase is very inefficient. Fewer than 25% of vasopressin precursor molecules were cleaved by signal peptidase. These studies *in vitro* suggest that inefficient cleavage by signal peptidase *in vivo* results in decreased levels of circulating vasopressin, thus contributing to the pathogenesis of CDI in these patients. Once again, a relatively simple amino acid change in a secretory protein signal sequence results in a serious metabolic problem.

SUMMARY

Cells have evolved an efficient system for directing proteins from their sites of synthesis to their sites of function inside and outside of the cell. This system depends on information incorporated into the proteins being targeted to their destinations. The cell is able to maintain its structural integrity by carefully interpreting the sorting signals carried by its proteins. In the case of signal sequences discussed here, the information is encoded in a wide range of amino acid sequences that appear to have little in common, yet they function by a common mechanism. This mechanism has been preserved throughout evolution and remains essentially the same today in organisms from bacteria to man. Now that the structure and function relationships of signal sequences are beginning to be understood, mutations in these sequences have been recognized that cause diseases in humans. Although it seems unlikely that mutations of this type can be corrected before gene therapy becomes a reality, identification of the mutations responsible will permit genetic screening in affected families. Studies of the effects of the naturally selected mutations will continue to contribute to the understanding of structure and function relationships in signal sequences and may suggest treatment methods that could provide symptomatic relief to affected individuals.

ACKNOWLEDGMENTS

The author acknowledges support from the National Institutes of Health, grant GM32861 and thanks Dr. Marco Racchi for consideration of the manuscript.

REFERENCES

Arnold, A., Horst, S.A., Gardella, T.J., Baba, H., Levine, M.A., & Kronenberg, H.M. (1990). Mutation of the signal peptide-encoding region of the preproparathyroid hormone gene in familial isolated hypoparathyroidism. J. Clin. Invest. 86, 1084–1087.

Brennan, S.O., Myles, T., Peach, R.J., Donaldson, D., & George, P. (1990). Albumin Redhill (-1 Arg, 320 Ala $\rightarrow$ Thr): A glycoprotein variant of human serum albumin whose precursor has an aberrant signal peptidase cleavage site. Proc. Natl. Acad. Sci. USA 87, 26–30.

Briggs, M.S. & Gierasch, L.M. (1986). Molecular mechanisms of protein secretion: The role of the signal sequence. Adv. Protein Chem. 38, 109–180.

Chou, P.Y. & Fasman, G.D. (1978). Prediction of the secondary structure of proteins from their amino acid sequence. Adv. Enzymol. 47, 45–148.

Daly, M., Bruce, D., Perry, D.J., Price, J., Harper, P.L., O'Meara, A., & Carrell, R.W. (1990). Antithrombin Dublin (-3 Val → Glu): and N-terminal variant which has an aberrant signal peptidase cleavage site. FEBS Lett. 273, 87–90.

Gerace, L. (1992). Molecular trafficking across the nuclear pore complex. Curr. Opin. Cell Biol. 4, 637–645.

Hartl, F.-H. & Neupert, W. (1990). Protein sorting to mitochondria: Evolutionary conservations of folding and assembly. Science 247, 930–937.

Inouye, M. & Halegoua, S. (1980). Secretion and membrane localization of proteins in *Escherichia coli*. Crit. Rev. Biochem. 129, 233–239.

Ito, M., Osio, Y., Murase, T., Kondo, K., Saito, H., Chinzei, T., Racchi, M., & Lively, M.O. (1993). J. Clin. Invest. 91, 2565–2571.

Jackson, C.M. & Nemerson, Y. (1980). Blood coagulation. Ann. Rev. Biochem. 49, 765–811.

Kaiser, C.A., Preuss, D., Grisafi, P., & Botstein, D. (1987). Science 235, 312–317.

Karaplis, A.C., Lim, S.-K., Baba, H., Arnold, A., & Kronenberg, H.M. (1995). Inefficient membrane targeting, translocation, and proteolytic processing by signal peptidase of a mutant preproparathyroid hormone protein. J. Biol. Chem. 270, 1629–1635.

Leytus, S.P., Foster, D.C., Kurachi, K., & Davie, E.W. (1986). Gene for human factor X: A blood coagulation factor whose gene organization is essentially identical with that of factor IX and Protein C. Biochem. 25, 5098–5102.

McLeod, J.F., Kovács, L., Gaskill, M.B., Rittig, S., Bradley, G.S., & Robertson, G.L. (1993). Familial neurohypophyseal diabetes insipidus associated with a signal peptide mutation. J. Clin. Endocrinol. Metab. 77, 599A–599G.

Melton, D.A., Krieg, P.A., Rebaliati, M.R., Maniatis, T., Zinn, K., & Green, M.R. (1984). Efficient *in vitro* synthesis of biologically active RNA and RNA hybridization probes from plasmids containing a bacteriophage SP6 promoter. Nucleic Acids Res. 12, 7035–7056.

Moses, A.M. (1985). In: Frontiers in Hormone Research 13. Diabetes insipidus in man. (Czernichow, P. and Robinson, A.G., eds.), pp. 156–175, S. Karger, Basel.

Müller, M. (1992). Proteolysis in protein import and export: Signal peptide processing in eu- and prokaryotes. Experientia 48, 118–129.

Nielsen, H., Engelbrecht, J., Heijne, G., & Brunak, S. (1996). Defining a similarity threshold for a functional protein sequence pattern: The signal peptide cleavage site. Proteins: Structure, Function, and Genetics 24, 165–177.

Nunnari, J. & Walter, P. (1992). Protein targeting to and translocation across the membrane of the endoplasmic reticulum. Curr. Opin. Cell Biol. 4, 573–580.

Perlman, D. & Halvorson, H.O. (1983). A putative signal peptidase recognition site and sequence in eukaryotic and prokaryotic signal peptides. J. Mol. Biol. 167, 391–409.

Racchi, M., Watzke, H.H., High, K.A., & Lively, M.O. (1993). Human coagulation factor X deficiency caused by a mutant signal peptide that blocks cleavage by signal peptidase but not targeting and translocation to the endoplasmic reticulum. J. Biol. Chem. 268, 5735–5740.

Randall, L.L. & Hardy, S.J.S. (1989). Unity in function in the absence of consensus in sequence: Role of leader peptides in export. Science 243, 1156–1159.

Rapoport, T.A. (1992). Transport of proteins across the endoplasmic reticulum. Science 258, 931–936.

Scheele, G. (1983). Methods for study of protein translocation across the RER membrane using reticulocyte lysate translation system and canine pancreatic microsomal membranes. Meth. Enzymol. 96, 94–111.

Silver, P.A. (1991). How proteins enter the nucleus. Cell 64, 489–497.

Visvikis, S., Chan, L., Siest, G., Drouin, P., & Boerwinkle, E. (1990). An insertion deletion polymorphism in the signal peptide of the human apolipoprotein B gene. Hum. Genet. 84, 373–375.

von Heijne, G. (1983). Patterns of amino acids near signal sequence cleavage sites. Eur. J. Biochem. 133, 17–21.

von Heijne, G. (1986). A new method for predicting signal sequence cleavage sites. Nucleic Acids Res. 14, 4683–4690.
von Heijne, G. (1990). The signal peptide. J. Membrane Biol. 115, 195–201.
Walter, P. & Blobel, G. (1983). Preparation of microsomal membranes for cotranslational protein translocation. Meth. Enzymol. 96, 84–93.
Watzke, H.H., Wallmark, A., Hamaguchi, N., Giardina, P., Stafford, D.W., & High, K.A. (1991). Factor $X_{Santo\ Domingo}$. Evidence that the severe clinical phenotype arises from a mutation blocking secretion. J. Clin. Invest. 88, 1685–1689.
Yamomoto, Y., Taniyama, Y., Kikuchi, M., & Ikehara, M. (1987). Engineering of the hydrophobic segment of the signal sequence for efficient secretion of human lysozyme by *Saccharomyces cerevisiae*. Biochem. Biophys. Res. Commun. 149, 431–436.

Chapter 9

Proteolysis in Protein Import and Export:
THE STRUCTURE AND FUNCTION OF SIGNAL PEPTIDASES

MATTHIAS MÜLLER

Principles of Medical Biology, Volume 7A
Membranes and Cell Signaling, pages 185–204.

ISBN: 1-55938-812-9

INTRODUCTION

Numerous proteins in pro- and eukaryotes have to cross biomembranes in order to reach their destination within the cell. These proteins are distinguished, and consequently sorted, from the bulk of cytoplasmic proteins by virtue of discrete sequence sections, termed signal sequences (synonyms are presequence, transit sequence, targeting sequence, leader sequence). A signal sequence is defined as a sequence which contains the information necessary to guide a protein bearing that sequence to a distinct cellular membrane, and initiate its transmembrane transport (which is referred to as translocation here).

A functional signal sequence is characterized by: (1) distinct structural features (discussed in Chapter 8) which when altered lead to a loss of function, (2) its selective occurrence in non-cytoplasmic proteins, and (3) its ability to function as a transport signal when artificially fused to a normally cytoplasmic protein. The latter holds true only for those cytoplasmic proteins whose structural conformation allows transmembrane transport. Signal sequences are in many instances NH_2-terminally located and cleaved during or shortly after the translocation process. The enzymes involved in the cleavage of signal peptides are the subject of this chapter.

PROTEIN TRAFFIC INVOLVING REMOVAL OF SIGNAL SEQUENCES OCCURS AT DIVERSE CELLULAR MEMBRANES OF PRO- AND EUKARYOTES

The cell envelope of Gram-negative bacteria such as *Escherichia coli*, is composed of an inner (plasma) membrane confining the cell body, an outer membrane, and an intermediate periplasmic space (Figure 1). After their synthesis in the cytoplasm many proteins are exported to one of the three envelope layers. The periplasmic and outer membrane proteins are proteolytically processed during export by the removal of their NH_2-terminal signal sequences, a process that is catalyzed by plasma membrane-located signal peptidases. The same applies to the cell surface proteins of Gram-positive bacteria which have no outer membrane layer.

In eukaryotic cells, the situation is more complex because of the different organelles present (Figure 2). There, signal sequences not only fulfill the function of targeting proteins to membranes but in addition must contain information that allows distinction between the various organellar membranes. In fact, differently

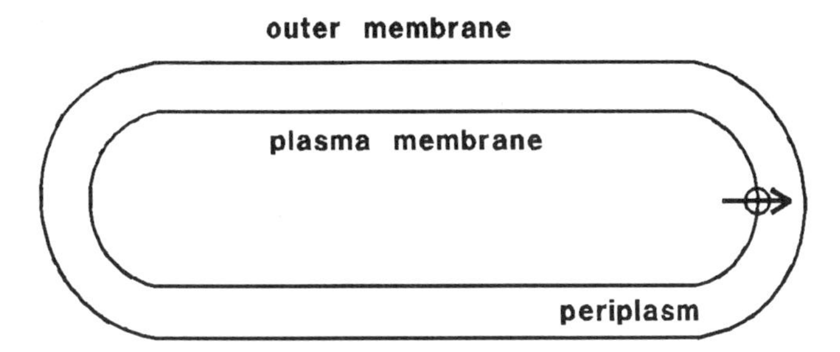

Figure 1. Protein export of Gram-negative bacteria. The arrow symbolizes translocation of precursor proteins across the plasma membrane. Precursors are proteolytically processed by signal peptidases residing in the plasma membrane (*circle*). The translocated proteins are destined for the periplasm and the outer membrane.

structured signal sequences occur. One group of proteins containing cleavable signal sequences encompasses secretory proteins, lysosomal proteins, house keeping enzymes of the endoplasmic reticulum (ER) and the Golgi apparatus, and a few membrane proteins. These proteins are initially targeted to the ER membrane by signal sequences which are hydrophobic in nature and which are cleaved upon entry into the ER cisternae by signal peptidases located on the ER membrane. These ER-targeting signal sequences are functionally interchangeable with those initiating translocation of the periplasmic and outer membrane proteins across the plasma membrane of Gram-negative bacteria.

A different class of cleavable signal sequences, mostly hydrophilic in nature, is required to target nuclear-encoded proteins to mitochondria and chloroplasts and to initiate their import into these organelles. The hydrophilic signal sequences have to be sufficiently specific to avoid errors in sorting between mitochondria and chloroplasts. Hydrophilic signal sequences guide proteins imported into mitochondria to the matrix of the organelle, and are removed by a matrix-located processing enzyme. Inner membrane and intermembrane space proteins are sorted from the pool of matrix-destined proteins by means of the information contained in additional signals. Some intermembrane space proteins thereby undergo a second proteolytic event, which is catalyzed by a signal peptidase of the inner mitochondrial membrane. An even more complex situation is found in chloroplasts. Protein traffic in them occurs at several levels: (1) import into the stroma of the organelle with the concomitant removal of signal sequences by a stromal processing peptidase, (2) targeting to the envelope layers, as seen in mitochondria, and (3) intraorganellar transport to the membrane and the lumen of the thylakoid. In the latter case, additional signal sequences are required which are cleaved off by a thylakoidal signal peptidase.

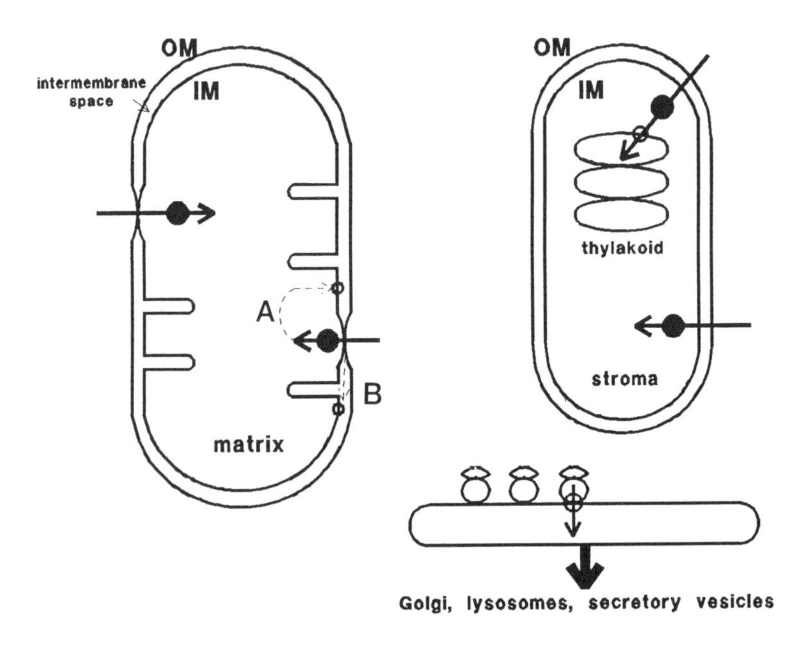

Figure 2. Protein traffic involving proteolytic processing of precursor proteins within a eukaryotic cell. Shown are transport processes at mitochondria, chloroplasts, and the endoplasmic reticulum (OM and IM, outer and inner membrane).

Nuclear encoded proteins are imported into the matrix of mitochondria through distinct translocation sites of both membranes (*upper left import site*). They are processed by a soluble processing peptidase (*dot*). Imported intermembrane space proteins (*lower right import site*) are processed by the same matrix peptidase. Afterwards, they are targeted to an inner membrane peptidase (*circle*) where they undergo a second cleavage step. Targeting to the intermembrane space (*dotted arrows*) might occur as re-export from the matrix (**A**) or by lateral transport within the inner membrane (**B**).

Similarly, the targeting sequence of proteins imported into the stroma of chloroplasts is removed by a soluble processing peptidase (*dot*). Proteins destined for the thylakoid contain an additional signal that is cleaved off by a thylakoidal peptidase (*circle*).

Proteins destined for the ER are cotranslationally translocated into the ER and processed by the ER-specific signal peptidase (*circle*).

THE CHARACTERISTICS OF MORE AND MORE INDIVIDUAL SIGNAL PEPTIDASES ARE BEING ELUCIDATED

Table 1 is a summary of the most extensively studied signal peptidases including some of their known characteristics.

Signal Peptidases of the Endoplasmic Reticulum (ER)

Most secreted, lysosomal, and some integral membrane proteins are targeted to the ER via NH_2-terminal signal sequences which are cleaved off upon entry into the ER by a signal peptidase (Figure 2). Two of these signal peptidases have been purified to homogeneity from mammalian tissue, canine pancreas signal peptidase complex (SPC) and hen oviduct signal peptidase (HOSP). Recently, the yeast enzyme has also been purified to near homogeneity. All three enzymes purify as complexes composed of several subunits.

Signal Peptidases of the Prokaryotic Plasma Membrane

Signal peptidases of the prokaryotic plasma membrane proteolytically process proteins destined for the periplasmic space and the outer membrane. This is seen in Gram-negative bacteria (e.g., *Escherichia coli*). In the case of Gram-positive bacteria (e.g., *Bacillus subtilis*) proteins with clearable signal sequences are destined for the extracellular milieu. Most of our present knowledge on bacterial signal peptidases stems from studies with the *E. coli* enzymes (Table 1). Signal peptidase I (SPase I), also termed leader peptidase, was the first signal peptidase to be purified (Wolfe et al., 1982) and its gene (*lepB*) was cloned and sequenced (Wolfe et al., 1983). SPase I was found to consist of a single polypeptide chain of 37kD molecular mass. It is essential for cell growth. The enzyme is anchored within the plasma membrane and its assembly is dependent on the same Sec proteins which are required in general for protein export from the cytoplasm of *E. coli*.

Typc I signal peptidases of bacteria other than *E. coli* show various degrees of overall homology with the *E. coli* enzyme. SPase I of *Salmonella typhimurium* is 92% identical to the *E. coli* peptidase at the amino acid level, whereas that of *Pseudomonas fluorescens* only 50%. Much less homology is shared between these enzymes and SPase I of the Gram-positive *B. subtilis*. However, distinct regions of high conservation can be identified even in this case (see below).

Prokaryotic plasma membranes contain another type of signal peptidase, designated signal peptidase II (SPase II) whose substrate specificities are quite different from SPase I. SPase II is specific for preproteins that undergo a distinct lipid modification, such as Braun's lipoprotein of *E. coli*. The precursor of this lipoprotein is first modified by the attachment of a diglyceride to the prospective NH_2-terminal Cys of the mature protein. Only the diglyceride-bearing precursor is accepted as substrate by the lipoprotein-specific SPase II which removes the signal

Table 1. Signal Peptidases

Organelle	*Organism*	*Nomenclature*	*Membrane Anchored*	*Subunits*	*Molecular Mass (kD)*
Endoplasmic reticulum	Dog pancreas	(Canine) signal peptidase complex (SPC)	+	5	12 18 21 22/23[a] 25
	Hen oviduct	Hen oviduct signal peptidase (HOSP)	+	2	19 22–24[a]
	Yeast	Yeast signal peptidase (ySP)	+	3–4	(13)[b] 18=SEC11 20 25
Mitochondria					
Matrix	*Neurospora crassa*	Matrix processing peptidase (MPP)	soluble	2	57 (α) 52 (β)
	Yeast		soluble	2	51 48

	Rat liver[c]		soluble	2	55 52
Inner membrane	Yeast	Inner membrane protease	+	2	21.4 (Imp1) 19 (Imp2)
Chloroplasts					
Stroma	Pea	Stromal processing peptidase	soluble	?	143/145[d]
Thylakoidal membrane	Pea	Thylakoidal processing peptidase	+	?	?[d]
Bacterial Plasma Membrane	*E. coli*[e]	Signal peptidase I (SPase I; leader peptidase)	+	1	37
		Signal peptidase II (SPase II)	+	1	18
	B. subtilis	SipS[f]	+	1	21
	Several Gram-negative & Gram-positive species	Prepilin peptidase[g]	+	1	~25

Notes: [a]Differently glycosylated forms of one polypeptide; [b]the association of the 13kD protein with the enzymatic activity is tentative; [c]two different matrix processing peptidases have been detected in rat liver (see text); [d]these have only partially been purified; [e]corresponding activities have been detected in several other Gram-negative bacteria (see text); [f]SipS, **s**ignal **p**eptidase of Bacillus **s**ubtilis; [g]for further details see text.

peptide immediately upstream of the modified Cys. Braun's lipoprotein is a prototype of many glyceride-modified proteins in Gram-negative and Gram-positive bacteria (reviewed in Wu and Tokunaga, 1986). The gene of SPase II (*lspA*) does not share sequence homology with the *lepB*-gene or genes encoding other known signal peptidases. Like SPase I, SPase II is essential for cell growth.

A third class of signal peptidases located at the plasma membrane of Gram-negative and Gram-positive bacteria are the so-called type IV prepilin peptidases. These were first identified as processing enzymes of type IV pili. Pili are rod-shaped fibers that emerge from the cell surface of Gram-negative bacteria and allow the bacteria to attach to other cell surfaces (reviewed in Pugsley, 1993). For instance, these pili are used by several pathogenic bacteria as adhesins to dock at their host cells. They are composed mainly of multiple copies of one polypeptide called pilin. Type IV pilins are synthesized as precursors with apparently normal signal peptides. However, they are aberrantly cleaved by a prepilin peptidase between the NH_2-terminal, positively charged section and the hydrophobic core instead of after the COOH-terminal region (see Chapter 8). In addition, prepilin type peptidases are required for the processing of a number of proteins somehow involved in either the secretion of proteins across both membranes of Gram-negative bacteria or in the transformation of Gram-positive bacteria (reviewed in Pugsley, 1993). Elucidation of the precise function and requirements these peptidases fulfill will have to await a detailed molecular analysis of the processes in which their substrate proteins are involved.

Mitochondrial Signal Peptidases

The majority of constituent mitochondrial proteins are synthesized in the cytoplasm and therefore have to be imported into the organelle. This process is driven by mitochondria-specific, hydrophilic signal sequences. Import is achieved by translocation machineries of the inner and outer membranes and involves cleavage of the hydrophilic presequence by a matrix-located processing peptidase (Figure 2). The enzyme consists of two subunits (Table 1) termed MPP (matrix processing peptidase) subunit α and β *Neurospora crassa* and *Saccharomyces cerevisiae*. The enzymes from *Neurospora crassa* and *Saccharomyces cerevisiae* have been obtained in pure form (Hawlitschek et al., 1988; Yang et al., 1988) and the nucleotide sequences of all four subunits have been determined. The subunits themselves are synthesized as precursors with hydrophilic signal sequences typical for matrix-directed import proteins. In contrast to the enzymes of the ER and the bacterial plasma membrane, these mitochondrial peptidases are soluble within the matrix. Under conditions in which cells depend on mitochondrial metabolism, the matrix processing peptidase is essential for growth. Furthermore, rat liver was reported to contain a second soluble enzyme, termed mitochondrial intermediate peptidase (MIP), which is required for some imported proteins in concert with the matrix processing peptidase to cleave in two steps the precursor to its mature size (Kalousek et al., 1988).

Some of the imported proteins which are directed to either the inner mitochondrial membrane or the intermembrane space are synthesized with a bipartite NH_2-terminal signal sequence, the NH_2-terminal part being a hydrophilic matrix-targeting signal, the COOH-terminal part resembling the hydrophobic signal sequences of exported bacterial proteins. Whereas the NH_2-terminal part of these signals is processed by the matrix-located peptidase to an intermediate-sized precursor, the COOH-terminal section is subsequently cleaved off by a distinct signal peptidase residing in the inner mitochondrial membrane. Currently, two views exist about the topography of the two-step processing event (Glick et al., 1992; Segui-Real et al., 1992). One hypothesis, called the conservative sorting hypothesis, postulates an initial translocation of the precursor into the matrix, specified by the mitochondrial targeting sequence. There, this signal is removed and re-export across the inner mitochondrial membrane ensues, mediated by the targeting information of the second, i.e., hydrophobic part of the signal (Figure 2A). The conservative trait would be the involvement of a mitochondrial export machinery in the re-export to the intermembrane space. According to the endosymbiont theory (see below) any mitochondrial export machinery is likely to be a direct descendant of that of the bacterial ancestor of mitochondria. An alternative view (Figure 2B) suggests that the processing-intermediates of intermembrane space proteins, resulting from cleavage by the matrix processing peptidase, have not crossed the inner mitochondrial membrane completely but rather remain anchored to it due to the stop transfer information of the COOH-terminal, hydrophobic part of the bipartite signal sequence, which is subsequently cleaved by the membrane-located signal peptidase. Recently, a unifying hypothesis has been put forward (Gärtner et al., 1995).

The enzyme which cleaves the second part of the bipartite signal sequence and which is different from the matrix peptidase has been characterized in detail. Interestingly, it also processes the precursor of a mitochondrially encoded protein destined for the inner mitochondrial membrane. This is inferred from the finding that a particular yeast mutant (*pet ts2858*) accumulates an incompletely processed precursor of an imported, intermembrane space protein (cytochrome b_2), as well as the precursor of the mitochondrially encoded subunit II of cytochrome oxidase (Pratje and Guiard, 1986). The protein affected in this mutant was shown to be a signal peptidase of the inner mitochondrial membrane of *S. cerevisiae*, designated inner mitochondrial protease (Imp1) (Schneider et al., 1991). A second subunit of this protease has been identified in *S. cerevisiae* (Imp2) which shows non-overlapping substrate specificities with Imp1 and which is required for the stable and functional expression of Imp1 (Nunnari et al., 1993).

Signal Peptidases of the Chloroplast and the Thylakoid of Cyanobacteria

The compartment structure of the chloroplast resembles that of mitochondria with the exception of the additional thylakoidal membrane system (Figure 2). Accordingly, the chloroplast requires processing peptidase activities similar to those of mitochondria, but in addition, a thylakoidal enzyme. Chloroplasts contain a processing peptidase of high molecular weight which is soluble in the stroma of the organelle. Little is known about the mechanism by which proteins of the chloroplast envelope (outer and inner membrane) are sorted. However, it was demonstrated that proteins destined for the thylakoid contain additional targeting information (summarized in Keegstra, 1989; Smeekens et al., 1990). In the case of thylakoid lumen proteins, the additional information resides in a second signal sequence immediately downstream of the one required for transport into the stroma. Hence, this class of proteins also contains a bipartite signal sequence, which is cleaved in two successive steps. After the NH_2-terminal signal specifying import into the stroma has been removed by the stromal processing peptidase, an integral thylakoid membrane peptidase cleaves the thylakoid-specific part of the signal sequence upon transport into the lumen of the thylakoid. Whether or not all signal sequence-bearing proteins which are destined for the thylakoid are subject to an obligate two step processing event is presently unknown. A peptidase activity from thylakoids has been partially purified. Presumably this enzyme also processes precursors of chloroplast-encoded thylakoid proteins such as cytochrome f.

Cyanobacteria contain a thylakoid membrane peptidase that is involved in proteolytic processing of proteins imported into the thylakoid. This enzyme has reaction specificities which are similar to that of the plant thylakoid peptidase. A thylakoid peptidase has been solubilized and partially characterized from the cyanobacterium *Phormidium laminosum*.

THE ENZYMES REMOVING SIGNAL SEQUENCES HAVE BEEN REMARKABLY CONSERVED DURING EVOLUTION

Substantial structural homologies exist between organelle-specific signal peptidases prepared from different organisms. Remarkably, this applies to enzymes of organisms as phylogenetically diverse as mammals and yeast (signal peptidases of the ER and matrix processing peptidases of mitochondria, respectively). The conservation in structure underlines the vital function of signal peptidases in all living organisms.

Moreover, structural and functional homologies are observed even between signal peptidases of different membrane systems. An extended sequence comparison between the signal peptidases of bacteria and the inner mitochondrial membrane, as well as distinct subunits of the peptidase complexes from the eukaryotic ER revealed five regions of remarkable similarity (Figure 3, A–E; van Dijl et al., 1992). These stretches represent a conserved pattern specific for type I signal

	A		B		C	
1	72	I V L I V	88	S G S M M P T L L	127	R G D I V V F
2	73	I V L I V	89	S G S M M P T L L	128	R G D I V V F
3	25	L A L L I	40	D G D S M Y P T L H	68	R G D I V V L
4	22	F L H I I	37	R G E S M L P T L S	65	K M G D C I V A L
5	37	I V V V L	42	S G S M E P A F Q	51	R G D I L F L
6	49	I V V V L	54	S G S M E P A F H	63	R G D L L F L
7	60	I V V V L	65	S G S M E P A F H	74	R G D L L F L

(★ marks the Ser column in B)

	D		E	
1	137	E D P K L D Y I K R A V G L P G D K	272	G D N R D N S A D S R
2	138	E D P K L D Y I K R A V G L P G D K	273	G D N R D N S A D S R
3	75	N G D D V H Y V K R I I G L P G D T	145	G D N R R N S M D S R
4	76	T D P N H R I C K R V T G M P G D L	130	G D N L S H S L D S R
5	75	E G K Q I P I V H R V L R Q H N N H	102	G D N N A G N D I S
6	88	E G R E I P I V H R V L K I H E K	115	G D N N A V D D R
7	99	E G R D I P I V H R V I K V H E K	126	G D N N E V D D R

Figure 3. Patterns of conserved amino acids in the deduced amino acid sequences of prokaryotic and eukaryotic signal peptidases. (1, *E. coli*; 2, *Salmonella typhimurium*; 3, *Bacillus subtilis*; 4, mitochondrial inner membrane protease 1; 5, Subunit SEC11 of the ER enzyme of yeast; 6,7, subunits 18 and 21 of the ER enzyme from canine pancreas). Identical or conserved patterns of amino acids are only boxed when present in at least two of the groups of peptidases. Box A represents the COOH-terminal end of the second transmembrane segment of the *E. coli* enzyme (cf. Figure 4A). The Ser residue (position 90 in the amino acid sequence of the *E. coli* enzyme) necessary for catalytic activity is marked with an asterisk. The sequence similarity shown here also extends to the second subunit (Imp2) of the mitochondrial inner membrane protease. (Reproduced from van Dijl et al., 1992, by permission of Oxford University Press).

peptidases since they are not found with lipoprotein-specific signal peptidases (type II signal peptidases) or mitochondrial processing peptidases. The occurrence of this consensus motif in the 18 and 21kD subunits of the signal peptidase complex from mammalian ER and in the 18kD subunit of the yeast enzyme (SEC11) identifies them as the polypeptides endowed with the catalytic activity.

The structural relatedness found for the inner mitochondrial membrane peptidase and signal peptidase I of the *E. coli* plasma membrane (Figure 3) is consistent with the finding that the COOH-terminal part of the bipartite signal sequence of mitochondrial intermembrane space proteins is very similar in structure to the signal sequence of bacterial export proteins. For example, the α-purple bacterium, *Rhodobacter capsulatus*, containing a cytochrome bc_1 complex highly homologous with that found in mitochondria, synthesizes cytochrome c_1 with a hydrophobic signal sequence which shows considerable similarity to the COOH-terminal part of the presequence of yeast cytochrome c_1 (summarized in Hartl and Neupert,

1990). The homologies of the signal sequences and the corresponding peptidases between bacteria and mitochondria are reflections of the endosymbiont theory proposing the origin of mitochondria from prokaryotic ancestors. Based on a sequence analysis of the 16S ribosomal RNAs it is the species of α-purple bacteria which is most closely related to the mitochondrial progenitors (Woese, 1987).

Similarly, chloroplasts are thought to have arisen from cyanobacterial progenitors after endocytosis by an ancestral plant cell (Gray and Doolittle, 1982). Cyanobacteria contain a thylakoidal membrane system. Transport of proteins into the lumen of this organelle involves cleavage of a hydrophobic signal sequence. The evolutionary relationship of chloroplasts and cyanobacteria is again suggested by the same type of hydrophobic signal sequence found on thylakoid proteins of both origins (Smeekens et al., 1990). No sequence data are as yet available on the thylakoidal signal peptidases of chloroplasts and cyanobacteria. Therefore the degree to which these peptidases are homologous with bacterial plasma membrane signal peptidases has not yet been determined. However, the reaction specificities of the thylakoidal enzyme of chloroplasts and *E. coli* SPase I are identical. This is inferred from the finding that SPaseI correctly and efficiently processes a precursor of a chloroplast-thylakoid lumen protein, while the thylakoidal processing peptidase cleaves a precursor with a bacterial signal peptide (Halpin et al., 1989).

Translocation of preproteins across the membrane of the mammalian endoplasmic reticulum and the prokaryotic plasma membrane involves cleavage of signal sequences which have a highly conserved secondary structure. The compatibility of both transport machineries in recognizing and processing the foreign substrate has also frequently been demonstrated both *in vitro* and *in vivo* (summarized in Saier et al., 1989). In addition, purified bacterial SPase I correctly processes a variety of eukaryotic precursor proteins. These results are in complete agreement with the sequence similarities between pro- and eukaryotic signal peptidases listed in Figure 3.

SOME SIGNAL PEPTIDASES HAVE AN OLIGOMERIC STRUCTURE

Signal peptidases of the endoplasmic reticulum and of mitochondria were found to consist of more than one subunit (cf. Table 1). In most cases, this was shown by the copurification of the various subunits with the catalytic activity. The activity of the mitochondrial matrix processing peptidase is lost if the two subunits reconstituting the enzyme are separated. A polymeric structure, however, has not been found for the bacterial signal peptidases which consist of a single polypeptide only. Interestingly, in those cases in which sequence data are available, it has become evident that different subunits of the same signal peptidase complex might show substantial sequence homology: this is true of the subunits 18 and 21 of canine signal peptidase, the two subunits of the mitochondrial matrix processing peptidases from yeast and *Neurospora crassa*, and the two subunits of the mitochondrial inner membrane protease. These similarities suggest that the homologous subunits have evolved from a common ancestor.

Why have some of the known signal peptidases a composite structure? The isolated subunit α of the mitochondrial matrix enzyme of *N. crassa* (MPP) contains low activity compared to the complex consisting of both subunits. This finding suggests that subunit α is the actual catalytic subunit whereas subunit β, which is inactive when isolated, has an enhancing function. Following the sequence comparison depicted in Figure 3, it is likely that the catalytic activity of all signal peptidases resides in distinct polypeptides as is the case of the monomeric signal peptidases of prokaryotes. Ancillary subunits like subunit β would then be activators of the catalytic subunit. This view is supported by the finding that subunit β which is about 15-fold more abundant in mitochondria than subunit α, is either identical (*Neurospora crassa*) or highly homologous (yeast) to the subunit I of the cytochrome c reductase (cytochrome bc_1 complex) of the inner mitochondrial membrane (Schulte et al., 1989).[1] Subunit I does not participate in electron transfer but is required for the reductase activity and assembly of cytochrome bc_1. Therefore, subunit β may have a stabilizing or even assembly-mediating function. Another reason for the occurrence of multiple subunits in signal peptidases might be a broadening of the range of substrate specificities. This is suggested by the finding that the two subunits Imp1 and Imp2 of the mitochondrial inner membrane protease have non-overlapping substrate specificities.

Translocation of preproteins into the ER and subsequent processing occur predominantly in a cotranslational manner. Therefore, the signal peptidase of the ER should be intimately associated with the pore through which nascent polypeptide chains traverse the membrane. The oligomeric structure of ER-signal peptidase may simply reflect an association of the catalytic protein with other subunits potentially involved in the translocation process or other related processes.

INTRACELLULAR TOPOGRAPHY OF SIGNAL PEPTIDASES

Signal peptidases of the ER, the bacterial plasma membrane and related enzymes of the mitochondrial inner membrane and the thylakoidal membrane of chloroplasts and cyanobacteria, are all integral membrane proteins requiring detergents for solubilization. In contrast, processing peptidases of the mitochondrial matrix and the chloroplast stroma are soluble enzymes.

A body of direct and circumstantial evidence has been accumulated showing that the integral membrane peptidases are located at the *trans*-side of the membrane with respect to the transport direction of the precursor proteins. This has been investigated most extensively for the *E. coli* signal peptidase I. This protein (Figure 4A) has a large (about two-thirds of the molecule) COOH-terminal domain facing the periplasm, while the NH_2-terminal part anchors SPase I within the plasma membrane. The anchor starts with an internal, i.e., uncleaved, signal sequence and a positively charged cytoplasmic domain. The considerably smaller signal pepti-

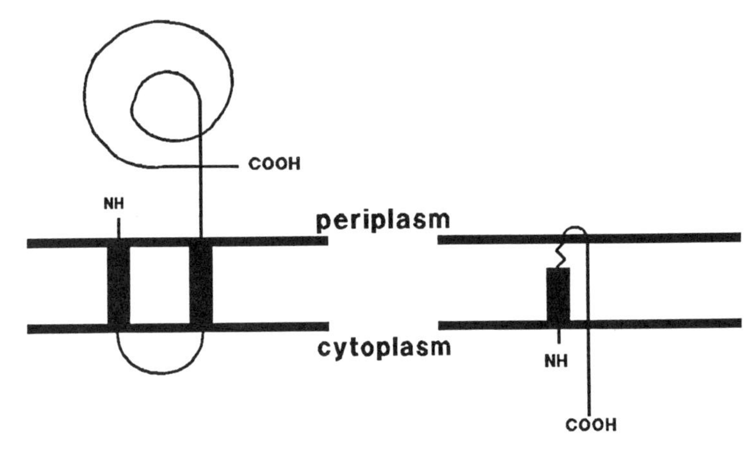

Figure 4. Panel A, topography of the *E. coli* SPase I (leader peptidase). The black boxes represent α-helical transmembrane segments, the left one functioning as internal (uncleaved) signal sequence of SPase I.

Panel B, hairpin loop structure of a precursor protein containing a hydrophobic signal sequence after it has initially inserted into the membrane. The signal sequence spans the membrane with its α-helical, hydrophobic core (*black box*) and its COOH-terminal, β-pleated part. A β-turn might function in exposing the cleavage site on the *trans* face of the membrane (periplasmic, in the case of Gram-negative bacteria, and lumenal, in the case of eukaryotic endoplasmic reticulum).

dase of *B. subtilis* exhibits a similar topography lacking the internal signal sequence and most of the cytoplasmic domain, however. Thus a single transmembrane segment anchors this peptidase to the membrane. This property is shared with the Imp 1 and the subunits Spc18, Spc21, and Spc22/23 of eukaryotic signal peptidases. Accordingly, deletion analysis of the *E. coli* SPase I revealed the dispensability of the first transmembrane segment and the cytoplasmic loop for an active enzyme.

The major part of SPase II is also thought to protrude into the periplasm. Four significantly hydrophobic domains are potential transmembrane segments, giving rise to a topography of SPase II such that 2 long hydrophilic loops extend into the periplasm while the NH_2-, the COOH-terminus, and a small loop between the second and third transmembrane domain are located at the cytoplasmic side of the membrane (Isaki et al., 1990). Both subunits of the inner membrane protease of yeast mitochondria have been found to be largely exposed to the mitochondrial intermembrane space and the thylakoid processing peptidase of chloroplasts faces the lumen of the organelle.

ER-signal peptidases are also oriented towards the lumen: in addition to the fact that some of the subunits are glycosylated (cf. Table 1), and that protease treatment of right-side out microsomal vesicles does not abolish signal peptidase activity, three of the five subunits of the canine signal peptidase complex were shown to be anchored to the ER membrane by a single NH_2-terminal transmembrane domain (Shelness et al., 1993; Kalies and Hartmann, 1996).

Extensive computer analyses of numerous pro- and eukaryotic, hydrophobic, NH_2-terminal signal sequences have unravelled the anatomy of a canonical signal peptide. Thus a typical hydrophobic signal sequence is 15–25 amino acids long. It has a tripartite structure consisting of a positively charged NH_2-terminal region, a central hydrophobic core probably arranged in an α-helix, and the more polar COOH-terminal part which probably forms a β-sheet including a β-turn around the cleavage site. This conformation is consistent with the hairpin loop structure (Inouye and Halegoua, 1980) proposed for the signal sequence after it has integrated into the membrane (Figure 4B): the NH_2-terminal part of the signal sequence remains in the cytoplasm, while the α-helical core and parts of the β-sheet span the membrane. The β-turn would then help to expose the cleavage site to the signal peptidase on the *trans*-side of the membrane with the remainder of the polypeptide chain running back through the bilayer. Such an array provides an explanation as to why the active sites of at least those peptidases recognizing hydrophobic signal sequences are exposed on the *trans*-side of the membrane. It has been suggested that the acquirement of the extended topography of this kind of signal sequences within their target membrane involves a looped insertion intermediate which is caused by a helix–breaking Gly or Pro in the middle of the hydrophobic core (Nouwen et al., 1994).

Studies of the structure-function relationship of the *E. coli* signal peptidase I have been carried out. This was based on examining the effects of mutant enzymes on the processing of signal peptides both *in vivo* and *in vitro*. The results obtained suggest that the second transmembrane segment of SPase I and the immediate downstream region protruding into the periplasm are directly involved in catalysis (Bilgin et al., 1990). This supports the notion that sections of high similarity between the known signal peptidases lie within this part of the molecule, including the conserved Ser 90 residue (cf. Figure 3).

SIGNAL PEPTIDASES YIELD TWO PRIMARY CLEAVAGE PRODUCTS

The reaction products of the signal peptidase-catalyzed reactions are the mature protein free of the signal sequence, and the signal sequence. Although proteolytic processing of precursor proteins by signal peptidase-containing membrane preparations has been demonstrated in countless experiments, usually the signal peptide is not recognized among the reaction products. Only under certain experimental conditions such as the use of purified peptidases, of protease inhibitors, or of synchronized *in vitro* systems can the signal peptide be detected. Therefore signal peptides normally do not accumulate because they are rapidly degraded. This process has been studied in more detail for the *E. coli* lipoprotein. Following cleavage of the precursor, a membrane-bound endoproteinase cuts the liberated signal peptide at various sites, while the resulting oligopeptides are degraded further by cytoplasmic proteinases. The physiological reason for the degradation of signal peptides is presumably to prevent inhibition of signal peptidases by the signal peptides generated, a phenomenon that has been demonstrated with purified enzymes.

WHAT IS THE REACTION MECHANISM OF SIGNAL PEPTIDASES?

Signal peptidases are endoproteinases. Little is known about the catalytic properties of these enzymes, mostly because no site-specific inhibitors are known. In agreement with the failure to inhibit signal peptide cleavage by conventional proteinase inhibitors, extensive site directed mutagenesis of SPase I of *E. coli* and *B. subtilis* did not disclose any essential Asp, Cys, or His residue: This would have been to be expected if these enzymes belonged to one of the four classes of proteinases, i.e., aspartyl proteinases, metalloproteinases, thiol proteinases, and classical serine proteinases (Black et al., 1992; Tschantz et al., 1993; van Dijl et al., 1995). The only critical amino acid residues found are a Ser, an Asp, and a Lys. Of those amino acids the Ser has been conserved throughout the known sequences as indicated in Figure 3 (marked with an asterisk). It has therefore been proposed that signal peptidases share the mechanism of catalysis with β-lactamases. Furthermore, a similarity between signal peptidases and LexA–like proteases has also been pointed out. The latter enzymes carry out catalysis by a Ser–Lys dyad.

The soluble processing peptidases of mitochondria and chloroplasts are inhibited by metal chelators and are therefore classified as metalloproteinases.

As pointed out earlier, peptidases cleaving hydrophobic signal sequences attack the precursor after it has inserted into the membrane, thereby forming a hairpin loop structure. Thus the conformation of the substrate to be cleaved is ultimately dependent on its molecular environment in the membrane. Furthermore, noncatalytic components of the known signal peptidase complexes could conceivably participate in the maintenance of a cleavage-competent structure of the precursor. In some instances, covalent modification of the precursor, such as heme attachment (Hartl and Neupert, 1990), is a prerequisite for cleavage, which is an additional indication that signal peptidases require a distinct tertiary structure of their substrates if they are to be active. Taken together, signal peptidases recognize structural epitopes of their substrates—as opposed to linear sequence motifs—and the concerted action of several factors helps to form these epitopes.

PHYSIOLOGICAL FUNCTION

The overwhelming majority of NH_2-terminal signal peptides that promote translocation of secretory proteins are cleaved off during or shortly after the translocation process. Why does proteolytic processing occur? Clearly, transmembrane translocation and signal sequence cleavage are independent processes. This has been repeatedly documented by different experimental approaches; these are: (1) a fraction of *in vitro* synthesized precursor proteins from *E. coli* consistently translocates into plasma membrane vesicles without being cleaved by SPase I, (2) mutations within the COOH-terminal region of signal peptides have been described which prevent cleavage but do not affect the translocation efficiency of the mutated precursor, (3) depletion of yeast cells of the mitochondrial matrix processing protease results in the expected failure to cleave mitochondrial precursor proteins but still allows import, and (4) many random sequences function as signals for transmembrane translocation of proteins into the ER and mitochondria. However, these surrogate signals are not proteolytically processed. Therefore, signal peptides are not removed for the sake of rendering membrane transport possible.

It has been proposed that the signal sequence functions as a modulator of the folding rate of a precursor molecule (Randall and Hardy, 1989). Since experimental data have accumulated demonstrating that stably folded precursors are translocation-incompetent, molecular mechanisms have to be postulated, that prevent premature folding of a precursor molecule. To what extent the signal peptide *in vivo* influences folding kinetics per se or to what extent it does so by interacting with recognition proteins that primarily or concomitantly function as molecular chaperones, has to await a complete elucidation of the protein transport mechanisms operating at the level of cellular membranes. After its complete translocation across a membrane, the conformation of a polypeptide chain appears to be controlled by the action of molecular chaperones, so that the signal peptide becomes dispensable in that respect also.

Presumably the most important aspect of signal sequence cleavage is the release of the translocated polypeptide chain from the membrane with which it otherwise remains associated via the signal sequence. The latter is demonstrated by studies in which site-directed mutagenesis has led to cleavage of a normally noncleavable signal sequence, thereby converting an anchor sequence to a classical signal sequence. Furthermore, repression of synthesis of *E. coli* signal peptidase I was shown to result in the membrane-anchorage of periplasmic and outer membrane proteins, clearly underlining the function of signal peptidases in completing the translocation process. It is evident that proteins which are destined for subcellular locations different from the compartment into which they have been initially translocated (i.e., outer membrane proteins in Gram-negative bacteria, truly secreted proteins, and so forth) have to be detached from the membrane across which they are translocated. Mis-sorting of translocated proteins by retaining them at the outer surface of the plasma membrane (lumenal surface of the ER-membrane, respectively) is a lethal event presumably due to blockage of the translocation machinery. Therefore it appears that the major reason for cleavage of signal sequences is to ensure that the translocated protein is liberated in order to reach further destinations and to unblock the export sites of the membrane.

SUMMARY

Numerous proteins in pro- and eukaryotes must cross cellular membranes in order to reach their site of function. Many of these proteins carry signal sequences that are removed by specific signal peptidases during, or shortly after, membrane transport. Signal peptidases have been identified in the rough endoplasmic reticulum, the matrix and inner membrane of mitochondria, the stroma and thylakoid membrane of chloroplasts, the bacterial plasma membrane, and the thylakoid membrane of cyanobacteria. The composition of these peptidases varies among one and several subunits. No general site-specific inhibitors are known for the majority of these enzymes. Signal peptidases recognize structural motifs rather than linear amino acid sequences. Recent evidence indicates that the reaction mechanism of peptidases cleaving hydrophobic NH_2-terminal signal sequences is similar to that of LexA–like proteases. Analysis of the reaction specificities and the primary sequences of several signal peptidases suggests that the enzymes of the endoplasmic reticulum, the inner mitochondrial membrane and the thylakoid membrane of chloroplasts all have evolved from bacterial progenitors.

ACKNOWLEDGMENTS

The work from the author's laboratory was supported by the DFG grant Dr 29/31-6C, the Sonderforschungsbereich 206, and the Fonds der Chemischen Industrie. I thank Dr. J. MacFarlane and E. Sambor for their help with the preparation of the manuscript.

NOTE

1. Interestingly, in potato both subunits of MPP are subunits of a decameric complex copurifying with the cytochrome c reductase activity (Braun et al., 1992).

REFERENCES

Bilgin, N., Lee, J.I., Zhu, H., Dalbey, R., & v. Heijne, G. (1990). Mapping of catalytically important domains in *Escherichia coli* leader peptidase. EMBO J. 9, 2717–2722.

Black, M.T., Munn, J.G.R., & Allsop, A.E. (1992). On the catalytic mechanism of prokaryotic leader peptidase 1. Biochem. J. 282, 539–543.

Braun, H.P., Emmermann, M., Kruft, V., & Schmitz, U.K. (1992). The general mitochondrial processing peptidase from potato is an integral part of cytochrome c reductase of the respiratory chain. EMBO J. 11, 3219–3227.

Gärtner, F., Bömer, U., Guiard, B., & Pfanner, N. (1995). The sorting of cytochrome b_2 promotes early divergence from the general mitochondrial import pathway and restricts the unfoldase activity of matrix Hsp70. EMBO J. 14, 6043–6057.

Glick, B.S., Beasley, E.M., & Schatz, G. (1992). Protein sorting in mitochondria. Trends Biochem. Sci. 17, 453–459.

Gray, M.W. & Doolittle, W.F. (1982). Has the endosymbiont hypothesis been proven? Microbiol. Rev. 46, 1–42.

Halpin, C., Elderfield, P.D., James, H.E., Zimmermann, R., Dunbar, B., & Robinson, C. (1989). The reaction specificities of the thylakoidal processing peptidase and *Escherichia coli* leader peptidase are identical. EMBO J. 8, 3917–3921.

Hartl, F.-U. & Neupert, W. (1990). Protein sorting to mitochondria: Evolutionary conservations of folding and assembly. Science 247, 930–938.

Hawlitschek, G., Schneider, H., Schmidt, B., Tropschug, M., Hartl, F.U., & Neupert, W. (1988). Mitochondrial protein import: Identification of processing peptidase and of PEP, a processing enhancing protein. Cell 53, 795–806.

Inouye, M. & Halegoua, S. (1980). Secretion and membrane localization of proteins in *Escherichia coli*. Crit. Rev. Biochem. 7, 339–371.

Isaki, L., Beers, R., & Wu, H.C. (1990). Nucleotide sequence of the *Pseudomonas fluorescens* signal peptidase II gene (lsp) and flanking genes. J. Bacteriol. 172, 6512–6517.

Kalies, K.U. & Hartmann, E. (1996). Membrane topology of the 12- and 25-kDa subunits of the mammalian signal peptidase complex. J. Biol. Chem. 271, 3925–3929.

Kalousek, F., Hendrick, J.P., & Rosenberg, L.E. (1988). Two mitochondrial matrix proteases act sequentially in the processing of mammalian matrix enzymes. Proc. Natl. Acad. Sci. USA 85, 7536–7540.

Keegstra, K. (1989). Transport and routing of proteins into chloroplasts. Cell 56, 247–253.

Nouwen, N., Tommassen, J., & de Kruijff, B. (1994). Requirement for conformational flexibility in the signal sequence of precursor protein. J. Biol. Chem. 269, 16029–16033.

Nunnari, J., Fox, T.D., & Walter, P. (1993). A mitochondrial protease with two catalytic subunits of nonoverlapping specificities. Science 262, 1997–2004.

Pratje, E. & Guiard, B. (1986). One nuclear gene controls the removal of transient pre-sequences from two yeast proteins: One encoded by the nuclear the other by the mitochondrial genome. EMBO J. 5, 1313–1317.

Pugsley, A.P. (1993). The complete general secretory pathway in Gram-negative bacteria. Microbiol. Rev. 57, 50–108.

Randall, L.L. & Hardy, S.J.S. (1989). Unity in function in the absence of consensus in sequence: Role of leader peptides in export. Science 243, 1156–1159.

Saier Jr., M.H., Werner, P.K., & Müller, M. (1989). Insertion of proteins into bacterial membranes: Mechanism, characteristics, and comparisons with the eucaryotic process. Microbiol. Rev. 53, 333–366.

Schneider, A., Behrens, M., Scherer, P., Pratje, E., Michaelis, G., & Schatz, G. (1991). Inner membrane protease I, an enzyme mediating intramitochondrial protein sorting in yeast. EMBO J. 10, 247–254.

Schulte, U., Arretz, M., Schneider, H., Tropschug, M., Wachter, E., Neupert, W., & Weiss, H. (1989). A family of mitochondrial proteins involved in bioenergetics and biogenesis. Nature 339, 147–149.

Segui-Regal, B., Stuart, R.A., & Neupert, W. (1992). Transport of proteins into the various compartments of mitochondria. FEBS Lett. 313, 2–7.

Shelness, G.S., Lin, L., & Nicchitta, C.V. (1993). Membrane topology and biogenesis of eukaryotic signal peptidase. J. Biol. Chem. 268, 5201–5208.

Smeekens, S., Weisbeck, P., & Robinson, C. (1990). Protein transport into and within chloroplasts. Trends Biochem. Sci. 15, 73–76.

Tschantz, W.R., Sung. M., Delgado-Partin, V.M., & Dalbey, R.E. (1993). A serine and a lysine residue implicated in the catalytic mecdhanism of the *Escherichia coli* leader peptidase. J. Biol. Chem. 268, 27349–27354.

van Dijl, J.M., de Jong, A., Vehmaanperä, J., Venema, G., & Bron, S. (1992). Signal peptidase I of *Bacillus subtilis*: Patterns of conserved amino acids in prokaryotic and eukaryotic type I signal peptidases. EMBO J. 11, 2819–2828.

van Dijl, J.M., de Jong, A., Venema, G., & Bron, S. (1995). Identification of the potential active site of the signal peptidase SipS of *Bacillus subtilis*. J. Biol. Chem. 270, 3611–3618.

Woese, C.R. (1987). Bacterial evolution. Microbiol. Rev. 51, 221–271.

Wolfe, P.B., Silver, P., & Wickner, W. (1982). The isolation of homogeneous leader peptidase from a strain of *Escherichia coli* which overproduces the enzyme. J. Biol. Chem. 257, 7898–7902.

Wolfe, P.B., Wickner, W., & Goodman, J.M. (1983). Sequence of the leader peptidase gene of *Escherichia coli* and the orientation of leader peptidase in the bacterial envelope. J. Biol. Chem. 258, 12073–12080.

Wu, H.C. & Tokunaga, M. (1986). Biogenesis of lipoproteins in bacteria. Curr. Top. Microbiol. Immunol. 125, 127–157.

Yang, M., Jensen, R.E., Yaffe, M.P., Oppliger, W., & Schatz, G. (1988). Import of proteins into yeast mitochondria: The purified matrix processing protease contains two subunits which are encoded by the nuclear *MAS1* and *MAS2* genes. EMBO J. 7, 3857–3862.

RECOMMENDED READINGS

Dalbey, R.E. & von Heijne, G. (1992). Signal peptidases in prokaryotes and eukaryotes—a new protease family. Trends Biochem. Sci. 17, 474–478.

Dev, I.K. & Ray, P.H. (1990). Signal peptidases and signal peptide hydrolases. J. Bioenerget. Biomemb. 22, 271–290.

Izard, J.W. & Kendall, D.A. (1994). Signal peptides: Exquisitely designed transport promoters. Mol. Microbiol. 13, 765–773.

Lively, M.O. (1989). Signal peptidases in protein biosynthesis and intracellular transport. Curr. Opin. Cell Biol. 1, 1188–1193.

Müller, M. (1992). Proteolysis in protein import and export: Signal peptide processing in eu- and prokaryotes. Experientia 48, 118–129.

Chapter 10

THE SORTING OF PROTEINS

Stephen S. Rothman

Principles of Medical Biology, Volume 7A
Membranes and Cell Signaling, pages 205–227.

ISBN: 1-55938-812-9

Biology seems complicated, ad hoc, unparsimonious, and I can't understand what it does, let alone why it does it.

Ann Finkbeiner, a science writer in a New York Times book review, 1994.

INTRODUCTION

Every student who has taken a biology course knows that the singular achievement of 20th century biological research has been the discovery that the nucleic acid molecule, DNA, carries our genetic heritage from generation to generation. But an equally important discovery was the realization that this genetic heritage, written in DNA, is comprised solely of the structures of various protein molecules. DNA contains the code for perhaps as many as 100,000 different proteins in higher organisms such as man, although an average cell only produces (translates) a small fraction of this number; perhaps 10,000 or so. That is, we seem to have learned that we and all other living things are defined wholly, at least in genetic terms, by these complex protein molecules. In this context, it seems that differences in the kinds of proteins expressed and the details of their molecular structure account for all observable (phenotypic) differences between individuals and species.

If this is true, then it follows that all of the functions of cells, indeed all of the functions of organisms, are traceable to the properties of the various protein molecules that comprise them. If these proteins (and DNA) alone distinguish living cells from the inanimate world, then it would seem a relatively simple matter to create a living cell. Just place the proteins together in a test tube, add all other necessary ingredients, such as the right salts, water, gases, small nutrient organic molecules, at the right concentrations, and will form a cell. That is, if we have each and every protein molecule, and the correct substrates and environmental preconditions so that they can carry out their physiological functions, then we should have created a living cell by definition.

But would such an enriched broth really be a cell? Most biologists today would agree that it would not. Yes, it would contain the ingredients necessary, prerequisite, for life, but these ingredients by themselves would not be sufficient. What is missing? Do we need to include other kinds of substances, perhaps still undiscovered, or must we invoke insubstantive forces? Curiously, what is missing is not material, but no less a part of the physical world for not being so. It is what makes biological cells far more complex than the incredibly complex mixture of protein-based chemical reactions that we have introduced into our hypothetical broth.

What is missing may already be obvious. It is *organization*; the organization of the components of the cell, particularly its protein molecules, in three dimensional space. That is, for these molecules to carry out their assigned functions they must be at the correct location or locations. Thus, if we could in some way specify protein location so that each molecule was in proper juxtaposition to every other molecule, then, assuming that modern biology is correct, a living cell would indeed be the outcome. This then is the subject of this chapter; how the biological cell produces topological order among its composite proteins. This is called the sorting problem, and is one of the outstanding problems of modern biology, some would say its central unsolved problem.

BACTERIA

Prokaryotes, bacteria, and certain algae, are the oldest known forms of life on the planet. The word "prokaryote" is taken from the Greek and means "taking the place (pro-) of one with nuts (-karyote), as compared to eukaryotes, which means "true (eu-) nut containing entities." The "nuts" are nuclei, absent in the former and present in the latter; the cells of higher organisms such as plants and animals. We will begin our consideration of the sorting of protein in the simple prokaryotic cell, and then move on to the far more complex situation in the cells of higher organisms. Paradoxically, many of the ideas that we will discuss about sorting in bacteria, were first developed in studies on eukaryotic cells. The simplest bacterial cells are comprised of the genome (the expressive DNA machinery that manufactures proteins), the proteins themselves and all associated metabolites. They are mixed together to form the cytoplasm, and are held in place by an enveloping single lipid bilayer membrane.

Given this description, one could say that protein sorting would not seem necessary in such simple structures. After all, everything is mixed together in the cytoplasm. This would be true, but for two crucial facts. Bacterial cells secrete proteins into the external medium and contain them in the lipid membrane that encloses the cell. These two sorting processes are required for cell survival. Digestive enzymes and toxins are secreted to obtain food, and membrane-embedded protein transporters scavenge the environment for needed organic and inorganic metabolites. Other membrane proteins protect the cell from the vagaries of the external environment and by acting as transducers (receptors) that allow the cell to respond to changes in their milieu. Without these processes, the cell would be totally isolated from its environment and at its mercy, and life as we know it would not exist.

Thus, we need two sorting processes even in this simple cell; a process for the secretion of proteins across its membrane and the insertion of proteins into it. Things are more complex for bacteria that have an additional (outer) membrane (for example, Gram-negative bacteria, such as *E. coli*). Here, secretion must not only include transport across the inner membrane, but through an intermembrane compartment, and across the external membrane. Moreover, in these cells, (non-secreted) proteins are also transported across the inner membrane to reside in the intermembrane space as well as to enter the external membrane.

In addition to these processes of sorting, cellular proteins turn over. That is, proteins are degraded into their substituent amino acids in the normal course of events, and are replaced with new protein molecules of the same or different type. Turnover rate varies greatly between different protein molecules, with half-times as brief as an hour for some, and as long as a week for others. A day or so is probably common. Protein turnover is a central regulatory process of cells and allows them to adapt to changing needs. That is, they are able to change their complement of proteins, in both type and amount, as environmental factors dictate. Turnover also allows for the removal of proteins that have become damaged due to cosmic radiation or other natural environmental forces. Of course, secreted proteins, at least once they are secreted, are not degraded by cellular processes. But membrane proteins, including those in the outer membrane and intermembrane space when present, are part of the cell's system of protein turnover. For them to be broken down, they must re-enter the cytoplasm where most degradative processes occur in bacteria. Thus, sorting for these proteins, is not only getting there, but coming back.

Membrane insertion, transport across membranes into an intermembrane space or the outer membrane when present, secretion, and movement back into the cytoplasm from sorted locations for subsequent degradation are the processes of protein sorting in bacteria. We have learned a great deal about these processes, but only recently. Even so, our understanding remains very incomplete. No successful general theory has yet emerged. In fact, thus far, the evidence runs counter to this hope. Sorting, even in the simple bacterium, seems to involve a variety of different processes, not a single universal mechanism.

Two Central Questions

Even though many different mechanisms may account for the sorting of its thousands of different proteins by a cell, protein sorting is conceptually quite simple. It involves the insertion of proteins into, and their deletion from membranes. That is, it involves two physical events; one in which proteins and membranes associate and the other in which they dissociate. In secretion, insertion is preferred at one surface and deletion at the other. For the turnover of membrane proteins insertion and deletion are favored at the same surface.

This brings us to the two questions that underlie all research in this area. What molecular mechanisms account for these events? And why do particular protein molecules enter and cross membranes, while others do not? The answer to the second question, may well be found in the answer to the first and vice versa. That is, the "why" of sorting may be part and parcel of the mechanism of insertion and transport. On the other hand, the two questions could have separate answers. Sorting involves two elements. One, that *identifies* the protein for targeting to a particular locus, and the other, the *mechanism* of insertion and transport that is needed to accomplish sorting. As scientists have tried to understand these processes, separate answers to the two questions have commonly been sought. What identifies a particular protein for sorting to a particular compartment? And what are the mechanisms that underlie the insertion and transport of proteins?

This dichotomy comes from the idea that sorting has an informational element. The protein must contain the necessary information to specify its eventual location. The location must be recognizable to the protein molecule and in turn recognize it as different from other proteins. This fact has led to attempts to discover the sorting code resident in the structure of each protein molecule; a code that specifies location. This code has been sought in great part in the primary sequence of the protein's constituent amino acids. For example, if we can determine that all proteins that reside in the intermembrane space of bacteria, contain the same particular sequence of amino acids, then this sequence must code for location. Such sequences are called *consensus* or *sorting sequences*.

The ability to genetically engineer proteins has been the driving force behind this approach. It is relatively easy to determine the amino acid sequence of any protein by cDNA replication and by then using computer algorithms to quickly seek sequences of amino acids that are common to different molecules. In addition, genetic engineering has made it possible to change the primary structure of any protein virtually at will. Having so done, it is then possible to determine whether or not this or that new form is properly sorted. From this information, inferences can be drawn about the nature of the sorting signal in proteins destined for particular compartments. If we change this or that amino acid or group of amino acids, and sorting is interfered with, then we conclude that we have identified the sorting signal. Consensus sequences for sorting have been uncovered, for example, for the nucleus. However, in other cases, such as the mitochondrion, there does not appear to be a consensus sequence.

In any event, before we accept the conclusion that because a particular sequence of amino acids is found in a variety of proteins slated for a particular compartment and is necessary for their proper sorting, that it is a code (information) that specifies location, we should consider whether such sequences could represent something else. For example, we often know that a particular sequence of amino acids is common to a family of proteins, and necessary for their common functions, but we do not think of this sequence as a "code" that allows these common functions. Usually we view these identities in structural terms. What is being coded for is structure, and the sequence in its own right, as a code of amino acids, has no meaning. Common amino acid sequences may tell us that proteins in this or that family contain such and such β-sheets, or histidine-containing active sites, or have other structures in common, but they are not a "code" that directly signifies a common function. The amino acid sequence leads to a particular structure and that structure to function.

To use another example, this time from transport physiology, we may know that a particular amino acid moves across a membrane by a particular transport system or "porter" by facilitated diffusion, and that both the amino acid and the porter display an affinity for each other. (A porter is a membrane protein or group of proteins to which transportable substrates bind with high affinity. As a result of this interaction the ligand is transferred across the membrane where it then dissociates. The process is called "facilitated diffusion," because this interaction facilitates transport as compared to simple or normal diffusion.) In our hypothetical system, we find that we can prevent transport in a variety of ways by altering the structure of the amino acid. We find also that simply by changing the amino acid from an L-isomer to a D-isomer transport is essentially prevented. Moreover, we find that this is not only true for a single amino acid, but for all the amino acids we examine. This is the actual case for amino acid transport in animal cells. Should we conclude from this information, that the effective L-isomer is the *code* for transport? We certainly can, because who can tell us what to call something? But is it really a code, or does it simply reflect chemical specificity between ligand and porter? The ligand has a high affinity for a particular site, an amino acid porter, when it has a specific structure, but not when the structure has been changed in a particular way.

Thus, simply because a sequence of amino acids is common to a group of proteins or because altering it, affects sorting, we cannot necessarily conclude that this sequence is a code that specifies sorting to a particular compartment, and that as such, is separable from the mechanism of insertion and transport. To understand these characteristics as being different, we must be able to separate the *coding* function from the *structural* features of the molecule related to its transport, as has been done with DNA coding for proteins.

The Paradox

During and subsequent to their manufacture, proteins released into the cytoplasm fold into what is known as the "mature" or folded state. Although our understanding of folding is still incomplete, in the end it follows a thermodynamically determined path which is at least in part a function of hydrophobicity or water avoiding amino acids trying to hide from the aqueous milieu, and hydrophilic or water seeking amino acids looking for the watery interface. For the most part, this leads proteins to ball up into spheres or as close to spheres as their particular amino acid sequence allows. Proteins today are understood to be amphiphilic or amphipathic molecules. The amino acid composition of proteins is a mixture, usually about 50-50, of hydrophilic and hydrophobic, water soluble and water insoluble, residues. How successful a particular amino acid is in locating itself in its desired environment depends on all of the other amino acids in the chain and the sequence they form. The movement of each amino acid is constrained by its place in the chain, and however hydrophobic or hydrophilic it may be, it may find itself in an environment that it would not seek left to its own devices. Thus, while protein molecules fold to form the most thermodynamically stable state in an aqueous medium, this state is a composite, an acceptable balance, in which many individual amino acids find themselves uncomfortably situated for the greater good of the whole (not that they have any choice in the matter). The proclivity to show its water seeking face, usually, although not always, makes proteins water soluble substances.

The lipid membrane for which both membrane and secreted proteins are destined, is, unlike the cytoplasm, a non-polar structure. Water avoiding aliphatic chains of membrane lipids form its structural base. These two facts—proteins tend to form hydrophilic balls in aqueous media and the structural base of the membrane is its non-polar lipid chains—taken together have stood as an immense barrier to our understanding of protein sorting; a barrier that lasted for more than half a century.

Although it has been known for a long time that cells secrete proteins and that their membranes contain proteins, the fact that proteins are in great part water soluble substances and that the lipid membrane presents an highly unfavorable hydrophobic world to be entered, led to the conclusion that proteins could not enter membranes, no less cross them. To do so would, it was thought, require solvation of large water-soluble substances (proteins) in non-polar environments (the lipid bilayer). Solubility measurements told us that this did not occur to any substantive extent. Most proteins simply would not go into solution in substantial amounts in a non-polar medium such as lipids.

And yet, it was known that proteins entered and crossed membranes in one way or another. Much of the struggle to understand the sorting of proteins by cells has had as a subtext attempts to deal with this paradox; that is, to try to understand how the processes of protein insertion and transport comport with our understanding of the physical properties of proteins and lipids and their potential interactions. Many of our ideas, and early models for these processes, were predicated on the view that such interactions were not possible, and that a way around the paradox had to be sought.

AN EARLY MODEL

The first model suggesting a specific mechanism to account for protein insertion and transport across biomembranes was proposed about 20 years ago. It attempted to overcome the problem of protein solvation in lipid in an ingenious way. Evidence for the hypothesis came primarily from eukaryotic cells, but the ideas were applied to bacteria as well. It is called the signal hypothesis and is really a series of related hypotheses that we can list as follows:

1. For a protein to be inserted into a membrane or transported across it, it has to contain an additional sequence of amino acids at its N-terminus, about 15–20 amino acids long, called the signal sequence. This sequence acts as the signal (information) signifying the need for these processes.
2. Protein insertion and transport only occurs during protein synthesis. This is called co-translational insertion and transport, because it occurs coincident with the translation of the protein.
3. Protein synthesis begins on ribosomes [a complex structure of more than 20 proteins and nucleic acids that carries out the synthesis of polypeptide chains] that are floating freely in the cytoplasm of the cell.
4. After its initiation, synthesis is halted by attachment to the ribosome of a group of proteins from the cytoplasm, called the signal recognition particle, that allows time for the ribosome to diffuse to the particular membrane being targeted without additional chain elongation (synthesis).

5. The ribosome attaches to the membrane and synthesis resumes.
6. Transport occurs only for the elongating peptide chain absent secondary or tertiary structure. If the chain is allowed to fold as a result of continued synthesis, transport is prohibited. This is the reason for stopping synthesis with the signal recognition particle while the ribosome is free in the cytoplasm.
7. The linear chain is advanced through a special pore in the membrane that is mobilized or opened only after the ribosome attaches to the membrane.
8. For proteins that cross the membrane the N-terminus of the peptide chain is cleaved to prevent further transport of the molecule across this or any other biomembrane.

The model's features were designed to overcome the problem of the unfavorable nature of protein/lipid interactions that insertion and transport seemed to require. Yes, protein insertion into and transport across membranes occur, but only under very special circumstances. Transport can only occur if the protein contains the signal sequence, it can only occur during protein synthesis, only for the linear amino acid chain, only through a special pore, and cleavage of the signal sequence would prevent recurrence. As such protein transport and insertion were one time only events; that is, they only occurred in one direction across a single membrane.

By only allowing transport of a linear chain, the problem of attempting to insert a whole protein into the lipid bilayer, with many surface hydrophilic groups, commonly of the order of 15–20, that would have to be simultaneously masked for entry to occur, could be avoided. Amino acids would enter one at a time into an aqueous channel. As an additional bonus, synthesis could provide the energy for transport, providing a ratchet like mechanism in which every time a peptide bond was formed the chain would be advanced through the channel.

The Signal Hypothesis and the Evidence

Since its original proposal some 20 years ago, a great deal of positive evidence has accumulated for the signal hypothesis; literally hundreds of scientific papers. Centrally, many proteins that enter or cross membranes have been found to have the predicted N-terminus addition as part of their DNA reading frame. Often cytoplasmic proteins do not have this sequence. Its removal or addition may prevent or allow membrane transport for a particular protein. Proteases that cleave the sequence have been discovered in the right locations. Recently, good evidence for the elusive pore has emerged, as well as substantive evidence for peptide transport through such a structure when ribosomes are attached to the membrane. On the basis of such evidence, the signal hypothesis can claim affirmation, and it has certainly earned the right to graduate from a hypothesis to a theory. Hypotheses need not have observational support. They can simply be mental constructs. A theory, however, must have a substantial observational base, and the signal "theory" has that.

But it was hoped by some that the signal hypothesis would provide a universal mechanism that applied to all protein transport across and insertion into membranes. However, if it is a universal hypothesis, then even one clear exception to it, must lead to its rejection. And unfortunately there have been many observations, perhaps equal in number to the affirmative ones, that present falsifying information for the hypothesis in particular cases. As such, the signal theory was clearly not universally applicable. Other mechanisms existed for protein insertion into and transport across membranes.

There were also other questions. If we think about bacteria for a moment, we can see major problems with the theory. For example, it argues that protein transport and insertion only occur during translation, but not subsequently. Considering a double-membrane bacterium, how does one explain secretion by these cells, in which two membranes must be crossed; the second presumably after the molecule is in its balled up state and lacks the signal sequence of amino acids (as well as its association with the ribosome). And how do we account for protein turnover by this mechanism? Membrane proteins have to dissociate from the inner membrane and proteins in the intermembrane space or outer membrane have to cross the inner membrane a second time, in order to enter the cytoplasm. The signal theory does not seem to be able to explain such processes.

It has also recently been discovered that the nascent (still forming) peptide chain in the membrane is about triple the length expected if it were in linear form. That is, the peptide being transported contains higher order structure and is folded back on itself in one way or another. And evidence of a channel is only observed subsequently. In addition, there is evidence that suggests that transport and synthesis may not truly be linked processes, but simply occur in proximity to each other in space and time. For example, if you eliminate an effective signal recognition event to halt synthesis, it continues to the point that secondary and tertiary structures form in the peptide chain. In this case, transport should not occur, and the cell, unable to sort its proteins, should not survive. However, such an experimental manipulation carried out genetically in yeast, does *not* lead to lethality, as predicted. That is, the cells survive under these conditions. This means that sorting must occur in other ways.

But perhaps the most problematic aspect of the signal theory has been the notion of the signal sequence itself. Today we know that many proteins released into the cytoplasm contain this extra sequence of amino acids. Some of these molecules subsequently enter or cross membranes after synthesis is complete and the folded structure is in place. That is, the presence of the signal sequence does not necessarily lead to the events listed above. Why not? Moreover, many protein molecules that lack the signal sequence enter and cross membranes in its absence.

Even more significant, it is not clear exactly what the signal sequence is? If it is a signal, a means of communication, what is the code? The easy availability of gene technology has provided us with hundreds, and soon thousands, of these signal sequences to examine and compare. They are not all the same, nor have any informational units been decoded similar to the genetic code. They vary greatly in sequence and content. The point of cleavage is not uniform. After much analysis, the central generalization has been that these sequences contain a relatively large number of hydrophobic amino acids. It is this feature, its hydrophobicity, that is thought to be responsible for its role in insertion and transport.

Of course, it would seem that a hydrophobic sequence of amino acids would reflect a structural feature of a protein, not an amino acid code. But more importantly, even this generalization about a hydrophobic sequence does not hold. Whole groups of proteins that are transported across membranes, such as across the outer membrane of the mitochondrion, contain amphiphilic signal sequences, or even hydrophilic ones. Putting the question of what the signal sequence is in further doubt, scientists have manufactured random sequences of the length of the usual pre- or signal sequence, and then tested these peptides for their ability to enter and cross membranes alone or as part of genetically engineered hybrid or fusion proteins. That is, these sequences were added to selected proteins and transport assessed. Some 20% of randomly synthesized sequences had the ability to promote transport. Hardly what one would expect for a conserved code. Moreover, if there was a common structural theme for the effective N-terminus peptides, it appeared to be their amphiphilicity—being a mixture of hydrophobic and hydrophilic amino acids, not their hydrophobicity. Thus, on this basis it seems doubtful that the signal sequence is a signal at all, at least in the sense of being an informational code. Instead, it seems to be a common structural prerequisite for protein insertion and transport; a structural prerequisite of variable character and perhaps variable function.

There is one final question about the signal theory to be addressed before we move on. The theory argues, as we have said, that transport occurs through pores in the relevant membranes. This is true of proteins that are transported across the membrane, but how about those that are inserted into it? Having entered the pore, how do they exit into the membrane proper? And how does the pore distinguish peptide chains that will cross the membrane from those that will remain. Many membrane proteins have multiple membrane spanning segments that are in intimate contact with membrane lipids. The signal theory proposes that such insertion is accounted for by the presence of a sequence of amino acids, called the stop-start sequence, that stops synthesis to allow time for the chain to enter the membrane, before it continues. But does this really solve the problem?

In a sense we still confront the same issue raised at the outset. How do we put a protein molecule into the lipid bilayer? If the molecule is itself hydrophobic, then why do we need this complex mechanism, and if not, how has it helped us? What we *do* know is that proteins exist in membranes and form bonds with membrane lipids and you have learned about this as the *fluid mosaic model* of the biological membrane. Membrane proteins are usually classified as being of two types. One is called *peripheral*. They are proteins that can be separated from the membrane by the addition of salts. This indicates that their association with the membrane is due to polar bonds, presumably with the head groups of phospholipids, such as phosphatidyl inositol or serine, or with other membrane proteins. The other group is called *intrinsic*. These proteins can only be separated from the membrane by detergent treatment. This indicates that they form hydrophobic bonds with the aliphatic chains of membrane lipids, or with other membrane proteins, and need a detergent to be interposed in order to free them. Thus, we know that proteins in membranes are associated by both polar and non-polar bonding to membrane lipids and to each other. A model for protein insertion and membrane transport that seeks ways of avoiding these interactions, that are known to occur in the end, cannot be viewed as being complete or satisfactory.

NEWER VIEWS

Post-Translational Transport

Perhaps the most significant of the new ideas about protein sorting is that the insertion of proteins into membranes and their transport across them is often, if not most frequently, an event that occurs after protein synthesis, or if coincident with it, not necessarily coupled to it. Although there is some evidence in bacteria that ribosomes attach to the cell membrane during translation to accommodate protein insertion, it seems that the majority of membrane and secreted proteins are released from ribosomes into the cytoplasm from whence sorting occurs by mechanisms that we shall now consider.

Protein/Lipid Interactions

It is clear today that other ways are available for protein insertion into and transport across membranes than that proposed in the signal theory. Perhaps most importantly, scientists understand that proteins and lipids can enter into high affinity associations with each other. That is, we can discard the paradox that we discussed initially concerning the impossibility of such associations. As we have noted, proteins are not simply water soluble substances unable to enter a non-polar lipid phase, but are amphipathic molecules capable of forming associations with lipids under the right circumstances. At least for certain proteins, interactions with membrane lipids alters their structure and proclivities, and makes their entry into the bilayer favorable.

And the biological membrane cannot be properly analogized to a bulk non-polar medium in which proteins must solvate. Indeed, it is not a bulk medium at all, nor are all membrane lipids simply non-polar substances. They are also amphipathic, much like proteins. For example, phospholipid molecules in the bilayer present their polar ends (such as inositol or serine) to the boundary with the aqueous medium, while their non-polar aliphatic chains face inward towards each other. The membrane is only two lipid molecules thick, not much wider than the diameter of a protein molecule, and in a highly mobile liquid crystal phase. Protein interactions with membrane lipids can alter or disrupt the bilayer structure, giving rise to the formation of intramembranous protein containing "micelles." In a micelle, the polar and non-polar elements of molecules are spatially segregated, such that non-polar elements may face internally and polar ones externally, much like the bilayer itself.

What is clear today is that the simple solubility of protein molecules in a lipid solution is not the issue. The question is whether interactions between proteins and lipids can alter the structure of either in a fashion that allows for more favorable secondary interactions; that is, interactions that lead to protein insertion into the membrane or transport across it. Some proteins, water soluble proteins, have been shown to be able to enter membranes spontaneously. Among these are a variety of bacterial toxins. Interactions with membrane lipids lead to changes in the protein's structure. These interactions are probably initially with polar head groups of phospholipids. As a result, the structure of the protein is altered so that its hydrophobic elements are more accessible. Hydrophobic bonds with the aliphatic chains of membrane lipids are then formed, and the protein is inserted into the membrane.

Chaperones

But proteins and lipids are not always, or perhaps usually, left to effect a union, or not, simply as they see fit, or more properly as they fit each other. We have begun to learn about a variety of mechanisms that exist to help in the consummation of this union. An important recent discovery is the existence of a special class of proteins, appropriately called chaperones, that bind other proteins, and as a consequence act, among other things, to assist in their transfer into the lipid membrane of the cell. Although what chaperones actually do and how they do it is not a settled issue, they seem to keep protein molecules in a loose configuration, preventing them from balling up after synthesis, or helping them unfold if the protein is already in its folded form. As such, non-polar (aromatic and aliphatic) amino acids, such as phenylalanine and leucine, that are normally water avoiding have more ready access to the surface of the molecule. This allows for the formation of secondary structures, such as α-helices, that may be relatively hydrophobic and more readily inserted into the lipid layer. Chaperones have ATP binding sites and ATP utilization is thought necessary for their action. Having been so prepared, these complexes diffuse to the surface of the membrane, and the loosely configured, more hydrophobic protein, interacts with membrane elements. The chaperone is then released for another cycle.

The Role of Membrane Proteins

It has become clear that in addition to interactions with membrane lipids and chaperones, many membrane proteins are involved in the insertion of other proteins into membranes, as well as their transport across them. One of the most well studied protein transport systems is the Sec system in the bacterium *E. coli*. It is responsible for the secretion of a variety of proteins by this organism. Unlike simple protein/ lipid interactions, at least six membrane proteins have thus far been shown to be part of this transport system, in addition to at least one cytoplasmic chaperone. Exactly what each protein does is not known. Some may function as membrane chaperones, others as recognition sites, and others of course as part of the transport system proper. It has recently been proposed that Sec proteins form a pore through which secreted proteins travel, with higher order structure in the protein apparently being allowed. Membrane pores that can accommodate proteins, are quickly becoming the most well established modality for protein transport across membranes. We will discuss these processes below as they are expressed in eukaryotes.

FROM THE SIMPLE TO THE INCREDIBLY COMPLEX

Not to suggest that bacteria in the modern world are not extremely complex organisms, they most certainly are, but sorting in the cells of higher organisms, eukaryotic cells, is a far more complex and therefore uncertain business. Centrally, unlike the bacterium, cells from animal and plant species contain "organelles," intracellular structures that partition proteins, and hence protein functions within the cell. These structures are usually separated from the cytoplasm by bilayer membranes. The presence in eukaryotic cells of this complex morphology, including the nucleus, mitochondrion, chloroplast, endoplasmic reticulum, Golgi apparatus (with four subdivisions—cis, medial, trans, and trans-Golgi network), secretion granules, assorted vesicle structures such as coated vesicles and small smooth-surfaced vesicles, as well as a wide range of other specialized structures found in one cell type or another, tells us in no uncertain terms that sorting is likely to be a complex process in such cells.

It no longer is just a problem of placing proteins in the cell membrane, secreting them across this membrane, or removing them from the membrane, as in the bacterium. Proteins must be sorted to all of these different intracellular structures, that is, their contents and membranes. Nuclear proteins must go to the nucleus, not the mitochondrion, and vice versa. Those slated for the lysosome, must not end up in secretion granules, and so forth. The cell sorts these products with high efficiency. But how?

Processes similar to those in bacteria in which direct membrane/protein interactions of the sort we have discussed that lead to sorting are important in eukaryotes, and we shall discuss them later. However, in eukaryotic cells a whole new class of processes is seen. In these processes, protein sorting occurs by means of membrane-enclosed vesicles, which contain the proteins to be sorted either within their internal space or as part of their membrane. These vesicles carry the substances to different loci in the cell. This process (actually an extremely complex series of processes, of which the signal hypothesis is one element) is known by the general term, *vesicle transport*.

Vesicle Transport

In vesicle transport, substances are moved from place to place within the cell as a result of the formation of a vesicle from a given compartment (budding), its movement to a target compartment, the fusion of the membrane of the vesicle to the target membrane, the formation of a hole in both membranes through which the products contained in the vesicle pass, and the subsequent inclusion of the contents of the vesicle, as well as its membrane in the new compartment. Our notions of vesicle transport have in great part been derived from the study of secretion in eukaryotic cells. The theory that grew from these studies has served as both an exemplar of sorting in eukaryotic cells, and as the framework for understanding sorting processes thought to involve vesicle transport in general.

The traditional view has been that the secretion of proteins by eukaryotic cells occurs by vesicle transport processes, and such processes alone. However, we know today that this is incorrect and that secretion by eukaryotic cells also occurs by membrane transport processes similar to those found in bacteria. We will discuss this briefly below, but for now, let us consider the standard vesicle theory of secretion.

The intellectual foundation of the vesicle theory is the same as the signal hypothesis. The existence of such processes was originally predicated as much on necessity, as on evidence. To some degree this still holds today. They were thought necessary because they were seemingly the only available way to overcome the paradox that although proteins cross membranes in all cells, such events are impossible because of the nature of proteins and lipid membranes, one polar, the other non-polar. Vesicle transport allows for the transfer of proteins between compartments, permitting them to cross membranes, without requiring that the membrane itself be permeable to these substances. It also explains the insertion of proteins into membranes, and their deletion from them, by either adding membrane to membrane, or deleting membrane from membrane, respectively. This avoids requiring that proteins partition into or out of these structure by themselves.

In the vesicle theory, the task of transport is taken away from the transported molecule. Its thermal motion, response to concentration gradients, ability to partition in phases, affinity for one or another other substance are all irrelevant. Instead, an external process takes care of business. It is akin to the difference between driving your own car and traveling by bus, along with many other passengers, and passively leaving the driving (timing, rate, and location) to the bus driver.

It was the vesicle theory of secretion that gave rise to the signal hypothesis, and they both, as we have noted, at least originally, held to the same notion of protein impermeability and the need to find a way around this impasse. Indeed, the signal hypothesis was proposed as a solution to a problem that arose from the vesicle theory. It seemed that whatever vesicle processes accounted for transport, in the beginning, that is, at the time of the synthesis of the protein, its movement across or into a biomembrane was required, however unlikely this seemed. This was because it was clear that proteins were manufactured on ribosomes that were suspended in the cytoplasm either freely or attached to membranes. Therefore, it seemed that proteins had to cross or enter at least one membrane by their own, non-vesicular, devices, i.e., by membrane transport, after their release from the ribosome, if they were to end up in any compartment other than the cytoplasm.

As we have discussed, the signal hypothesis' proposed solution to this problem was the first specific model for membrane protein transport. It hypothesized that synthesis and transport were coupled and that the growing peptide chain, in linear form, crosses membranes. In this way, the proteins entered structures from which vesicles could be derived. In eukaryotic cells this appears to be exclusively the cisternal (internal) spaces of the endoplasmic reticulum, or its membrane (the endoplasmic reticulum is also called the rough-surfaced endoplasmic reticulum or RER to distinguish it from smooth reticula that do not have ribosomes appended to their surface). It is from this source that all sorting begins, whether for the matrices of organelles or the various membranes of the cell, including the cell membrane itself, or for secretion.

Subsequently, small membrane enclosed vesicles bud from the RER, to carry the products, and specified membrane components, to the Golgi apparatus. The Golgi apparatus, like the RER, is comprised of elongated membrane-bounded sacs organized in layers. Unlike the RER, however, it is thought that the different Golgi sacs carry out different functions. The Golgi apparatus has been divided into four components thus far, primarily on the basis of differences in their protein composition as determined immunocytochemically (that is, by the binding or hybridization of antibodies to specific Golgi proteins, visualized by the addition of microscopic markers, such as fluorescent molecules, to these antibodies). These four compartments are the cis-, medial and trans-, and trans-Golgi network, each of which seems to be involved in the post-translational processing of proteins. Post-translational processing refers to the chemical modification of proteins subsequent to their synthesis; for example, the addition of polysaccharides, their sulfation, or peptide cleavage.

Proteins destined for secretion move through these sacs, one after the other, carried from one to the other in small vesicles that bud from one sac, fuse to the next, and so on, in sequence. Finally, small vesicles bud from the trans-Golgi network and carry the product to secretion granules, the final repository of substances secreted by the cell. Secretion itself occurs by another vesicle transport process called *exocytosis*, in which the membrane of the secretion granule fuses with the cell membrane, and the contents are released into the extracellular environment. The membrane proteins of the secretion granule are sorted as well, and carried in the membrane of the small vesicles that carry the secreted proteins.

Sorting via the Secretory Pathway

What I have just described is the conventional or standard secretory pathway. Initially, it was thought to describe not only how secretion occurred, but more generally how the cell sorts its proteins otherwise. Central to this theory is the idea of membrane flow. Membrane is moved from the RER to the Golgi to secretion granules, and eventually into the plasma membrane. The plasma membrane is then recycled by *endocytosis*, in which process the cell membrane buds internally, like phagocytosis, and forms *endosomes* which contain fluid from the external milieu. Originally, it was thought that these vesicles return to the RER from which they were formed. Today such endocytic vesicles are thought to have a variety of destinations, centrally the Golgi, and we will touch on this below.

In this view, proteins destined for a variety of intracellular loci follow the same pathway as products to be exported. They are packaged in small vesicles that form from the RER and travel to the Golgi apparatus. It is at the Golgi that their paths diverge. Secretory products are transferred to secretion granules in vesicles derived from the Golgi, whereas vesicles carrying products to intracellular loci travel to those specified sites instead. An example of intracellular sorting by small Golgi derived vesicles is for lysosomal proteins; that is proteins, mainly proteases, that are involved in the degradation of other proteins in lysosomes. Similarly, membrane components that do not reach the plasma membrane but are sorted to the membranes of intracellular organelles, such as the lysosomal membrane, are recycled internally; that is, without first going to the plasma membrane.

First, let us consider the sorting of proteins contained *in* vesicles, not in their membranes. What is required for sorting in this case? Of course, the proteins first have to be transported into the cisternal spaces of the RER, and the signal hypothesis provides a mechanism for this step in the signal sequence of amino acids. Where the paths of different products separate, there must be a means of sorting them at these branch points. For example, it would not do to secrete lysosomal proteins, proteins that function in the RER or Golgi cisternae, nor would sorting secretory proteins to intracellular sites do either. That is, vesicles formed from the Golgi must be specific in terms of their contents. If they are to travel to the cell surface, for example, they would have to contain only location appropriate proteins, rejecting all others.

It is not clear how all this happens, but let us run through one scenario. In this case, the protein is included in the vesicle because it is able to bind to a membrane protein at the internal face of the vesicle membrane. Proteins that do not form such associations are excluded. In this model, the vesicle would contain no soluble proteins—that is, proteins not attached to the membrane directly or indirectly, excluding them by unspecified mechanisms (perhaps you have some ideas about this, or is it not possible?). Certain membrane proteins containing domains that face the cytoplasm specify the port for the particular vesicle. The vesicle then moves to its targeted site, where it deposits its contents as a result of membrane fusion.

The problem is somewhat more complex for membrane proteins themselves. Like secreted proteins and those targeted to the matrix of organelles, it is thought that the synthesis and insertion of proteins destined for the various membranes of the eukaryotic cell takes place on the RER. That is, according to this view we should find all of the various membrane proteins of the cell in the membranes of the RER to be sorted along with the contents in the way just described. Although this is widely believed to be true, evidence supporting this supposition is lacking and it is clearly incorrect in regard to at least certain membrane proteins; for example, those of the mitochondrion.

Before much was known about the chemical composition of membranes, it was thought that the membranes of all of the compartments in the secretory pathway were compositionally the same and that a simple model of membrane flow could be applied; in which case RER membrane would become Golgi membrane and so forth, as one membrane flows one into the other, as a result of vesicles forming and fusing. But as we have learned more about the chemistry of membranes, it has become clear that this hypothesis must be incorrect. The chemical composition, in terms of both proteins and lipids, of each membrane in the secretory pathway is not only different, but indeed diagnostic. That is, we are able to distinguish one type of membrane from any other on the basis of their chemical composition.

Thus, just as with the contents of vesicles, their membrane proteins (and lipids) must be sorted, separated from each other. In some presently unknown fashion, the different membranes are presumably segregated in the secretory pathway. Both proteins and lipids put in place in the endoplasmic reticulum, are separated during the budding of vesicles. A vesicle going to the lysosome contains only lysosomal membrane proteins (and lipids), those going to a secretion granule contain only those marked for this locus. It is thought that this sorting takes place primarily in the Golgi.

Perhaps most confusing is the fact that even the membranes of some structures in the secretory pathway that are thought to be essentially identical objects, have different, distinct protein and lipid composition. "Condensing granules" provide a good example. These objects are thought to be identical to secretion granules. That is, they are secretion granules that have not as yet been completely filled with product. But the protein contents of their membranes is quite different from that of the mature (filled) secretion granule. They contain proteins that are absent in the membrane of the mature granule, and lack others that are present in it.

Transcytosis

As complex as the hypothetical system we have just described may seem, the reality appears even more complex. For example, at least certain proteins destined for the apical membrane of polarized epithelial cells, appear to leave to the Golgi only to reappear in the basolateral membrane at the other end of the cell. Evidence suggests that it is only from here that they are transported to the correct location in the apical membrane. Endosomes formed from this membrane, called *late endosomes*, carry the protein by a process called *transcytosis*; namely, vesicle transfer from one membrane surface of a cell to another in polarized epithelia.

Turnover

Well, we have, if with some difficulty, carried many, but as we shall discuss in a moment, not all, of the protein products to their proper homes within the cell. But as with prokaryotes, sorted proteins, excepting those that are secreted, turn over. That is, in the normal course of events they are degraded. How is this accomplished? The vesicle theory proposes that protein turnover occurs by means of vesicle mechanisms. Regions of membrane from various cellular compartments bud from the home compartment, for example, plasma membrane proteins are endocytosed from particular regions of the cell membrane, transported to lysosomes, with which they fuse, and it is there that the components are degraded. How this actually occurs is not understood.

There are many unknowns in this story, but one fact introduces a major complexity that any successful theory must account for. As noted above, all proteins in cells turnover at different rates relative to each other. That is, not all the proteins, say, in the basolateral plasma membrane, turnover at the same rate. As a result, at the end of their life cycle when proteins are made available for degradation, they must be withdrawn from the home compartment at different rates. How is this accomplished? The cell could segregate each protein in different vesicles. Or alternatively, a mechanism might exist to remove them from the membrane in the same vesicle but at different proportions relative to each other, to reflect their turnover rates, not the absolute amount of each present in the membrane. There are other scenarios that we might think about, but however it may be accomplished, we must also think about its *basis*, presumably the structural basis, for selecting one or another protein. How this would be accomplished is not known.

The Non-Vesicular Sorting of Proteins in Eukaryotes

To this point then it would seem that sorting in eukaryotes is quite different from that in prokaryotes. That is, in the latter, molecules move as the result of forces that directly involve them, molecule by molecule, whereas in eukaryotes hundreds, thousands, even millions of molecules are moved *en masse* in vesicles. In recent years it has become clear that such a view is incorrect. Not only are mechanisms like those found in bacteria also found in eukaryotic cells, but such processes are central elements in protein sorting in eukaryotes just as they are in prokaryotes. Indeed, sorting to arguably the two most important compartments of the cell, the nucleus and mitochondrion (or chloroplasts in plants) does not involve vesicle processes. In much the same way that we described for bacteria, proteins destined for these compartments are released from ribosomes into the cytoplasm, and cross organellar membranes subsequently and individually.

It was realized that protein traffic into and out of the nucleus occurs by non-vesicular, membrane transport mechanisms even before such things were thought possible. This was because the membrane of the nucleus, alone among organelles, seemed to contain pores large enough to accommodate proteins. We now know that many cellular structures contain large pores through which proteins can travel, though they are not as large nor often as numerous as in the nucleus. Nuclear pores are very large indeed compared to the size of proteins, about 0.1 microns in diameter versus about 50 Å for an average protein molecule. They cannot only easily accommodate proteins, but much larger structures, such as nucleic acids and ribosomal subunits. The problem of permeability if anything seems reversed here. How does one keep proteins out of the nucleus that have no business there? We do not have a good answer to this question yet, but it is clear that many proteins that travel through the nuclear pore do so by means of specific transport mechanisms, and modification of protein structure can prevent proper sorting to the nucleus. Indeed, as mentioned above, it is thought that there are consensus sequences for nuclear targeting, without which, however large the pore, such proteins do not enter. It has become clear today that what looked simply like large holes in the membrane are really very complex structures made up of many membrane proteins that carry out specific identification and transport functions.

Sorting to the mitochondrion is perhaps the most interesting, because while much is known, simple rules do not seem to apply. Proteins that go to the mitochondrion often, but not always, contain the N-terminus presequence referred to as the signal sequence. Sometimes this sequence is cleaved pursuant to transport and sometimes not. But perhaps most importantly, this presequence has not been possible to characterize in any simple fashion. It is not hydrophobic on most proteins, and a wide range of sequences, that are most often amphiphilic. In addition, mitochondrial proteins are targeted to four compartments; outer membrane, intermembrane space, inner membrane and inner matrix. To reach the inner membrane the protein must of course cross the outer membrane. Thus, such proteins must contain the information, or simply have the proclivity, required for passage into or across both membranes.

The Non-Vesicular Secretory Pathway

Thus, it is clear today that non-vesicular or membrane transport modalities are important for sorting in eukaryotes, as they are in prokaryotes. However, it is widely held that for processes where vesicle based sorting mechanisms are thought to occur, that such vesicle mechanisms, and such mechanisms alone, are responsible. That is, sorting either occurs by vesicles or membrane transport, but not both for a particular protein. As a result, observations of sorting by membrane transport processes to effect protein secretion in eukaryotic cells, has been controversial, because secretion in eukaryotes is widely viewed as being synonymous with the vesicle process outlined above. Today we know that this view is incorrect, and that membrane transport processes also exist in eukaryotic cells for proteins that are secreted. The most well studied of these processes is for the secretion of some 20 different protein molecules, digestive enzymes, by cells in the pancreas of mammals. It is interesting that this is the same cell in which evidence for the vesicle theory was first developed.

Questions about whether vesicle processes alone could account for secretion first arose from observations that the proportions of different proteins in secretion can vary due to transport related events. Their proportions were not necessarily constant, nor were the proportions found in secretion always the same as the proportions in the source compartment, particularly the secretion granules, from whence they in great part came. This was called "non-parallel" secretion. This phenomenon could have been explained within the vesicle construct if different secretion granules carried different proteins, protein x in granule x and protein y in granule y, and so forth, but the evidence was that each granule contained all of the proteins. In this case, how could the proportions be altered, if secretion occurred solely by exocytosis, in which circumstance all of the contents of the granule would be released together into the extracellular medium? One obvious explanation was that the membranes of the secretion granule and cell were permeable to these substances and that they crossed them in much the same way as in bacteria; molecule by molecule.

This was found to be the case, and protein secretion in this, and other cells, occurs by both vesicle and membrane transport processes. It is not clear what different physiological purpose the two processes serve, but it has been suggested that vesicle transport is useful for moving large amounts of material quickly, such as at the onset of digestion or during synaptic transmission, whereas membrane transport processes are better suited to modulation, where one might wish, for example, to alter the amount of a substance, or the relative amounts of different substances, secreted by the cell over longer durations. It is known that during digestion the secretion of one enzyme may be favored over another as the process proceeds. Digestive need varies depending upon the relative efficiency of the different enzymatic processes as a function of time and the contents of the ingested meal.

Constitutive and Regulated Secretion

Things may be even more complex, in that in addition to the standard vesicle pathway and an alternative non-vesicular route, it has been proposed that there are multiple forms of vesicles that undergo exocytosis. Two vesicle processes have been proposed, one that involves the secretion granule, and that is responsive to stimuli such as hormones, and the other, that involves small vesicles, presumably the small vesicles we discussed above, derived from the Golgi, that are thought to carry products within the cell. These vesicles, it has been proposed, also release their products by exocytosis. But in this case, the event is not responsive to external stimuli. The latter process is called constitutive secretion because it is thought to occur continuously and in the absence of stimuli, whereas the former is called regulated secretion because secretion via this route only occurs in response to stimuli. This having been said, whether the small objects in the cell thought to be transport vesicles indeed function as such, and undergo exocytosis at the cell membrane; that the constitutive pool is vesicular, not cytoplasmic; that secretion from the constitutive pool cannot be augmented by appropriate stimuli, and that secretion from the regulated pool only occurs in the presence of a stimulus, remain hypotheses.

CYSTIC FIBROSIS: A SORTING DISORDER?

The skeptical student headed for a career of helping people, may wonder whether it makes much difference how proteins are sorted. He or she might accept its importance as fundamental biology, but question its utility for the health of the patient. When the author was a student, there were no diseases of protein sorting to be studied. Indeed, as I have pointed out, the whole subject was enveloped in a cloud labeled "necessary but impossible." Although one cannot yet write a learned monograph describing a myriad of sorting disorders, it seems likely that such disorders will be found and such monographs written.

I would like to end this discussion with what may be, at least in part, an example of a sorting disorder. Cystic fibrosis (CF) is a genetic disease, that appears to involve a single membrane protein, CFTR, or the Cystic Fibrosis Transmembrane conductance Regulator. The disease is devastating; with the severe impairment of digestion, and continuous bouts of equally severe pulmonary infection. Until recently children born with this disease died in the first years of life. With a variety of interventions to improve nutrition and help overcome pulmonary disease, CF children are living into adulthood. It has been discovered that CFTR is a chloride channel, and although other actions of the protein may well be found, at present it is thought to be a disorder of water and ion transport. Most research has naturally focused on attempts to determine what is wrong with the channel, with the hope of being able to fix it or provide alternative means of accomplishing what it cannot. Recently one investigator wondered whether this protein was actually where it was supposed to be in the cell. That is, could the problem be not so much that the channel is ineffective, or that this alone causes the disease state, but that the protein never gets to the proper location in the cell. That is, could cystic fibrosis be at least in part a sorting disorder? Recent studies in some experimental systems suggest that this may well be the case. Some abnormal forms of CFTR do not leave the RER, and as a result are not sorted to the plasma membrane, as they are normally.

FINAL COMMENT

I have outlined the major themes in this area, but there is also much that I have not covered, for example, how the immune system transports peptides for presentation or the presence of proteins in multiple compartments in cells, not merely a single home. In any event, I hope that I have been able to pique your interest in this important subject. Because our understanding of mechanisms remains unsettled, even if a particular view dominates thinking at a given time, the subject may seem, as the book reviewer said about biology in general, in turn, "complicated, *ad hoc*, and unparsimonious." I have tried to expose, not avoid, these uncertainties in our understanding, and focus your attention on the questions that the subject raises, and not simply provide you with the present popular answers. You can find these answers, at various levels of simplification, in any traditional textbook that describes the biology of the cell. To understand the temporal nature of textbook truths, find, if you can, books that cover this subject that are just ten years old, not fifty or a hundred. The subject of membrane protein transport, that is, the passage of proteins across membranes, save vesicle mechanisms, will probably be nowhere in sight because such processes were until recently thought impossible. Future textbooks, will no doubt present this subject as if we had always known that protein molecules pass through biological membranes. Here I have tried to focus on the questions that underlie current research, not the answers. The answers often change with changing evidence and vogue, but the questions, ah the questions, they remain constant. In my view, it is in truly understanding the nature of the basic questions in an area, not the answers, that one gains lasting knowledge.

REFERENCES

Blobel, G. & Dobberstein, B. (1975). Transfer of protein across membranes. II. Reconstitution of rough microsomes from heterologous components. J. Cell. Biol. 67, 852. *The first experimental evidence for the signal hypothesis.*

Denning, G.M., Ostedgaard, L.S., & Welch, M.J. (1992). Abnormal localization of cystic fibrosis transmembrane regulator in primary cultures of cystic fibrosis airway epithelia. J. Cell Biol. 118, 551–559. *A description of the abnormal sorting of CFTR.*

Jamieson, J.D. & Palade, G. (1967). Intracellular transport of secretory protein in the pancreatic exocrine cell. II. Transport to condensing vacuoles and zymogen granules. J. Cell Biol. 34, 597–615. In vitro *studies that confirm the original observations of Siekevitz and Palade.*

Isenman, L., et al. (1995). Membrane Protein Transport—A Paradigm in Transition. JAI Press, Greenwich, CT. *A recent critical review on the membrane transport of proteins.*

Rothman, S.S. (Ed.) (1995). Membrane Protein Transport. JAI Press, Greenwich, CT. *A compendium of articles on membrane transport in a variety of cells.*

Rothman, S.S. (1985). Protein Secretion: A Critical Evaluation of the Vesicle Theory. Wiley, New York. *An extensive discussion of the evidence for the vesicle theory and evidence suggesting non-vesicular mechanisms of secretion.*

Rothman, S.S. & Ho, J.J.H., eds. (1985). Nonvesicular Secretion. John Wiley, New York. A *monograph on non-vesicular secretion in a variety of cells.*

Rothman, S.S., et al. (1991). Nonparallel transport and mechanisms of secretion. Biochim. Biophys. Acta 1071, 159–173. *A recent review concerning the non-parallel transport of proteins by cells and the implications of this phenomenon for mechanisms of secretion.*

Siekevitz, P. & Palade, G.E. (1960). A cytochemical study on the pancreas of the guinea pig. V. *In vivo* incorporation of leucine-l-C^{14} into the chymotrypsinogen of various cell fractions. J. Biophys. Biochem. Cytol. 7, 619–630. *A paper from the original series of experiments that led to the modern vesicle theory.*

RECOMMENDED READINGS

Alberts, B., et al., eds. (1985). Molecular Biology of the Cell. 2nd edition, Garland Press, New York, chap. 8. *An extensive exposition on this subject that gives a different view of these processes than this chapter.*

Palade, G. (1975). Intracellular aspects of the process of protein synthesis, Science 189, 347–358. *George Palade's Nobel prize speech describing the vesicle theory.*

Rothman, S.S. (1975). Protein secretion by the pancreas. Science 190, 747–753. *A contemporaneous critique of the vesicle theory.*

Chapter 11

Molecular Genetics and Evolution of Voltage-Gated Ion Channels

LAWRENCE SALKOFF and TIMOTHY JEGLA

Principles of Medical Biology, Volume 7A
Membranes and Cell Signaling, pages 229–244.

ISBN: 1-55938-812-9

INTRODUCTION: THE BRAIN IS AN ELECTRONIC DEVICE DESIGNED TO FUNCTION IN WATER

The brain is essentially an electronic device designed to function in an aqueous environment. Like all of organic evolution, the evolution of biophysical processes in the nervous system took place in the unique environment of molecular water. Water is a special solvent with polar qualities that define, order, and constrain the interactions of macromolecules in a way that appears essential for the evolution of life. Life began in the sea probably because the aqueous environment was uniquely permissive of the complex molecular interactions required for metabolism. In an aqueous environment macromolecules are coated with a sticky shell of constantly exchanging water molecules. This "hydration shell" surrounds virtually all macromolecules in solution and plays a large part in their interactions. Water is so essential to the chemistry of life that most metabolic processes cease when water is removed. When life left the sea to venture on land it brought encapsulated within it a facsimile of the sea's environment, an aqueous environment rich in sodium chloride. This is the environment in which the brain must operate.

Because hydrated ions (and not electrons) carry an electrical current in water, the whole environment within and surrounding the brain and nervous system has the properties of a conductor. Special problems are presented for an electronic device required to function in such an environment. For the same reason that an electrical appliance fails in the bathtub, the brain can not have a central power source; the current would be shorted out into the bath before reaching the circuit components of the device. Instead, nature has designed each unit of the brain's circuitry (each neuron) to function as its own generator and battery. Thus, each individual cell in the nervous system generates and maintains an electrical gradient across its plasma membrane by expending energy in the form of ATP. Creating this battery in every cell is the work of the Na^+-K^+-ATPase, which produces an ion gradient between the inside and outside of the cell. This generates an electrical gradient of about sixty to ninety millivolts with the inside of the cell negative relative to the outside. This electrical gradient is known as the resting potential. During the conduction of an electrical impulse called an action potential, each cell expends energy stored in its individual battery to propagate the impulse across the length of the cell. Thus, energy is stored throughout the brain; it needs no external energy source to function, except for nutrients necessary for the production of ATP.

ION CHANNELS: ORGANIC TRANSISTORS THAT GATE AN ELECTRIC CURRENT IN AN AQUEOUS ENVIRONMENT

A fundamental component of all electronic devices is a rapidly activating switch to turn current flow on or off. In transistor radios and computers the device most commonly used is the transistor, which gates (switches on and off) a current flow carried by electrons. However, ions in solution carry electrical currents in the brain, and a special switching device is necessary to control this current. The ion channel evolved in living systems just for this purpose. When open, ion channels allow specific ions to flow across the otherwise impermeable cell membrane. Like transistors, many ion channels open in response to changes in voltage. Thus the ion channel is truly the "transistor" of the brain.

Since extracellular fluids have a high sodium ion concentration, and intracellular fluids are high in potassium ions, sodium and potassium ions carry most of the electrical current in the brain. It is no accident then, that the two most common types of voltage-gated ion channels are selectively permeable either to sodium or potassium ions. *Voltage-gating* means that the channels open in response to a change in the sixty to ninety millivolt resting potential of the cell. This occurs, for example, during the spread of depolarization when the action potential travels along the cell membrane. Sodium channels gate an ion current that flows into the cell, following the concentration gradient for that ion. In contrast, potassium channels gate an electrical current that flows from the inside to the outside of the cell, following the concentration gradient for potassium ions. Both sodium and potassium ions carry a net positive charge. The effect of the sodium current is to carry positive charge into the cell interior making it more positive (this is called depolarization). The sodium current is the predominant current during the upstroke of the action potential. The potassium current carries a net positive charge out of the cell, leaving the inside of the cell more negative. This occurs during the repolarization phase of the action potential. The action potential itself is a wave of depolarization, initiated by the sodium current, that spreads along the cell membrane. This is the main mechanism for transmitting electrical signals over distances in the nervous system. The action potential is "active" current flow, powered by the energy stored in every cell in the form of the resting potential. In the aqueous environment of the nervous system, the action potential mechanism insures that a signal arrives at its destination with undiminished intensity. The fundamental components that make this possible are the voltage gated ion channels.

MOLECULAR GENETICS HAS REVOLUTIONIZED THE STUDY OF ION CHANNELS

Since the early 1980s the study of the protein structures of ion channels has been sharply advanced by the application of molecular-genetic techniques. The genome (the total DNA) of any living organism is a catalog of genes coding for all of the proteins required to build a complete organism. One of the breakthroughs of modern biology has been the cracking of the genetic code, enabling us to read these protein-coding gene "files." Molecular-genetic techniques now allow us to study the structures of proteins simply by analyzing the DNA sequence of their genes. Because the genetic code is virtually universal among all life forms, gene sequence information integrates all fields of biology and medicine which have a focus on proteins and their functions.

Not only does all life share a common genetic code, but the basic molecular machinery of life is also, in large part, shared by all life forms. This commonality is present because the vital innovations of molecular evolution occurred early in the history of life. Ion channels were one of these early innovations and are ubiquitous in modern life forms. They have been exploited most extensively by animals, whose nervous systems require them in myriad forms to properly gate complex ionic currents. Many of these special animal ion channels evolved prior to the divergence of vertebrate and invertebrate phyla (more than 500 million years ago), and remain conserved today in most species. Exploiting this fact has been a great help in elucidating the molecular structure of these proteins.

Gene cloning and sequencing techniques have thus been adopted by neurobiologists to investigate the protein structures of ion channels. Recombinant DNA technology applied to these problems achieved a breakthrough in the early 1980s when the acetylcholine receptor from torpedo electroplax (Numa et al., 1983) was first cloned and sequenced. Soon to follow was the voltage-gated sodium channel from electric eel (Noda et al., 1984). Protein purification and sequencing experiments had been able to provide partial amino acid sequence data from these channels which was used to synthesize oligonucleotide probes to select for the channel cDNAs from a cDNA library (a cDNA library contains the sequences of all the individual mRNAs from a specific tissue source copied into DNA, which is technically easier to work with). The sequencing of these cDNAs yielded the complete deduced amino acid sequence for these channels. The cloning and sequencing of a calcium channel from the sarcoplasmic reticulum of rabbit muscle was achieved in a similar way (Tanabe et al., 1987).

MUTANT ANALYSIS OF THE *DROSOPHILA SHAKER* GENE: CLONING THE POTASSIUM CHANNEL

The approaches discussed in the preceding section worked partially because neurotoxins were available which bound to these channels with high affinity. These toxins aided in the isolation of small amounts of protein which yielded bits of peptide sequence information. For other channel types, however, where no toxins were available for channel protein isolation, the task of cloning presented more formidable obstacles. The approach that worked for cloning potassium channels was less direct and involved the exploitation of the fruit fly, *Drosophila melanogaster*, an animal system that can be manipulated genetically.

A variety of behavioral mutants of *Drosophila* had been produced which had the promise of being useful for neurobiological studies (Benzer, 1973; Suzuki, 1974). Some of these had behavioral defects that were most likely due to mutational alterations in membrane excitability. Physiological studies of one of these mutations, *Shaker*, suggested that the mutation altered a gene coding for a potassium channel. *Shaker* mutations produce a behavioral phenotype of poor coordination and violent shaking upon exposure to ether.

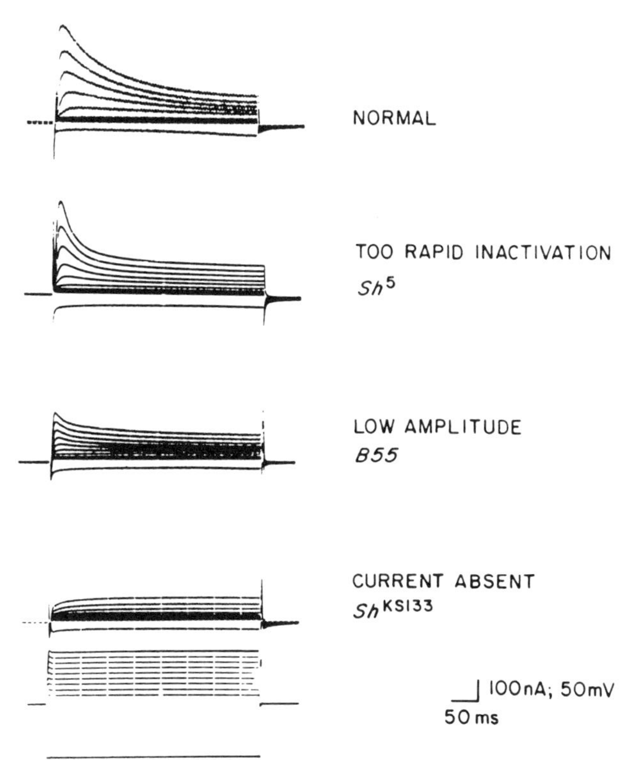

Figure 1. Voltage clamp experiments showing the fast transient potassium current in wild-type (normal) and *Shaker* mutant flies. The potassium current is shown as a deflection of the current trace in the upward direction. In Sh^{5} the current is present but inactivates too rapidly. The chromosomal breakpoint *B55* which falls in the *Shaker* region of the salivary chromosome map causes a marked reduction of current. Mutants like Sh^{ks133} completely eliminate the transient potassium current. All records shown are from the 72 h stage of pupal development when the voltage-gated transient potassium current is mature in normal wild-type flies. (See Salkoff, 1983.)

Direct evidence that *Shaker* mutations affected a potassium current required the voltage clamp technique. The effects of the *Shaker* gene were most clearly observed by voltage clamp studies of *Drosophila* flight muscles (Salkoff and Wyman, 1981; Salkoff, 1983) (see Figure 1). These studies directly compared the potassium currents in mutant and wild-type muscles, and found that the mutation affected one particular type of potassium current, a rapidly activating and inactivating potassium current called the *A-current* (Connor and Stevens, 1971; Neher, 1971).

The subsequent cloning of the *Drosophila Shaker* locus by three independent groups finally revealed the primary structure of this voltage-gated potassium channel (Papazian et al., 1987; Kamb et al., 1988; Pongs et al., 1988). These laboratories employed the chromosome *walking* strategy. The entry point into the chromosome was a cloned segment of DNA mapping near the *Shaker* gene that was fortuitously available from a project unrelated to *Shaker*. This cloned DNA was used as a hybridization probe to isolate other larger cloned DNA sequences from a DNA library that contained randomly overlapping clones representing the entire *Drosophila* genome. Overlapping clones were then chosen which extended the greatest distance to the left and right along the chromosome. The left and right ends of these new clones were again chosen as new hybridization probes and the selection of new overlapping clones repeated. Thus, the map of cloned DNA was extended further to the left and right. The process was repeated until the entire genomic region was cloned and the actual *Shaker* gene identified. The identification of the *Shaker* gene was made by the molecular mapping of *Shaker* mutations that fell within the cloned area and by isolating and sequencing cDNAs that mapped close to the mutations.

A SUPERFAMILY OF GENES INCLUDES SODIUM, CALCIUM, AND POTASSIUM CHANNELS

Molecular and genetic evidence now suggests that most voltage-gated ion channels have a similar protein structure and probably a common evolutionary origin. Before any of the primary sequence data from cloning experiments was known, Bertil Hille (1984) suggested that potassium, calcium, and sodium channels evolved from a common origin, an ancestral cation channel in the earliest eukaryotes; thus he predicted common structural features. The hypothesis of a common evolutionary origin for these voltage-dependent channels is supported by similarities in their biophysical properties. For example, all of these channels are sharply responsive to voltage once a sufficient level of depolarization is reached, but show virtually no activation in response to voltage changes that remain near the resting potentials of most cells. Considering similarities like these, it was not unexpected that molecular cloning experiments eventually revealed a common structure and molecular mechanism of voltage-dependence for all of these channels.

Cloning and sequencing experiments have shown that both the sodium and calcium channels are composed of a long polypeptide having four internal repeated domains which are homologous but not identical (Figure 2A) (Salkoff et al., 1987; Barchi, 1988; Catterall, 1988). Each of these homology domains in Figure 2 is represented as a pie-shaped wedge. Each homology domain contains six hydrophobic segments which are proposed to be alpha-helical transmembrane segments (the six transmembrane segments comprising a single homology domain are not indicated in Figure 2). The hydrophilic linker segments which connect the homology domains are postulated to be cytoplasmic while the four homology domains are hypothesized to contain all of the portions of the channel residing within the membrane. Thus the channels are composed of $4 \times 6 = 24$ membrane spanning domains surrounding a central ion-conducting pore.

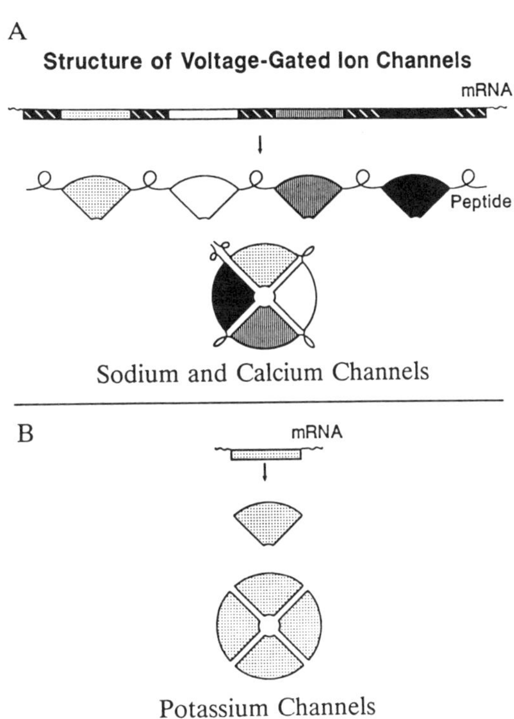

Figure 2. Diagrammatic representation of voltage-gated ion channels.

A. Sodium and calcium channels. A single mRNA (top) codes for a single long polypeptide (middle) which has four repeated domains which are homologous but not identical. Each of the four domains codes for six hydrophobic regions which are proposed to be transmembrane segments. This single polypeptide folds into a bagel-like structure (bottom) which has an ion selective pore in its center. Thus, both the voltage-gated sodium and calcium channels are composed of 24 membrane spanning segments arranged in a pseudosymmetrical configuration in four groups of six. The structure is said to be pseudosymmetrical (Noda et al., 1984) rather than symmetrical because the four homology domains are not identical.

B. Potassium channels. A single mRNA (top) codes for a smaller polypeptide (middle) which resembles a single repeated domain of the sodium or calcium channels. This has led to the speculation that the *Shaker* family potassium channels are composed of four small polypeptides (bottom) (Stevens, 1987; Agnew 1988; Timpe et al., 1988) and is, thus, a symmetrical structure.

Potassium channels differ in one important way: the gene product is a smaller polypeptide resembling only a single one of these homology domains (Figure 2B) containing six conserved hydrophobic regions. This has led to the speculation that the tertiary structure of the potassium channel is composed of four of these smaller polypeptides (Stevens, 1987; Agnew, 1988) and is, thus, analogous to the sodium and calcium channels. Numerous experimental tests of this channel structure have not been definitive, but strongly support a tetrameric structure for potassium channels.

DIVERSITY OF POTASSIUM CHANNELS: POTASSIUM CHANNEL GENES HAVE SUBFAMILIES

Potassium channels are virtually ubiquitous among the cells of eukaryotes and are the most heterogeneous of the voltage-gated cation channels (Hille, 1984; Rudy, 1988). They are involved in controlling the height, duration, and rate of repolarization of the action potential. In cells that beat like clocks, potassium channels control the rate of beating. In cells that fire bursts of action potentials at regular intervals, potassium channels are involved in regulating both the interburst interval, as well as the duration of the burst. Because potassium channels fine-tune almost all aspects of membrane excitability, considerable functional diversity is needed. Since it was observed that alternative RNA processing of the *Shaker* gene produced a variety of related peptides, it was initially assumed that this might be the mechanism for producing the great diversity of potassium channel types (Schwarz et al., 1987; Agnew, 1988). Indeed, using the *Xenopus* oocyte expression system, it was found that the alternative *Shaker* peptides do produce potassium channels having different kinetics (Timpe et al., 1988; Iverson et al., 1988). The observed kinetic differences, however, are not sufficient to account for the great diversity of potassium channel types seen in *Drosophila* muscle (Salkoff, 1985; Wu and Haugland, 1985; Zagotta et al., 1988) and nerve cells (Solc and Aldrich, 1988; Byerly and Leung, 1988).

More recently, it was observed that a greater range of potassium channel diversity results from an extended gene family coding for homologous proteins; the *Shaker* gene is but one of at least four homologous potassium channel genes in *Drosophila* (Butler et al., 1989; Wei et al., 1990). The proteins encoded by all of these genes, *Shaker*, *Shal*, *Shab*, and *Shaw*, share a similar organization in that each resembles one of the four repeated domains of the sodium or calcium channels as in (Figure 2B).

EXPRESSION OF HOMOLOGOUS GENES SHOWS A WIDE RANGE OF BIOPHYSICAL PROPERTIES

The extended family of *Shaker*-like genes codes for potassium channels with a diverse range of biophysical properties. This was shown by expression of the genes in the *Xenopus* oocyte system as shown in Figure 3 below. The expressed currents differ with regard to all voltage-sensitive and kinetic properties. The *Shaker* current activates and inactivates very rapidly while the *Shaw* current activates very slowly and does not inactivate; the *Shal* current is intermediate between the two. The *Drosophila Shab* gene has the properties of the delayed rectifier. The delayed rectifier was the first potassium current ever to be characterized (Hodgkin and Huxley, 1952). A more detailed description of the biophysical properties of these

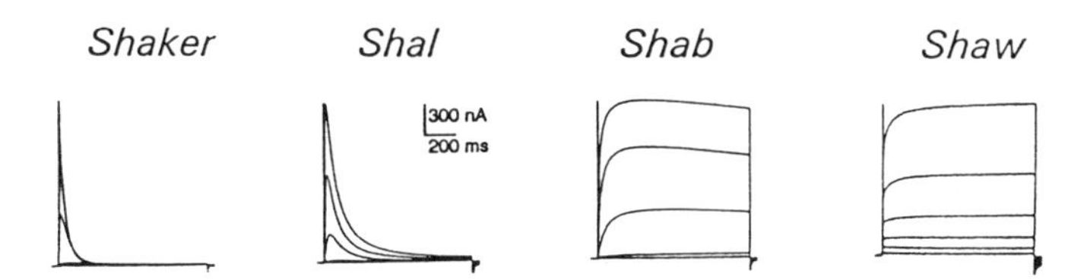

Figure 3. Expression of potassium currents in *Xenopus* oocytes injected with *ShakerH37* (*Shaker*), *Shal2* (*Shal*) and *Shaw2* (*Shaw*) cRNAs. a, b, and c, Outward currents recorded in response to one second voltage step depolarizations ranging between –80 and +20 mV in 20-mV steps from a holding potential of –90 mV. While the *Shaker* and *Shal* currents turn on sharply at about –40 mV, the *Shaw* current has a shallower activation response to voltage and turns on at a more hyperpolarized potential (–60 mV). See Wei et al. (1990).

expressed channels is given in Wei et al. (1990). It appears that the multigene family of potassium channels evolved in order to provide a much broader range of voltage sensitivity and kinetic diversity than can be produced by alternative splicing of the *Shaker* gene.

MORE EXTENSIVE GENE SUBFAMILIES EXIST IN MAMMALS

Current indications are that each of the *Drosophila* potassium channel genes, *Shaker*, *Shab*, *Shaw*, and *Shal* represents a subfamily of genes in mammals. It is not presently known how many members of each subclass there are, but there are at least eight genes of the *Shaker* class expressed in mammalian brain (see McKinnon, 1989, for a description of two). There are presently known to be four separate genes expressed in brain which have deduced protein features similar to *Drosophila Shaw*, and two genes each similar to *Drosophila Shab* and *Drosophila Shal*. The number of homologous potassium channel genes is almost certain to be larger. It appears that gene duplication has produced subfamilies of voltage-gated sodium channels and voltage-gated calcium channels in mammals as well, although they are not as large. Apparently extensive duplication of highly conserved ancestral channel types (followed by specialization of structure) has produced the great diversity of voltage-gated ion channels needed by complex mammalian nervous systems.

A UNIVERSALLY CONSERVED MECHANISM OF VOLTAGE-DEPENDENT GATING

Voltage-dependent activation of a voltage-gated channel is a property encoded by a specific structural feature which senses changes in transmembrane potential and

Sodium Channels

1626	**R** V I **R** L A **R** I G **R** I L **R** L I **K** G A **K** G I	RAT Brain I
1439	**R** V I **R** L A **R** I G **R** I L **R** L I **K** G A **K** G I	RAT Muscle I
1413	**R** V V **R** V F **R** I G **R** I L **R** L I **K** A A **K** G I **R**	FLY
1417	**R** V I **R** L A **R** I A **R** V L **R** L I **R** A A **K** G I **R**	EEL

Potassium Channels

362	**R** V I **R** L V **R** V F **R** I F **K** L S **R** H S **K** G L Q	FLY *Shaker*
362	**R** V I **R** L V **R** V F **R** I F **K** L S **R** H S **K** G L Q	RAT *BK1*
295	E F F S I I **R** I M **R** L F **K** V T **R** H S S G L **K**	FLY *Shaw*

Calcium Channel

883	**K** I L **R** V L **R** V L **R** P L **R** A I N R A K G L K	RABBIT Muscle

Notes: Positive residues presumed to be gating charges are bold type and labeled with a "+." Single letter abbreviations for the amino acid residues are: A, Ala; C, Cys; D, Asp; E, Glu; F, Phe; G, Gly; H, His; I, Ile; K, Lys; L, Leu; M, Met; N, Asn; P, Pro; Q, Gln; R, Arg; S, Ser; T, Thr; V, Val; W, Trp; and Y, Tyr.

Figure 4. Conservation of gating charge strings from all voltage-gated ion channels.

initiates a conformational change in response to them, resulting in the opening of the channel pore. This structure appears to be a region of regularly spaced positive charges in the fourth membrane spanning region (S4) common to all voltage-gated channels. Each string consists of a repeated pattern of a positively charged amino acid residue alternating with two uncharged residues (Figure 4). The positions of these positive charges are conserved between voltage-gated ion channel gene families and across species.

Empirical and experimental observations suggest that these charges are, indeed, responsible for gating. As a general rule, it is possible that the more positive charges a channel has, the faster the channel opens in response to a voltage change. Sodium channels and the *Shaker* potassium channel, which have the largest numbers of S4 gating charges, respond quickly to changes in voltage and have steep activation curves. That is, a population of these channels transits from the closed to the open state over a narrow voltage range. The *Shaw* channel, which has a lower number of gating charges than either the *Shaker* potassium channel or the sodium channel, is slowly activating with a shallow activation curve, as predicted for a channel with a low number of gating charges. Site directed mutagenesis experiments have shown that the elimination of charges from the S4 of the *Shaker* potassium channel alters the voltage-gating properties of the channel (Papazian et al., 1991; Liman et al., 1991). Similar mutagenesis of a sodium channel (Stuhmer et al., 1989) has also shown that channel gating is affected in a way that strongly implicates these charges in channel gating.

GENE DUPLICATION: NATURE MAKES THE MOST OF A GOOD THING

It has been suspected for some time that membrane proteins mediating transmission of information in the brain would fall into genetically related families (Hille, 1984; Stevens, 1987). This has now been proven to be the case for ligand-gated as well as voltage-gated channels. Thus, the acetylcholine receptor is a member of a superfamily of channels that includes both the GABA and glycine receptor subfamilies. It should be pointed out that gene duplication has produced not only the separate channel subfamilies, but also the specialized subunits within each subfamily. The nicotinic acetylcholine receptor in muscle is a pentamer composed of the products of four separate, yet, homologous genes (one of the subunits, *alpha*, is used twice in each channel). In the mammalian brain, the molecular structure of the acetylcholine receptor is probably similar, but the subunits are encoded by additional genes which are expressed only in neural tissue (Galzi et al., 1991). Thus, in mammals there are a large number of separate genes coding for nicotinic acetylcholine receptor subunits. The same pattern of molecular evolution involving gene duplication is seen in the supergene family coding for rhodopsin/beta-adrenergic/muscarinic receptors.

For voltage-gated ion channels, again, a similar evolution has occurred; there is a superfamily of genes encoding all voltage-dependent cation channels (sodium, calcium and potassium), and gene duplication has occurred within each category to produce subfamilies. Figure 5 summarizes the probable evolutionary line of descent of modern voltage-sensitive channels from a common origin. As mentioned previously, this common ancestor was originally proposed to be a primitive nonselective cation channel which itself probably evolved from earlier genes encoding peptide segments corresponding to regions S1 through S6. New channel sequences from primitive eukaryotes (Jegla and Salkoff, 1994) and bacteria (Milkman, 1994), however, suggest that the common ancestor may have been potassium selective. Perhaps potassium selective channels were important for processes such as osmotic regulation before the evolution of electrical excitability. In their evolution, sodium and calcium channels possibly underwent two rounds of intragenic duplication; the end result is the modern form of the genes containing four homology domains within a single large gene. Hille suggested that calcium selectivity may have evolved prior to sodium selectivity because of the vital role that calcium plays in cell metabolism, but this is, at present, only a hypothesis.

Potassium channels evolved along a different pathway: they retained the ancestral gene structure of a single S1–S6 domain. Thus, each potassium channel gene encodes only a single homology domain, and four of the gene products are required to form a complete channel. This has allowed a further opportunity for the production of potassium channel diversity. It has been demonstrated in mammals that channel subunits from different genes within the Shaker subfamily can form

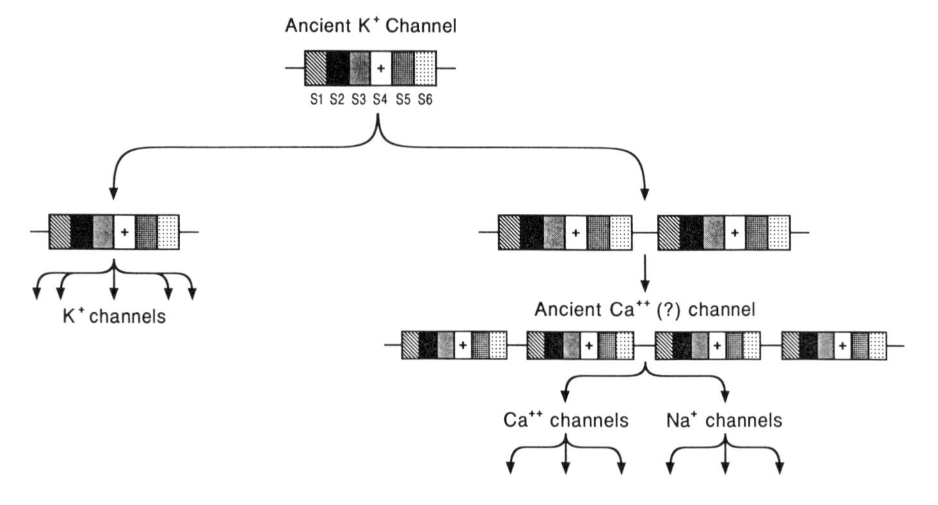

Figure 5. Possible evolutionary descent of voltage-gated cation channels.

heteromultimeric channels (Sheng et al., 1993; Wang et al., 1993). This mechanism of producing potassium channel diversity is limited since it appears that heteromultimers can only be formed by subunits from genes within the same subfamily (Covarrubias et al., 1991); that is, channel subunits from one *Shaker* gene can mix with subunits from another *Shaker* gene, but not with subunits from *Shal*, *Shab*, or *Shaw* genes.

The voltage-gated ion channel superfamily of genes is vast, and extends beyond the voltage-gated sodium, calcium and potassium channels. The cloning of calcium-activated potassium channels (Atkinson et al., 1991; Butler et al., 1993) and cyclic-nucleotide-gated cation channels (Kaupp et al., 1989; Goulding et al., 1992), has revealed that these channel types also belong to the superfamily of voltage-gated ion channels. Although they are not primarily gated by voltage, these channels are voltage-sensitive and contain the charged S4 motif. Another gene that belongs to the voltage-gated ion channel superfamily is *ether-a-go-go* (*eag*), which encodes a channel permeable to both potassium and calcium that is gated by voltage (Warmke et al., 1991; Bruggemann et al., 1993). Like voltage-gated potassium channels, all of these channel genes encode a single homology domain and most likely function as tetramers.

Over and over again nature has used gene duplication for expanding and modifying the functional roles of a rare and uniquely valuable protein motif. Examples of convergent evolution in molecular evolution are rare; most enzymes and proteins that perform similar tasks are the products of genes with a common evolutionary origin. Molecular evolution rarely reinvents the wheel.

THE EVOLUTIONARY ORIGINS OF VOLTAGE-GATED ION CHANNELS

Since both *Drosophila* and mammals have the same highly conserved genes coding for both sodium and potassium (and probably calcium channels), it is almost certain that the genes evolved to virtually their modern form prior to the separation of vertebrates and invertebrates. This means that even the individual subfamilies of the potassium channel gene family had evolved their specialized features by the late Precambrian period, about six hundred million years ago.

It now appears that these potassium channel gene subfamilies are as old as the nervous system itself. We have used a PCR-based screen (utilizing primers to the most conserved regions of voltage-gated potassium channels) to isolate clones of two *Shaker* homologs and a *Shal* homolog from a jellyfish (Jegla et al., 1995), which is a member of the most primitive surviving animal phylum to contain an organized nervous system (Cnidaria). Jellyfish differ radically from more advanced animals (or metazoans) in several ways; most notably, they are diploblasts, having only two embryonic germ layers (humans and most other animals are triploblasts and have three) and are radially symmetric (triploblasts are bilaterally symmetric). Virtually the only structure shared between the diploblastic jellyfish and triploblasts is an organized nervous system. The nervous system most likely evolved in the last common ancestor of diploblasts and triploblasts, nearly one billion years ago. A sodium channel with high homology to *Drosophila* and mammalian sodium channels has also been isolated from jellyfish (Anderson et al., 1993), and it is thus likely that highly conserved jellyfish calcium channels will be found as well.

Although voltage-gated potassium channels are found outside the animal kingdom, particularly in the electrically excitable protozoa, it appears that these channels are only distantly related to the voltage-gated channels found in the Metazoa. We have screened extensively for homologs of *Shaker*, *Shal*, *Shab*, and *Shaw* in *Paramecium*, a ciliate protozoan with several potassium channel types (Saimi and Kung, 1987), but instead of finding these, our highly sensitive screen led to the isolation of two novel and divergent potassium channel genes that bear little resemblance to *Shaker*, *Shal*, *Shab*, or *Shaw*. These genes appear to be derived from an early radiation of eukaryotes predating the metazoans. Nevertheless, they are clearly members of the voltage-gated ion channel gene superfamily (Jegla and Salkoff, 1995). Our screen has also failed to find *Shaker*, *Shal*, *Shab*, or *Shaw* homologs in flagellate protozoans, which may be more closely related to metazoans, fungi (yeast) and plants. Two plant potassium channels that have been cloned (Anderson et al., 1992; Sentenac et al., 1992) are also apparently derived from an earlier eukaryotic radiation and are most similar to the *Paramecium* potassium channels. This is a strong suggestion that *Shaker*, *Shal*, *Shab*, and *Shaw* may represent a set of voltage-gated potassium channels specifically designed for the propagation of electrical signals in metazoan nervous systems. It will be interesting to see if more primitive metazoans such as sponges, which do not contain nervous systems but can propagate electrical signals from cell to cell, contain homologs of these voltage-gated potassium channels.

CONCLUDING REMARKS

With regard to the large number of similar yet distinctive ion channel genes, a challenge remains to piece together an understanding of how all types fit into the pattern of animal behavior and cell metabolism. For more than two decades following the pioneering work of Hodgkin and Huxley (1952) nearly all of our understanding of what ion channels are and how they work was obtained by biophysical investigations. During this period the horizon of our understanding of membrane electrical properties barely extended beyond one single type of sodium and potassium channel. With the great increase in knowledge that molecular genetic techniques has added, a substantial challenge is presented: How are all these different gene products used in customizing membrane electrical properties in the nervous system, thereby generating the enormously complex and sophisticated patterns of animal behavior?

ACKNOWLEDGMENT

Supported by the NIH, the Muscular Dystrophy Association of America, and G. D. Searle, Inc.

REFERENCES

Agnew, W.S. (1988). A rosetta stone for K channels. Nature 331, 114–115.

Anderson, J.A., Huprikar, S.S., Kochian, L.V., Lucas, W.J., & Gaber, R.F. (1992). Functional expression of a probable Arabidopsis thaliana potassium channel in Saccharomyces cerevisiae. Proc. Nat. Acad. Sci. USA 89, 3736–3740.

Anderson, P.A.V., Holman, M.A., & Greenburg, R.M. (1993). Deduced amino acid sequence of a putative sodium channel from the scyphozoan jellyfish *Cyanea capillata*. Proc. Nat. Acad. Sci. USA 90, 7419–7423.

Atkinson, N.S., Robertson, G.A., & Ganetzky, B. (1991). A component of calcium-activated potassium channels encoded by the *Drosophila slo* locus. Science 253, 551–555.

Barchi, R. (1988). Probing the molecular structure of the voltage-dependent sodium channel. Ann. Rev. Neurosci. 11, 455–495.

Benzer, S. (1973). Genetic dissection of behavior. Sci. Amer. 229, 24–37.

Bruggemann, A., Pardo, L.A., Stuhmer, W., & Pongs, O. (1993). *Ether-a-go-go* encodes a voltage-gated channel permeable to K^+ and Ca^{2+} and modulated by cAMP. Nature 365, 445–448.

Butler, A., Wei, A., Baker, K., & Salkoff, L. (1989). A family of putative potassium channel genes in *Drosophila*. Science 243, 943–947.

Butler, A., Tsunoda, S., McCobb, D.P., Wei, A., & Salkoff, L. (1993). *mSlo*, a complex mouse gene encoding "maxi" calcium-activated potassium channels. Science 261, 221–224.

Byerly, L. & Leung, H.-T. (1988). Ionic currents of *Drosophila* neurons in embryonic cultures. J. Neurosci. 8, 4379–4393.

Catterall, W. (1988). Structure and function of voltage-sensitive ion channels. Science 242, 50–61.

Conner, J.A. & Stevens, C.F. (1971). Voltage clamp studies of a transient outward membrane current in gastropod neural somata. J. Physiol. 213, 21–30.

Galzi, J.-L., Revah, F., Bessis, A., & Changuex, J.-P. (1991). Functional architecture of the nicotinic acetylcholine receptor: from electric organ to brain. Ann. Rev. Pharmacol. 31, 37–72.

Goulding, E.H., Ngai, J., Kramer, R.H., Colicos, S., Axel, R., Siegelbaum, S.A., & Chess, A. (1992). Molecular cloning and single-channel properties of the cyclic nucleotide-gated channel from catfish olfactory neurons. Neuron 8, 45–58.

Hille, B. (1992). Ionic Channels of Excitable Membranes, 2nd. ed. Sinauer Associates. Sunderland, MA.

Hodgkin, A.L. & Huxley, A.F. (1952). Currents carried by sodium and potassium ions through the membrane of the giant axon of *Loligo*. J. Physiol. 116, 449–472.

Iverson, L.E., Tanouye, M.A., Lester, H.A., Davidson, N., & Rudy, B. (1988). A-type potassium channels expressed from *Shaker* locus cDNA. Proc. Nat. Acad. Sci. USA 85, 5723–5727.

Jegla, T. & Salkoff, L. (1994). Molecular evolution of K^+ channels in primitive eukaryotes (in *Molecular Evolution of Physiological Processes*, D.M. Fambrough, Ed., pp. 211–222).

Jegla, T. & Salkoff, L. (1995). A multigene family of novel K^+ channels from Paramecia. Receptors and Channels 3, 51–60.

Jegla, T., Grigoriev, N., Gallin, W., Salkoff, L., & Spencer, A. (1995). Multiple *Shaker* potassium channels in a primitive metazoan. J. Neurosci. 15, 7989–7999.

Kamb, A., Tseng-Crank, J., & Tanouye, M.A. (1988). Multiple products of the *Drosophila Shaker* gene may contribute to potassium channel diversity. Neuron 1, 421–430.

Kaupp, B.U., Niidome, T., Tanabe, T., Terada, S., Bonigk, W., Suhmer, W., Cook, N., Kanagawa, K., Matsuo, H., Hirose, T., & Numa, S. (1989). Primary structure and functional expression from complemetary DNA of the rod photoreceptor cyclic GMP-gated channel. Nature 342, 762–766.

Liman, E.R, Hess, P., Weaver, F., & Koren, G. (1991). Voltage-sensing residues in the S4 region of a mammalian K^+ channel. Nature 353, 752–756.

McKinnon, D. (1989). Isolation of a cDNA clone coding for a putative second potassium channel indicates the existence of a gene family. J. Biol. Chem. 264, 8230–8236.

Milkman, R. (1994). An Escherichia coli homologue of eukaryotic potassium channel proteins. Proc. Natl. Acad. Sci. USA 91(9), 3510–3514.

Neher, E. (1971). Two fast transient current components during voltage clamp on snail neurons. J. Gen. Physiol. 58, 36–53.

Noda, M., Shimizu, S., Tanabe, T., Takai, T., Kayano, T., Ikeda, T., Takahashi, H., Nakayama, H., Kanoak, Y., Minamino, N., Kangawa, K., Matsuo, H., Raftery, M.A., Hirose, T., Inayama, S., Hayashida, H., Miyata, T., & Numa, S. (1984). Primary structure of *Electrophorus electricus* sodium channel deduced from cDNA sequence. Nature 312, 121–127.

Numa, S., Noda, M., Takahashi, H., Tanabe, T., Toyosato, M., Furutani, Y., & Kikyotani, S. (1983). Molecular structure of the nicotinic acetylcholine receptor. CSHSQB *XLVIII*, 57–69.

Papazian, D.M., Schwarz, T.L., Tempel, B.L., Jan, Y.N., & Jan, L.Y. (1987). Cloning of genomic and complementary DNA from *Shaker*, a putative potassium channel gene from *Drosophila*. Science 237, 749–753.

Papazian, D.M., Timpe, L.C., Jan, Y.N., & Jan, L.Y. (1991). Alteration of voltage-dependence of *Shaker* potassium channel by mutations in the S4 sequence. Nature 349, 305–310.

Pongs, O., Kecskemethy, N., Muller, R., Krah-Jentgens, I., Baumann, A., Kiltz, H.H., Canal, I., Llamazares, S., & Ferrus, A. (1988). *Shaker* encodes a family of putative potassium channel proteins in the nervous system of *Drosophila*. EMBO J. 7, 1087–1096.

Rudy, B. (1988). Diversity and ubiquity of K channels. Neuroscience 25, 729–749.

Saimi, K. & Kung, C. (1987). Behavioural genetics of *paramecium*. Ann. Rev. Genet. 21, 47–65.

Salkoff, L.B. & Wyman, R.J. (1981). Genetic modification of potassium channels in *Drosophila Shaker* mutants. Nature 293, 228–230.

Salkoff, L. (1983). Genetic and voltage-clamp analysis of a *Drosophila* potassium channel. Cold Spring Harbor Symp. Quant. Biol. 48, 221–231.

Salkoff, L. (1985). Development of ion channels in the flight muscles of *Drosophila.* J. Physiol. Paris 80, 275–282.

Salkoff, L., Butler, A., Wei, A., Scavarda, N., Baker, K., Pauron, D., & Smith, C. (1987). Molecular biology of the voltage-gated sodium channel. TINS 10, 522–527.

Schwarz, T.L., Tempel, B.L., Papazian, D.M., Jan, Y.N., & Jan, L.Y. (1988). Multiple potassium-channel components are produced by alternative splicing at the *Shaker* locus in *Drosophila.* Nature 331, 137–142.

Sentenac, H., Bonneaud, N., Minet, M., Lacroute, F., Salmon, J.-M., Gaymard, F., & Grignon, C. (1992). Cloning and expression in yeast of a plant potassium transport system. Science 256, 663–665.

Sheng, M., Liao, Y.J., Jan, Y.N., & Jan, L.Y. (1993). Presynaptic A-current based on heteromultimeric K^+ channels detected *in vivo.* Nature 365, 72–75.

Solc, C.K. & Aldrich, A.W. (1988). Voltage-gated potassium channels in larval CNS neurons of *Drosophila.* J. Neurosci. 8, 2556–2570.

Stevens, C. (1987). Channel families in the brain. Nature 328, 198–199.

Stuhmer, W., Conti, F., Suzuki, H., Wang, X., Noda, M., Yahagi, N., Kubo, H., & Numa, S. (1989). Structural parts involved in activation and inactivation of the sodium channel. Nature 339, 597–603.

Suzuki, D. (1974). Behavior in *Drosophila melanogaster*, a geneticist's point of view. Can. J. Gen. Cytol. 16, 713–735.

Tanabe, T., Takeshima, H., Mikami, A., Flockerzi, V., Takahashi, H., Kangawa, K., Kojima, M., Matsuo, H., Hirose, T., & Numa, S. (1987). Primary structure of the receptor for calcium channel blockers from skeletal muscle. Nature 328, 313–318.

Timpe, L.C., Schwarz, T.L., Tempel, B.L., Papazian, D.M., Jan, Y.N., & Jan, L.Y. (1988). Expression of functional potassium channels from *Shaker* cDNA in *Xenopus* oocytes. Nature 331, 143–145.

Wang, H., Kunkel, D.D., Martin, T.M., Schwarzkroin, P.A., & Tempel, B. (1993). Heteromultimeric K^+ channels in terminal and juxtaparanodal regions of neurons. Nature 365, 75–79.

Warmke, J., Drysdale, R., & Ganetzky, B. (1991). A distinct potassium channel polypeptide encoded by the *Drosophila eag* locus. Science 252, 1560–1562.

Wei, A., Covarrubias, M., Butler, A., Baker, K., & Salkoff, L. (1990). K^+ current diversity is produced by an extended gene family conserved in *Drosophila* and mouse. Science 248, 599–603.

Wu, C.-F. & Haugland, F.N. (1985). Voltage clamp analysis of membrane currents in larval muscle fibers of *Drosophila*: Alteration of potassium currents in *Shaker* mutants. J. Neurosci. 5, 2626–2640.

Zagotta, W.N., Brainard, M.S., & Aldrich, R.W. (1988). Single-channel analysis of four distinct classes of potassium channels in *Drosophila* muscle. J. Neurosci. 8, 4765–4779.

Chapter 12

Molecular Biology of Voltage-Gated Ionic Channels:

STRUCTURE-FUNCTION RELATIONSHIPS

DALIA GORDON

Principles of Medical Biology, Volume 7A
Membranes and Cell Signaling, pages 245–305.

ISBN: 1-55938-812-9

INTRODUCTION

Ion channels are integral membrane proteins that form ion conductive pathways across cell membranes. The channel mediated movement of ions into and out of cells underlies a variety of cellular functions such as cell volume regulation, muscle contraction and the production of electrical signals in the nervous system. In order to fulfill their many different roles the ion channels are a diverse class of proteins. Some ion channels open in response to a change in membrane electric potential, and others open in response to the binding of specific ligands. After being opened some ion channels are not very selective and allow many different ions to pass, but others are highly selective and allow the conduction of only one kind of ion present in physiological solutions.

The voltage-gated ionic channels that serve as the topic of this chapter are plasma membrane proteins responsible for the generation and propagation of electrical signals in excitable cells, such as nerve, muscle and endocrine cells. These channels are characterized by the steep voltage dependence of their opening behavior, so called activation (Hille, 1992). Their response to changes in membrane potential is by forming water-filled pores in the cell membrane through which specific ions can diffuse passively down their pre-established gradients. The large increase in permeability to specific ions is regenerative, mostly on a time scale of milliseconds. Within this class of channels there are large differences in kinetics, voltage dependence of activation and inactivation and ion selectivity. The voltage-gated channels are named after the principal ion permitted to pass: K^+, Na^+ and Ca^{2+}. Potassium (K) channels are responsible to maintain the resting membrane potential (negative inside), by permitting selectively K^+ to flow out from the cell. The sodium (Na) and calcium (Ca) channels tend to drive the voltage to the positive direction when allowing Na^+ and Ca^{2+} ions to flow inside the cell.

Voltage-gated ionic channels have three fundamental functional properties that enable them to fulfill their role in cellular electrical signaling: selective ion conductance, voltage dependent activation that results in opening of the ion conductive pore, and inactivation, that terminates the ion flux. Accordingly, these channels share several structural similarities as well as distinct characteristics. These similarities are underlined in this presentation and the homologous structural elements of the different channels, shown to be directly related to their common functional properties are presented in the figures.

The amino acid sequence of a growing number of Na, Ca, and especially K channels has been determined. The voltage activated Na, Ca, and K channels are members of the same gene superfamily in terms of structural homology at the amino acid level. They are distinguished from other channels by having multiple of six putative transmembrane segments. Most of these channels have been functionally expressed in *Xenopus* oocytes by microinjection of complementary RNAs (cRNA) derived from each channel cDNA, confirming their identity. The channels expressed from the cDNAs closely resemble those from native membranes in their properties, including sensitivity to toxins, the rate at which the current turns on in response to a depolarizing voltage step (activation), the decay of the current during a maintained depolarization (inactivation) and the single channel conductance.

The next sections provide a brief introduction to each of the main types of voltage-gated ionic channels. The reader is referred to review articles that include the original works that are not cited here.

Sodium (Na) Channels

Na channels mediate the rapid increase in membrane Na^+ conductance responsible for the rapidly depolarizing phase and propagation of the action potential in many excitable cells (Hille, 1992). This transient increase in Na^+ conductance lasts only milliseconds and then spontaneously inactivates, and the membrane can be repolarized as K^+ permeabilities are augmented by K channels activities. All Na channels that have been purified and biochemically characterized contain a large glycoprotein α subunit of 240–280 kD. In some tissues and species the α subunit is non-covalently associated with a smaller subunit of 30–40 kD. In mammalian brain, the Na channels consist of a heterotrimeric complex of α (260 kD), $\beta1$ (36 kD) and a disulfide linked $\beta2$ (33 kD) subunits (reviewed by Catterall, 1988, 1992; Barchi, 1988; Cohen and Barchi, 1993).

Reconstitution of purified Na channels into defined phospholipid vesicles demonstrated that the protein components of each of the isolated channels are sufficient to mediate ion fluxes with the ion selectivity, voltage dependence and pharmacological properties expected from the native channels. In rat brain Na channels, the $\beta2$ subunit can be removed without any functional effect, while removal of $\beta1$ subunits causes loss of all functional activities. The purified electroplax Na channel is functional with only a single α subunit, indicating that it contains most, if not all, of the functional domains of the channel.

The availability of purified, functional preparations of Na channels provided the basis for the isolation of cDNA clones encoding the primary structure of the principal α subunit of the Na channels from various species and tissues. Oocytes injected with cRNA that encodes for a single isoform of the α subunit express functional voltage-gated Na channels, confirming that all the functional properties of the channel are contained within this principal subunit. Co-injection of cRNAs for α and $\beta 1$ subunit from rat brain (Isom et al., 1992) resulted in expressed Na channels that rapidly inactivate, suggesting a stabilizing function for the smaller subunit (Isom et al., 1994).

Calcium (Ca) Channels

Voltage-gated Ca^{2+} channels that are present in excitable cells are activated by depolarization of their plasma membrane. The Ca^{2+} that enters the cell through this pathway is important for triggering excitation-secretion coupling and neurotransmitter or hormone release, or excitation-contraction coupling and muscle contraction, and a large number of other intracellular functions (Catterall, 1988, 1991). The activity of voltage-activated Ca^{2+} channels is modulated by a wide variety of ligands acting on cell surface receptors and through G proteins and protein phosphorylation. Voltage-sensitive Ca channels are also the sites of action for several clinically important drugs (for review see Catterall, 1991; Miller, 1992; Snutch and Reiner, 1992).

Ca channels are a heterogeneous group of proteins. It is well established that there are at least three distinct classes of voltage-activated Ca channels: The low threshold, rapidly inactivating T-type and the high-threshold, dihydropyridine (DHP) sensitive L-type, present in most cell types, and the N-type and P-type present only in neurons (reviewed in Bean, 1989; Miller, 1992; Snutch and Reiner, 1992).

The structural information about Ca channels has come from work on skeletal muscle, which has an unusually high density of voltage-gated Ca channels in the transverse tubule membrane, as detected by the high concentration of DHP binding sites. Solubilization and purification of L-type Ca channels from this source have shown that these channels have a multisubunit complex structure. L-type Ca channels consist of a central transmembrane α_1 (175 kD) subunit, thought to be associated with four other polypeptides. The disulfide-linked α_2 (143 kD) and δ (27 kD) subunits are transmembrane glycoproteins as is the γ (30 kD) subunit, which is noncovalently associated with the others. The hydrophilic β subunit (54 kD) is located on the intracellular side. cDNAs coding for all of these proteins have been purified and sequenced. The sequence of the α_1 subunit revealed very high similarity in its proposed structure and organization to that proposed for the α subunits of the voltage-gated Na and K channels.

Expression of the α_1 subunit by itself from cDNA or the corresponding RNA indicated that it may be able to form a functional Ca channel, but with some altered properties as compared to native channels. Coexpression of the other subunits restored many of the functional defects of the α_1 subunit expressed alone. Each of the α_2/δ, β, and γ subunits can interact directly with the α_1 subunit and can modify α_1 expression and functional properties. However, the type of α_1 determines the function of the channel (for review and references see Catterall, 1991; Miller, 1992; Snutch and Reiner, 1992).

Potassium (K) Channels

Voltage-gated K channels represent an extraordinary diverse class of ionic channels that are essential for setting the resting membrane potential, the duration of action potentials and for the generation and tuning of firing patterns (Hille, 1992). They are remarkably diverse in their conductance and gating mechanisms and most cells express several types of K channels. Many invertebrate and vertebrate voltage-gated K channels have been cloned and their primary sequence shown to be remarkably similar despite the functional diversity of K channels (for review, see Jan and Jan, 1990; Pongs, 1992). The voltage-dependent K channels can be broadly subdivided on the basis of their biophysical and pharmacological properties. One type is the delayed rectifier K selective channel which is practically non-inactivating and is found in most nerve and muscle cells. Its primary function is to repolarize the action potential. The other type is the rapidly inactivation transient A-channels that are important in shaping the action potential. Many voltage-gated K channels exist with intermediate properties and consist of several subtypes. Other K channels are opened or closed by second messengers (for references see Rudy, 1988; Pongs, 1992).

The primary structure of many K channel forming proteins are known because it has been possible to clone cDNAs encoding for K channel subunits. The *Shaker* behavioral mutant of *Drosophila* has been observed to encode K channels, followed by the cloning and characterization of the gene products. The central *Shaker* gene encoded a membrane protein that had structural similarities to the primary structures of Na and Ca channel proteins. Several different mRNAs encoding at least 10 different K channel subunits are synthesized from the *Shaker* gene by alternative splicing and alternative transcription. However, distinct genes defined separate K channel subfamilies in *Drosophila* and in vertebrates; this contributes to the great diversity of K channels.

Comparison of the sequences and electrophysiological properties of cloned K channels suggests that minor sequence alterations sufficed to build a variety of diverse voltage-gated K channels and on a molecular level, these highly selective channels are the best characterized among all ionic channels (for review, see Stuhmer, 1991 and Pongs, 1992).

GENERAL DESCRIPTION OF THE STRUCTURE

Voltage-gated ionic channels share many functional and structural features. The primary subunits of voltage gated Na and Ca channels are single polypeptides of about 2000 amino acid residues. Na and Ca channels comprise contiguous repeats of four internally homologous domains, called repeats I–IV. Each domain has six putative transmembrane segments (Figure 1). The presence of four internal repeats suggests that the Na and Ca channel evolved by duplications of an ancestral gene (Hille, 1992). To examine whether individual repeats or their combinations can

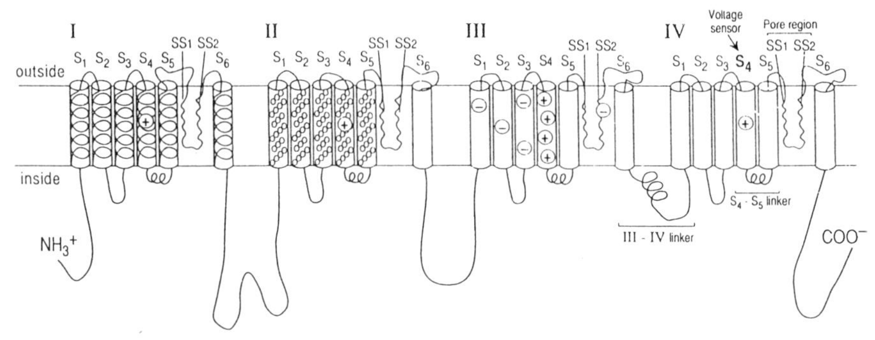

Figure 1. Diagrammatic view of a generalized voltage-sensitive ionic channel. The general transmembrane folding pattern in this model shows six homologous segments (marked S_1–S_6) in each repeat domain (I–IV) that fold in transmembrane α helix, illustrated by the spiral line in repeat I. The amino acid residues that form the helixes are illustrated by circles in repeat II, to visualize the structure. In repeat III the conserved charges, present in transmembrane segments in each repeat, are schematically indicated. The most highly conserved α helix is segment S4, present in each repeat, which contains a unique motif of positively charged amino acid residues followed by two non polar residues that repeat four to eight times in the helix. This structure is present in a comparable location in each of the Na, Ca, and K voltage-sensitive channels and is suggested to function as the voltage sensor of the channel. S4 segments are indicated by (+) sign.

In repeat IV the putative functional domains of the channels are indicated. Transmembrane α helixes are presented by cylinders, while the putative β sheet hairpin conformations of the short segments SS1 and SS2, are illustrated as waving lines. According to this model, these short segments of each repeat form the ion conducting pathway, indicated as the pore region. Presently, experimental data suggest the presence of two functional domains on the intracellular side of the channel: the sequence linking repeats III and IV is suggested to play an important role in the Na channel inactivation mechanism (III–IV linker); the short segment linking transmembrane S4 and S5 in each repeat (S4–S5 linker) is suggested to play an important role in activation as well as in inactivation mechanisms (Guy and Seetharramulu, 1986; Guy, 1988; Durell and Guy, 1992).

form functional channels, Stuhmer et al. (1989) prepared mRNAs encoding single repeats or severe contiguous repeats of Na channels by transcription *in vitro* of the corresponding cDNAs. mRNAs encoding only for one repeat or combinations of three separate repeats did not give rise to sodium currents that could be measured using the two electrode voltage clamp. By contrast, significant expression of sodium currents was observed with mutants having cuts between the repeats or deletions in the N- or C-terminal region. These results suggest that all four repeats are required for the formation of functional Na channels.

Typical voltage gated K channels are encoded as monomers homologous to each of the four Na or Ca channel domains. Independent subunits of about 600 residues symmetrically assemble to form homomeric or heteromeric K channels (Figure 2)

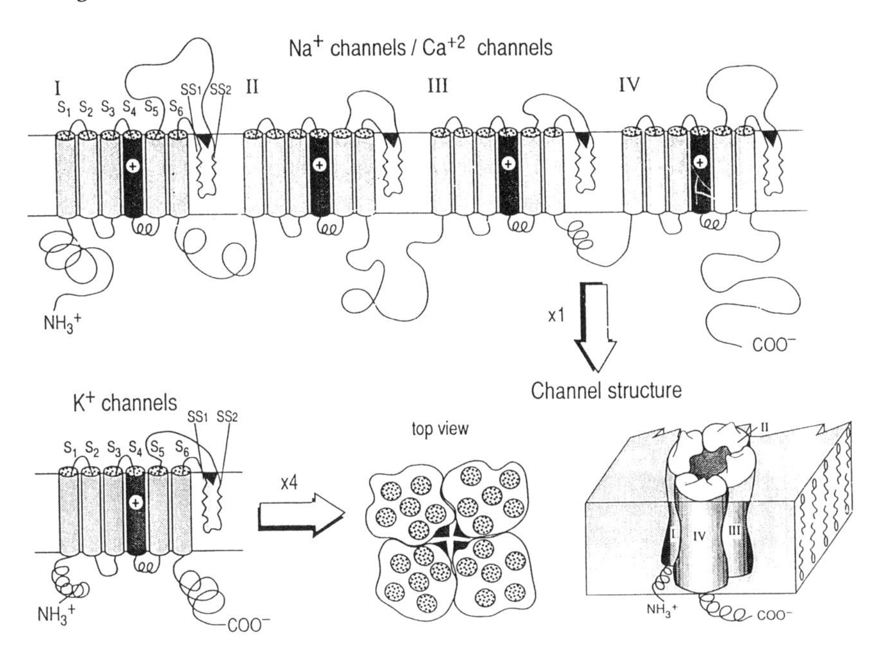

Figure 2. Schematic representation of the transmembrane arrangements of the principal subunits of Na, Ca, and K channels. Sodium and calcium channels share the general structure of four homologous repeat domains (I–IV) that are thought to act as 'pseudo' subunits coming together in the membrane to form a central ion pore. The potassium channel is composed of four peptides, each homologous to one repeat of the Na or Ca channel domains. Four subunits form a functional K channel. The short segments suggested to form a β hairpin in the membrane (SS1 and SS2, indicated by a dark triangle) from each repeat or subunit form the ion conducting pore.

(Catterall, 1988; Jan and Jan, 1990; MacKinnon, 1991). Some K channels may be homopolymers that comprise four identical subunits since injection of mRNA for one subunit into *Xenopus* oocytes produces functional channels; however, heteropolymers can also be formed by coinjecting different mRNAs or injecting mRNAs that produce covalently linked dimmers of two different subunits (Timpe et al., 1988; Isacoff et al., 1990; Christie et al., 1990; Ruppersberg et al., 1990).

An estimation of the subunit stoichiometry of the K channel has been based on engineering of constructs containing 2, 3, 4, or 5 channel subunits linked together in a single open reading frame. When four copies of the coding sequence for K channels were ligated contiguously and translated *in vitro*, the resulting RNA encodes four covalently linked subunit domains. Injection of this RNA into *Xenopus* oocytes resulted in the expression of voltage dependent K currents. Channels expressed by a tetrameric construct did not incorporate free monomer subunits, and a single subunit that expressed only as part of a heteromultimer contributed to the expression of functional channels when coexpressed with the trimeric construct but not a tetrameric construct, strongly suggesting that the channel is a tetramer (Hurst et al., 1992; Liman et al., 1992) (see Figure 2).

Recent structural models of voltage-gated ionic channels have been based on sequence comparisons of the growing number of cloned channels available (Guy and Conti, 1990; Durell and Guy, 1992). The rationale for the comparative approach has been that the conserved sequence regions among different channels are likely to represent regions responsible for common channel functions. The variable regions might correspond to sequences that are either not directly relevant to the function of the channel or that produce the unique characteristics of a particular channel type or subtype.

Structural models of voltage-gated ionic channels have been proposed in which the channel is formed by a symmetrical association of four subunits of the voltage-gated K channels or by a pseudosymmetrical assembly of the four repetitive homologous domains (pseudo-subunits) of Na channel (Figure 2) (Noda et al., 1984, 1986; Guy and Conti, 1990). Each Na channel repeat contains six potential transmembrane segments; two hydrophobic segments (S5 and S6) that have no charge residues, three hydrophobic segments (S1, S2, and S3) that have a few charge residues, and one segment (S4) that contains positively charged residues at every third position and usually hydrophobic residues at the remaining positions (see Figure 1). Each of these segments is long enough to span the membrane as an α-helix. The segments preceding and following each repeated domain are postulated to be on the cytoplasmic side of the membrane (Catterall, 1988, 1992). The largest degree of homology exists in the putative transmembrane regions, with a few charged residues in S2, S3, and S4 being strictly conserved among all voltage dependent channels. The S4 is the most characteristic structure shared by the members of all this superfamily and is thought to serve as a transmembrane voltage sensor for channel gating (Guy and Conti, 1990).

The structural model predicts that the polypeptide segment between domains S5 and S6 forms a deep invagination into the membrane to form part of the lining of the channel pore. Experimental support for the latter is provided by constructing chimeric channels in which the putative pore region has been exchanged between channels with different ion conductance properties. In each case, the properties of the ion conductance followed the S5-S6 channel donor (Hartmann et al., 1991; Stocker et al., 1991). Moreover, channels in which amino acids in this region are changed by site-directed mutagenesis, have altered ion permeability and sensitivity to block by non-permeable ions or neurotoxins (see below).

THE PORE REGION

A Model for the Ionic Pore

It is generally believed that the ionic pathway through voltage-gated channels has an "hour-glass" shape, with large entrance on each side of the membrane and a narrow region in the middle that controls the ion selectivity (see Figure 3) (Armstrong, 1981). Guy and Seetharramulu (1986) first noticed that two short segments (SS), located between S5 and S6 (called SS1 and SS2) are highly conserved in each of the ionic channel structures. They suggested that these segments form part of the conducting pathway and may also span part of the membrane as β-structure hairpin (Figure 3). These segments may span the narrow section of the hour-glass, where negative charges on SS2 may contribute to the cation selectivity.

Estimations of the conductance of a single channel imply physiological ion transport across membranes at rates approaching their rates of diffusion through free solution. In addition to being unusually rapid, ion conductance by voltage-gated ion channels is very selective: K^+ is about 8% as permeable as Na^+ through Na channels, and Ca channels are 1000 times more selective to Ca^{2+} than to any other ion. Hille measured the conductance of ionic channels to many metal and organic cations and estimated the narrowest region of the ion-conducting pore, called the ion selectivity filter, to be approximated in Na channels by a 0.31×0.51 nm rectangular orifice. Electrophysiological studies on K channels suggested that the movement of ions through K channels is kinetically best described if ions pass through the channels pore in a single file, having multiple binding sites (Hille, 1992).

Mutagenesis experiments have provided an excellent idea of the region of the peptide that constitutes the pore lining in K and Na channels. The proposed pore (P) region (called also H5 in K channels) in the S5-S6 linking sequence, consists of two short segments SS1 and SS2. Such a region has counterparts in equivalent positions, in the S5-S6 linkers in each pseudosubunit; within each ion channel class these sequences are highly conserved (Figures 4 and 5). Among K channels, P is the region with the highest degree of sequence identity. Many mutations in P did not yield currents detectable by two electrode voltage clamp (Yool and Schwartz, 1991; Pongs, 1992). Mutations in this region in K and Na channels alter the sensitivity of the channel to drugs and selective neurotoxins that block ionic currents as well as alter the ion conductance and selectivity (see below).

Elucidation of the Pore Structure by the Use of Selective Blockers of Ionic Currents

Several ligands known to interact specifically with the ionic pathway of Na and K channels, such as tetrodotoxin (TTX) and saxitoxin (STX), that block the Na^+ current in various Na channels, and charybdotoxin (CTX) and tetraethylammonium ions (TEA), that block K^+ current in many K channels, have been used as tools to study the amino acid residues involved in the formation of the pore.

In general, amino acids suspected to take part in the pore structure according to the channel models have been specifically substituted and the mutant channels expressed in the suitable expression system. The mutant channel functional activity

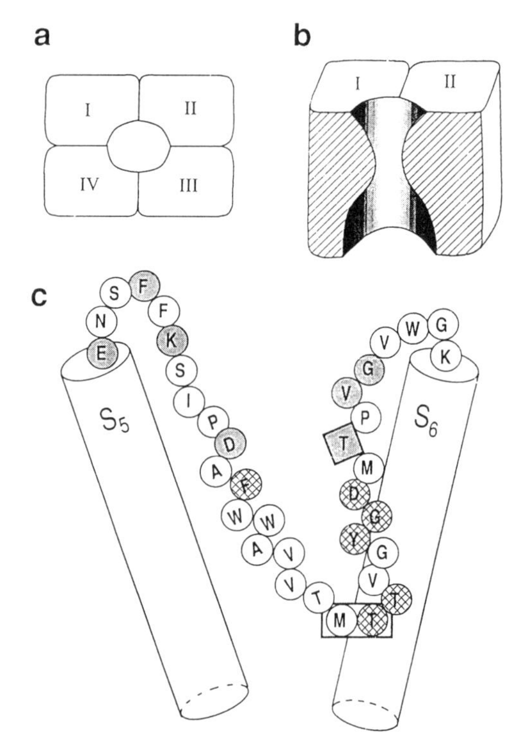

Figure 3. Diagram of the ionic channel pore. Four repeat domains (I–IV) of Na and Ca channels or subunits of K channels are symmetrically arranged in the membrane around a central pore. (*a*) Top view. (*b*) Section through the middle of the channel pore. The wider intra- and extracellular entrances to the channel pore form an hour-glass shape. The ion conducting pore itself with the selectivity filter is suggested to be in the narrowest part. (*c*) The residues forming the pore region in the loop between transmembrane segments S5 and S6 in the *Shaker* K channel (see Figure 4) are shown in standard 1 letter code. Residues are diagrammed as hairpins to illustrate the putative structure of the pore (P) region. Assuming a K channel tetramer with fourfold symmetry, four β hairpins would form an 8-stranded β barrel (Durell and Guy, 1992). The two prolines set the bounds for membrane entry and exit of a hairpin.

The Tyr residue (Y15 in Figure 4) is suggested to be in the narrowest part of the pore, thus dividing the pore region into the extracellular vestibule. Mutations right to it influence external TEA sensitivity, and mutations to the left influence internal TEA. This Tyr is completely conserved in all the known clones of K channels (Pongs, 1992). The square-boxed residue indicates the critical amino acid position (T449 in *Shaker*), substitution of which alters external TEA and CTX binding, as well as inward ion conductance. Hatched circles indicate amino acid residues, substitution of which alters ion selectivity of K channels. The boxed circles indicate amino acid residues in the most inner part of the pore, substitution of which alters internal TEA binding. Shaded residues indicate amino acids involved in external CTX binding. See Figure 4 for details and references.

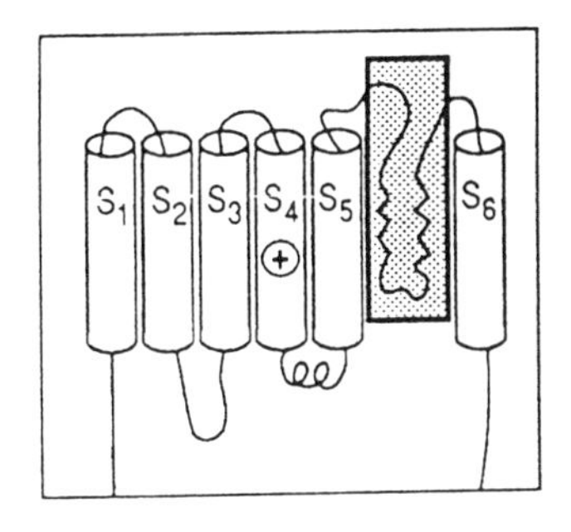

	422 … 430	<----SS1----> 440 <--SS2--->	449	
P region	-8 … -1<	1 … 10 … 15	19	> +1
Sh K channel	e nS f **F** k S**IP** d	A F **W** W A v V T **M T T** V **G Y G D** M	t	**P** v G
a reduce sensitivity to external TEA	K (at d)		K/R, V	
b reduce sensitivity to CTX (external)	K (at e), R (at f), K, N, E (at d)		K, N, Y	K R
c increase sensitivity to external TEA	G (at f)		Y/F	
d increase sensitivity to CTX (external)	G (at f), E (at k)			E
e reduce channel conductance			K/R, Q	
f reduce sensitivity to internal TEA		I S (at M T)		
g alter ion selectivity		S, Y (at F); S S (at T T); - - E (at G D M)		
h alter inactivation			E/K, V/Y	
Ca^{+2}- I	n **FDN** I	l **F** A i **L T V** F **Q** C **I T M** - - **E G**	**W**	**T** D i
Ca^{+2}- II	N **FD**T **F**	**P** A A I l T **V F Q** I **L T G** - - **E D**	**W**	**N** a **V**
Ca^{+2}- III	h Y**DN** V	L w **A** L L t **L F T V S T** G - - **E G**	**W**	**P** m V
Ca^{+2}- IV	**N F**RT **F**	l Q A L M **L L F R** S **A T G** - - **E A**	**W**	H e **I**

Figure 4. Map of site-directed mutagenesis experiments on *Shaker* K channel that revealed the pore (P) region, and their functional effects. Primary amino acid sequence of *Shaker* K channel protein between segments S5 and S6 that includes the pore (P) region, is presented in 1 letter code. The P region is indicated between < > and the amino acids within the region are given numbers 1 to 19 for discussion and those flanking the region are given negative numbers on the N-terminal side and (+) sign on the C-terminal side of P. Numbers above the amino acid sequence row correspond to the amino acid residue position in the *Shaker* K channel protein (422–452) (Temper et al., 1987; Durell and Guy, 1992). The SS1 and SS2 segments are marked according to Guy and Conti (1990). Identical residues among all voltage-gated K channels are in bold upper case, the conserved residues (no more than three substitutions) are in plain upper case, and the non-conserved residues are in lower case (according to Durell and Guy, 1992). *(continued)*

Figure 4. (Continued) Amino acid substitutions introduced by site-directed *in vitro* mutagenesis into *Shaker* K channels are given in rows designated a through h, according to their functional effect on the mutated and expressed channel behavior, as indicated. Deletions are presented by (–) under the deleted amino acid (row g). Mutations that resulted in more than a single functional effect are cited at each of the relevant rows.

The homologous pore region amino acid sequence of the four repeat domains of Ca channels is given for comparative purposes. Sequence alignment with K channel is according to Heginbotham et al. (1992). Ca channel sequence is that of the rabbit brain III Ca channel (BIII) and is compared to all the Ca channel known sequences (see Fujita et al., 1993). Bold upper case letters correspond to identical residues in each domain of rabbit brain Ca channels (BI, BII, BIII); rabbit cardiac and skeletal muscle DHP-sensitive Ca channels and human brain DHP-sensitive Ca channel α1D (see Tanabe et al., 1987; Mikami et al., 1989; Mori et al., 1991; Niidome et al., 1992; Williams et al., 1992b; Fujita et al., 1993 for references). The conserved residues (with one substitution among all the Ca channels) are in upper case letters. Lower case corresponds to non conservative residues (two or more substitution among all Ca channels).

The rows marked a–h present the *in vitro* amino acid substitutions that resulted in the following functional modifications of the expressed mutated channels: (*a*) Reduce external TEA binding (1D/K; 19T/K,R,V) (MacKinnon and Yellen, 1990). (*b*) Reduce external CTX binding (1D/N,K,E; 19T/K,Q,Y); +1V/K; +3V/R; F425R; –8E/K) (MacKinnon et al., 1990; Goldstein and Miller, 1992). (*c*) Increase external TEA sensitivity (19T/Y,F; –5F/G; F425G) (Kavanugh et al., 1991; Hurst et al., 1992; Heginbotham and MacKinnon, 1992; Goldstein and Miller, 1992). (*d*) Increase sensitivity to CTX (–3K/E; –6F/G) (MacKinnon et al., 1990). (*e*) Reduce single channel conductance (19T/K<R<Q) (MacKinnon and Yellen, 1990). (*f*) Reduce internal TEA binding (M440I; T441S) (Yellen et al., 1991; Choi et al., 1993). (*g*) Affect ion selectivity: T441S and F433S—increase permeability to big cations with no change in selectivity to Na^+. Deletion of two amino acids, Tyr and Gly at positions 15 and 16 in P (YG) results in channels having little selectivity among the alkali metal ions and abolishes K^+ selectivity. Mutation of the deletion mutant at 17D/E produced channels with high affinity to Ca^{2+} (Yool and Schwartz, 1991; Heginbotham et al., 1992). Mutation of the equivalent position in the Na channel results in Ca channel like ion conductance properties (see Figure 5). (*h*) Altered inactivation kinetics: by replacement of Thr449 by Glu or Lys (T449E,K) leads to a 100-fold acceleration of inactivation. In contrast, placing Val at this position results in channels that do not inactivate (Lopez-Barneo et al., 1993).

Darkly shaded amino acids represent residues that line the internal part of the pore; mutations at these residues affect ion selectivity, or/and reduce sensitivity to internal TEA. Residues shown to affect ion selectivity of the pore are underscored twice. Shaded residues represent those amino acids suggested to be located at the extracellular entrance of the pore that affect sensitivity to external TEA and/or CTX. The Thr residue at the critical position 19 of the pore that affects single channel conductance and gating, as well as sensitivity to external blockers is shaded and boxed. See also Figure 3.

and sensitivity to known ligands is monitored and analyzed using electrophysiological methods. The methodological aspects are beyond the scope of this chapter.

TEA and CTX—Selective Blockers of K Channels

Although the K channels are extraordinarily diverse with respect to voltage-dependent gating, the ion-conduction properties of K channels are quite invariant. Most K channels share the common features of very high K^+ selectivity, multi-ion occupancy, and block by the impermeant tetralkylammonium ions. Ammonium ions can pass through most K channels, whereas its quaternary derivative TEA cannot, resulting in the blockade of most voltage-activated K channels. The efficacy of extracellular TEA in blocking K channels in different preparations is quite variable, while the effect of intracellularly applied TEA is relatively uniform among different voltage-activated K channels (Hille, 1992). These well conserved functional properties suggest that the amino acid sequence and structure of the ion conduction pore may be highly conserved among the voltage-activated K channels.

Many K channels are inhibited by scorpion toxins like charybdotoxin (CTX), that has been studied in detail and the mechanism of inhibition is well understood (MacKinnon and Miller, 1989). CTX is a 37-residue basic peptide derived from scorpion venom which inhibits a variety of K channels at nanomolar concentrations (Park et al., 1991).

The TEA and CTX affinities depend on amino acids belonging to a sequence of 19 amino acids that lie between transmembrane segments S5 and S6 and, with small variations, is found in all voltage gated K channels cloned so far. When the region between S5 and S6 has been transplanted from a CTX-sensitive or TEA-sensitive channel to a CTX-insensitive channel, or vice versa, the resulting chimeric channels become as sensitive to CTX or TEA as the donor channel (Stocker et al., 1991; Hartmann et al., 1991). Mutagenesis experiments on K channels demonstrated that residues near the beginning and end of the P region affect binding of TEA and CTX to their extracellular sites (see Figures 3 and 4 shaded residues).

External domains of the pore revealed by TEA and CTX binding sites. Most K channels fall into one of two classes with respect to external TEA sensitivity: those that are sensitive (inhibition constant Ki < 1 mM) and those that are relatively insensitive (Ki > 20 mM) (Hille, 1992). Two amino acids located in the putative pore forming region of a K channel (position 1 and 19 in P region, Figure 4) were shown to fulfill a critical role in determining the sensitivity to external TEA and CTX. External TEA ion competes with CTX for channel blockade, suggesting that at least part of the peptide blocker is bound close to the ion conduction pore (MacKinnon et al., 1990). Substitution of the aspartate residue at position 1 of P by either an asparagine or lysine (mutants D431N and D431K in *Shaker*, Figures 4, b) renders the channel insensitive to the toxin and reduced TEA sensitivity by about a factor of 2 (mutant D431K) (Figures 4, a). Interestingly, a conserved substitution by

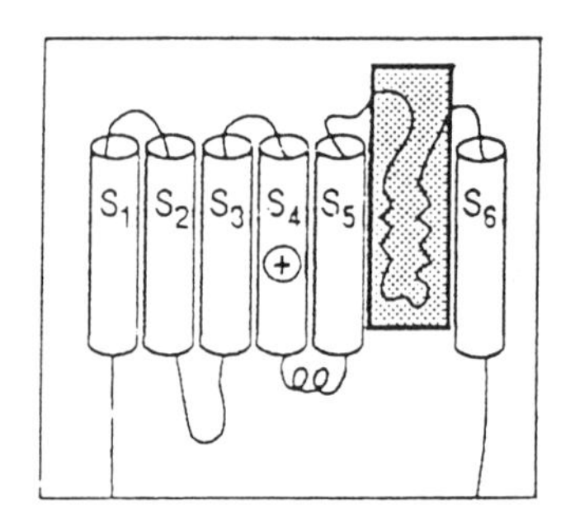

Figure 5. Map of site-directed mutagenesis experiments on Na channels that revealed the pore region, and their functional effects. Amino acid sequences of the putative pore region in the Na channel repeat domains, compared to the homologous region in *Shaker* K channels and Ca channels which are presented in 1 letter code. Numbers above residues correspond to the amino acid position in rat brain Na channel I, the sequence of which is presented. Letters in bold upper case correspond to identical residues present in rat brain Na channels I, II, and III (Noda et al., 1986; Kayano et al., 1988), eel electroplax (Noda et al., 1984), Drosophila *Para* locus Na channel (Loughney et al., 1989), rat heart (TTX-resistant Na channel) (Rogart et al., 1989) and rat skeletal muscle (Trimmer et al., 1989) sodium channels. Upper case letters represent highly conserved amino acids (one substitution) among the Na channels and lower case correspond to non conservative residues. Alignment of repeat domains according to Terlau et al. (1991). Sequence of skeletal muscle Ca channel is presented and compared with all Ca channel sequences, as in Figure 4. The *Shaker* K channel sequence is as in Figure 4.

Shaded residues in Na channel sequence correspond to amino acids affecting TTX/STX sensitivity (*row a*). Boxed residues represent charged residues involved in TTX/STX sensitivity and ion conductance or selectivity. The boxed residues in Ca channel sequence represent the corresponding residues in Ca channel that are suggested to be involved in ion selectivity and were mutated accordingly in Na channel domains III and IV (*see row c*). The boxed and shaded Thr residue in K channel sequence corresponds to position 19 in P region (Figure 4), shown to be critical for ion selectivity and sensitivity to external TEA and CTX. Double underscore indicates residues that alter ion selectivity and/or sensitivity to divalent ions. Sequence-directed antibodies (*Ab*) used for the localization of scorpion toxin binding sites in Na channels are indicated above the Na channel sequences in domains I, III, and IV (Thomsen and Catterall, 1989; Gordon et al., 1992).

The rows marked a–d present the *in vitro* amino acid substitutions in the indicated Na channel repeat domains (I–IV) that resulted in the following functional modifications of the expressed mutated channels: (*a*) Reduce significantly TTX/STX sensitivity (I-a: D384N; E387Q,S,Y; W386Y; F385C/Y. II-a: E942Q; E945Q,K. III-a: K1422E; M1425Q/K. IV-a: A1714E; D1717N,Q,K) (Terlau et al., 1991; Satin et al., 1992). (*b*) Reduce single channel conductance (I-b: E387Q; Q383E,K; D384N; W386Y; E387Q,S,Y. II-b: E945Q,K. III-b: M1425K. IV-b: D1717Q,K). (Terlau et al., 1991). (*c*) Alter ion selectivity: III-c: K1422E and IV-c: A1714E - convert Na selective to Ca selective channel (Heinemann et al., 1992b). (*d*) Alter sensitivity to divalent ions: F385C increases sensitivity and N388R reduces sensitivity to Cd^{2+} and other divalent ions (Satin et al., 1992; Heinemann et al., 1992a; Backx et al., 1992). (*continued*)

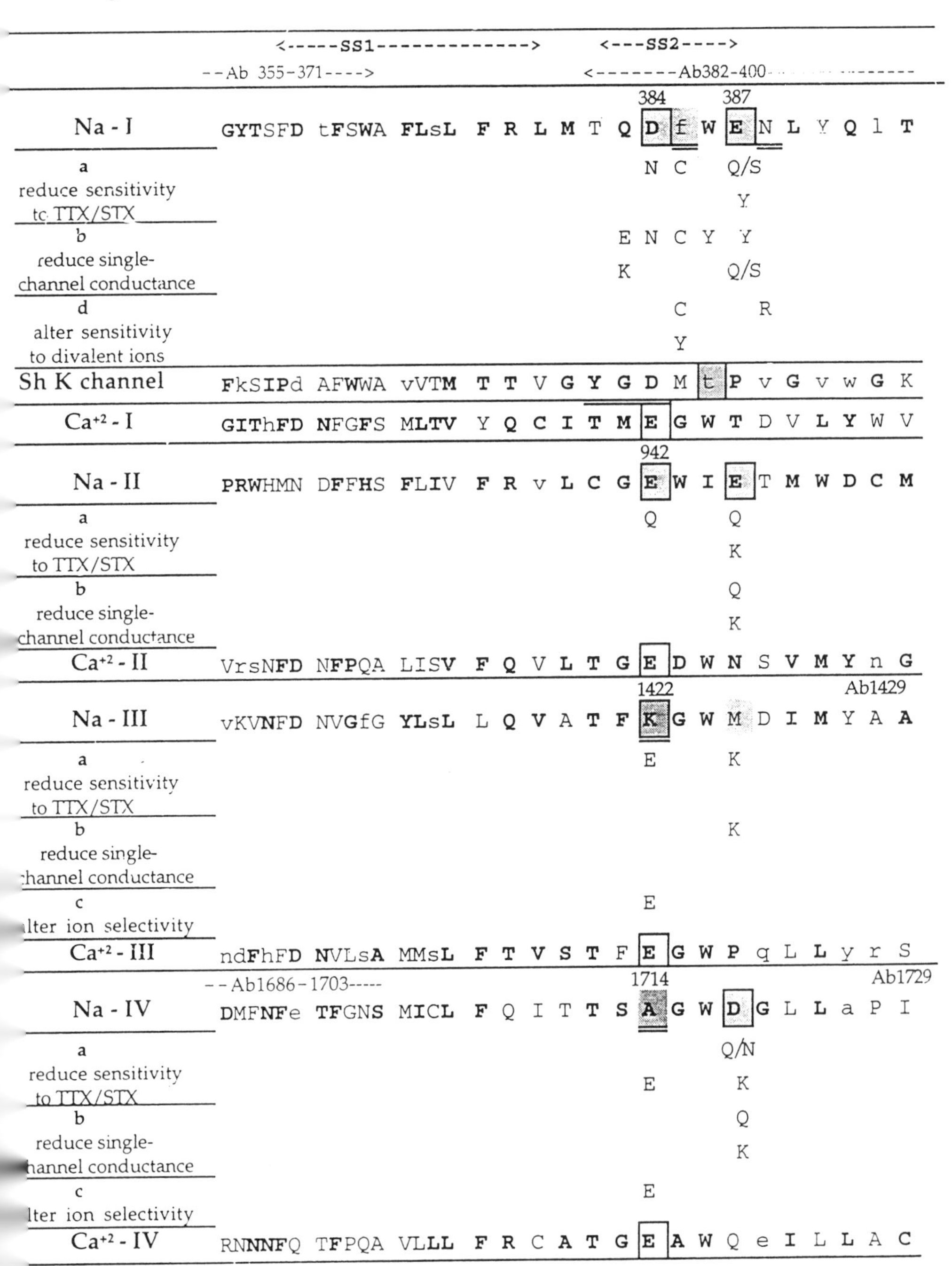

Figure 5. (Continued)

glutamate, which conserved the negative charge, but its side chain is about 0.2 nm longer, reduces the sensitivity to CTX by 3-fold, indicating a direct interaction between the acidic residue and the cationic toxin molecule (MacKinnon et al., 1990).

Replacing the amino acid threonine at position 19 of P (mutant T449K) with lysine effectively abolished the inhibition by TEA, producing a 1000-fold change in the inhibition constant (MacKinnon and Yellen, 1990). Substitution of arginine or valine for the threonine also reduced binding of TEA dramatically (Figure 4, a). Comparable to the external TEA block, mutating threonine at position 19 to lysine or asparagine completely abolished the channel sensitivity to CTX; in contrast to its effect on TEA (Figure 4) tyrosine at that position greatly reduced the sensitivity to the toxin (Figure 4, b) (MacKinnon et al., 1990). Substitutions at this residue have also been shown to dramatically alter inactivation of K channels (Lopez-Barneo et al., 1993), as discussed in a later section.

An aromatic residue at the 19 position of P appears to be a requirement for high affinity TEA blockade (Figure 4, c). Tyrosine substitution, that produces no change in charge, increased the sensitivity of the channel to TEA by about 50 fold (MacKinnon and Yellen, 1990). All naturally occurring channels that are very sensitive to TEA (Ki around 0.5 mM) have tyrosine in this position, and the only substitution shown to increase sensitivity is phenylalanine (Kavanaugh et al., 1991). Tyrosine and phenylalanine residues have in common a six-membered aromatic ring as the predominant structure of their side chain.

Constructing tandem dimmers, produced by two subunits of *Shaker* K channels, enabled production of hybrid channels in which 0, 2, or 4 tyrosines at the 449 positions were present. Two tandem dimmers are expected to participate in the formation of a tetrameric channel. The free energy for TEA binding to its external binding site scales linearly with the number of subunits containing a tyrosine at residue position 449. Heginbotham and MacKinnon (1992) and Hurst et al. (1992) have found that four aromatic residues, one from each subunit, participate in the formation of the high affinity site. Moreover, TEA interacts with all 4 of these aromatic residues at once, as if they "coordinate" the blocker in the pore (Kavanaugh et al., 1992). These 4 residues must lie near the central axis of the pore, where the four subunits come in contact (see Figures 2 and 3), assuming that indeed TEA inhibits by occluding the pore. The weak voltage dependence of external TEA blockade would indicate that TEA binds at a superficial point along the axis of the pore, near the external entryway. Thus, the findings suggest that amino acid 449 is located at the pore entryway on the extracellular side (Figures 3 and 4). In support of this notion, mutation of lysine, arginine or glutamine at the same position 19 of P dramatically reduced the single channel conductance, while substitution of tyrosine had no effect (Figure 4, e) (MacKinnon and Yellen, 1990).

These results indicate that the two positions 1 and 19 of the P region in the primary structure of the K channels appear to be located in or near the outer mouth of the ion conduction pore. These effects are probably produced by a specific alteration in the region of the conduction pathway that bind TEA. In support of this model, a point mutation located in an analogous region in the Na channel protein abolishes inhibition by TTX.

Other external domains of K channels have also been revealed using CTX. In addition to its binding sites within the P region ends (Figure 4), CTX was shown to interact with other amino acid residues in the segment flanking the ion conductance pore in the S5-S6 linker, but not in the putative external loops connecting S1, S2, and S3, S4 (MacKinnon and Miller, 1989). When a negatively charged glutamate residue at position 422 (located at position-8 on the N-terminal side of P, Figure 4, b) is replaced by a glutamine (neutral) or lysine (positive), the channel's sensitivity to the cationic toxin is progressively reduced. Replacing a positively charged lysine residue, located 5 amino acids further (427 in *Shaker*) by neutral (asparagine) or negative (glutamate) residues, makes the channel progressively more sensitive to the toxin inhibition (the inhibition constant (Ki) for the K427E mutant is 0.15 nM, compared with 3.7 nM for the wild-type *Shaker* channel). Mutating the phenylalanine residue at position 425 to glycine (mutant F425G, Figure 4) resulted in a large increase in affinity of the *Shaker* K channel to CTX due to an effect on toxin off-rate, that was decreased over 1800-fold in the mutant (Figure 4, c). This demonstrated that the phenylalanine at position 425 destabilizes CTX in the binding site in the CTX receptor structure that exists in the *Shaker* channel (Goldstein and Miller, 1992).

The C-terminal end of the S5-S6 linker is also involved in the binding of CTX. Toxin inhibition depends critically upon the amino acid substitution of valine at position 451 (+1 of P, V451 K, Figure 4, b), replacement of which by lysine renders the channel insensitive to CTX. Mutation at an adjacent position also influenced the toxin-channel interaction: glycine (G452) substitution to arginine (positive) decreased while substitution to glutamate (negative) increased sensitivity to toxin (Figure 4). These results support the notion that the electrostatic potential at the toxin binding sites stabilizes the toxin-channel complex (MacKinnon and Miller, 1989: MacKinnon et al., 1990).

The mutations that alter toxin inhibition by electrostatic and thus a local mechanism are located at both ends of the highly conserved P region. When four subunits are put together to make a channel, P forms the K selective pore, and the regions flanking P form the pore outer vestibule (Durell and Guy, 1992) (Figure 3). The K^+ currents carried out by all the mutant channels mentioned above are very similar to those carried by wild type channels, suggesting that in each case, the channel structure is largely intact and that the mutations act at a local manner, perhaps by sterically interfering with toxin binding.

The structure of CTX has been determined, and its large size (dimensions of 1.5 × 2.5 nm) (Bontems et al., 1991) is compatible with the notion that the toxin may interact with residues that are 2.5 nm away, perhaps at the very outer rim of the vestibule that leads down to the ion conduction pathway (MacKinnon et al., 1990) (Figure 3).

It has been shown that entering of K^+ ions from the internal side into the highly selective conduction pathway destabilizes TEA or CTX bound to its receptor in the external mouth of the channel. The interaction of channel bound CTX with pore-associated K^+ is mediated exclusively by a single positively charged residue on CTX, lysine 27, most likely by direct electrostatic repulsion between the e-amino group of lysine-27 and a K^+ ion bound to an ion conductance site in the channel (Park and Miller, 1992). The nature of the site at which K^+ exerts its destabilizing influence on CTX and TEA is not yet known. However, only permeant ions such as K^+ and Rb^+, but not Na^+, interact with the toxin in this way. Accordingly, this site is located in the conduction pore and is in an essential element of the channel's K^+ selectivity mechanism (Park and Miller, 1992).

Thus, several lines of evidence indicate that the regions flanking the SS1-SS2 are located externally (Figure 3). First, the interaction of the channel with CTX, which blocks the external mouth of the pore, is influenced by mutations in these regions. Second, external TEA blockade is affected by mutations at 431 and 449 positions. The 449 position is especially critical in determining external TEA sensitivity; substitution of different amino acids at this position results in inhibition constants that vary over a 500-fold range (MacKinnon and Yellen, 1990).

Internal TEA site. Internal application of TEA prevents the conductance of K^+ ions through voltage activated K channels. TEA is a good probe of the internal entryway to the ion conduction pore of K channels since several lines of evidence indicate that the internal TEA inhibition site is located 15% of the way into the transmembrane electric field and acts by plugging the pore (Yellen et al., 1991).

Mutating two highly conserved residues at the middle of the P region (Figure 4, positions 10 and 11), alters the internal TEA sensitivity dramatically (Figure 4, f) without any effect on the external TEA sensitivity (Yellen et al., 1991; Choi et al., 1993). The internal TEA sensitivity is decreased by 13-fold for the mutation T441S, and by 54-fold for the mutation M440I.

The effect on TEA binding of the two substitutions in the middle of the P region, M440I and T441S, is reduced or eliminated when the TEA ion has an alkyl tail of increasing chain length, suggesting that this site, unlike the external site, contains a hydrophobic region, shown to be located within the S6 transmembrane segment. Using a series of alkyl-tetraethylammonium (alkyl-TEA) derivatives to probe the internal binding site it has been shown that the blockade is state dependent and requires the channel to open (Choi et al., 1993). The affinity for binding of the longer blockers can be affected by varying the hydrophobicity of one residue, threonine 496 within the transmembrane segment S6, by its substitution for valine. This results in an increase in the affinity for alkyl-TEA by 27-fold but alters TEA affinity by only 2-fold, in a manner consistent with a direct interaction between the side chain at the S6 site and the alkyl chains of the blocker. Mutations at this position appear to modify specifically the binding of more hydrophobic blocks, opposite to the effects of mutations in the P region (Choi et al., 1993). These results suggest that part of the S6 segment may lie near the inner mouth of the pore.

The receptor site for verapamil, a probable intracellular pore blocker of Ca channels, involves residues immediately on the intracellular side of transmembrane segment S6 in repeat IV (Striessing et al., 1990). Such results support the idea that the ends of S6 segments are close to or form part of the intracellular openings of the transmembrane pore (Catterall, 1992; Choi et al., 1993).

Both internal and external TEA cause voltage dependent block in a manner consistent with the receptor site located about 15–20% into the membrane field, and the block is antagonized by K^+ on the opposite side of the membrane as though increased K^+ influx can sweep the quaternary ammonium ion out of the pore. The ability of a permeant ion to relieve blockade produced by open channel blockers has been observed for a variety of K channel types. Increasing internal K^+ can speed the dissociation of externally applied CTX (MacKinnon and Miller, 1989). It is suggested that electrostatic repulsion between K^+ and the positively charged blockers leads to the increased exit rates of the blockers (Newland et al., 1992).

Results showing antagonism between TEA and K^+ from opposite sides of the membrane provide strong evidence that the site lies within the aqueous pore, by interacting within the ion permeation pathway. A location of TEA binding sites within the pore is also suggested by the finding that amino acids strongly influencing block by TEA are found in the same 19 amino acid sequence (P region) as other amino acids that influence ion selectivity (Figure 4, g).

To summarize, the prevailing view of the structure of the ion conduction pathway in K channels is that it consists of a narrow pore opening at either end into the wider mouth (Figure 3). Mutation of conserved residues in the middle of the P region alters the sensitivity to internal, but not external TEA (Figure 4, f). In contrast, two residues, one at each end of the P region, significantly alter external TEA and CTX sensitivity (the mutations D431K and T449Y had completely normal sensitivity to internal TEA block (Figure 4, a, b) (MacKinnon and Yellen, 1990; Yellen et al., 1991). These findings support the notion that the two receptor sites for TEA are distinct, and agree with the idea that the P region spans the membrane twice, namely, folds into the membrane and lines the channel (Figure 3) (Durell and Guy, 1992; Yellen et al., 1991; Kirsch et al., 1992).

TTX and STX—Selective Blockers of Na Channels

Voltage sensitive sodium channels are responsible for the fast depolarizing phase of the action potential in most electrically excitable cells (Hille, 1992). As a critical element in excitability, a variety of neurotoxins affect many different properties of the sodium channel, including conductance, ion selectivity, activation, and inactivation (reviewed by Catterall, 1986, 1992). TTX was the first neurotoxin shown to block Na channels specifically. Because the toxin has a guanidino moiety and guanidine is a permeant ion of the Na channel, it has been proposed that the toxin enters the extracellular opening of the transmembrane pore of the Na channel and lodges there, however, see Gordon, 1996.

Na channels consist of four major symmetrical domains that are suggested to form a central pore (Figure 2). In further support of the notion that the domain between S5 and S6 contributes to formation of the ion conductance pore (as shown in K channels, see above and Figure 3), several studies have found that point mutations in an analogous region to the P region in the voltage gated Na channel, alter both TTX inhibition and reduces single channel conductance (Figure 5).

Initially, Noda et al. (1989) have shown that a single mutation (E387Q) in rat brain Na channel that neutralizes the negative charge of the conserved glutamic acid residue 387 to glutamine, renders the channel insensitive to TTX and STX and reduces five times its single-channel conductance, but is still highly selective for Na^+ over K^+ and its macroscopic current properties are only slightly affected. In fact, mutations of charged residues symmetrically located at homologous sites in the S5-S6 loops in each of the four repeat domains were found to dramatically reduce TTX binding (Figure 5, shaded residues) by making the IC_{50} for TTX and/or STX more than 100 times larger than the wild-type values. Charge mutations at adjacent positions in that region in each domain or mutations without a change in the net charge at positions 384 (D384E) and 1425 (M1425Q), and at other positions of the SS2 segments had minor or insignificant effects on toxin sensitivity (Terlau et al., 1991). These suggest that portions of all four S5-S6 loops may contribute to the formations of the TTX/STX binding site.

All of the mutations involving a decrease in net negative charge that strongly reduced toxin sensitivity also caused a marked decrease in single-channel conductance (Figure 5, shaded residues). The mutants K1422E and A1714E which have an increased negative charge, showed nearly no reduction in single-channel conductance, whereas their toxin sensitivity was strongly reduced. These indicate that the TTX/STX interaction cannot be solely explained by electrostatic attraction between the positively charged toxin guanidinium groups and negatively charged acidic groups in their Na channel binding site (Satin et al., 1992). In some mutants (Q383E and Q383K), single-channel conductance was affected without significant alterations in toxin sensitivity (Figure 5). These support the notion that this region participates in forming the extracellular mouth and/or the lining of the ion-conducting pore itself, as modeled by Guy and Conti (1990) (see Figure 3).

Although TTX and STX may share several amino acid residues in their binding site (Figure 5), some mutations affected TTX and STX sensitivity in varying degree, as exemplified by the mutant D1717N, which is essentially insensitive to STX but has an IC_{50} for TTX which increased only 20-fold. These results suggest that the two toxins that have different structures and carry 1 and 2 positive charges, respectively, may have overlapping binding sites within the extracellular pore region (Terlau et al., 1991; for a review see Gordon, 1996).

To summarize, the sensitivity to TTX and STX of the sodium channel is strongly reduced by mutations of specific amino acids residues in the SS2 segment equivalently positioned in the 4 internal repeats. Most mutants reducing net negative charge in these amino acids markedly diminish single-channel conductance (Figure 5). These results, together with the observation that the gating kinetics of all the mutants presented in Figure 5 appeared normal, provide evidence that the predominantly negatively charged residues line part of the extracellular mouth and/or the pore wall.

Two classes of Na channels subtypes have been distinguished by their sensitivity to block by TTX and STX and external divalent cations. Sodium channels in brain and innervated skeletal muscles that are TTX-sensitive are blocked by nanomolar concentrations of TTX and are relatively resistant to block by Zn^{2+} and Cd^{2+}, whereas TTX-resistant Na^+ channels in heart and denervated skeletal muscle are blocked by micromolar concentrations of TTX but have increased sensitivity to blockage by divalent cations. These divalent cations also competitively inhibit TTX and STX binding.

The differences between the TTX sensitivity of the two Na channel subtypes can be attributed mainly to the single amino acid difference, located in the SS2 segment of repeat I, where the TTX-sensitive rat brain Na channel has a phenylalanine residue whereas the TTX-resistant channel has a cysteine residue. The mutation F385C leaves the channel virtually insensitive to TTX and STX, and also results in a drastic increase in sensitivity to Cd^{2+} and Zn^{2+}. Mutation of the nearby asparagine residue (N388R) causes a reduction in sensitivity to all of the divalent cations but it exerts only a minor effect on the block by TTX (Heinemann et al., 1992a; Satin et al., 1992) (Figure 5). These mutations probably make structural changes in the binding site for TTX, STX, and Cd^{2+} that specifically alter only corresponding Na^+ channel functional properties because both substitutions occur naturally in other Na^+ channel isoforms and do not alter other properties of Na^+ current.

The similarity of external blockage of Na^+ channels by TTX and STX and of K channels by CTX and TEA suggests that these agents occupy structurally homologous binding sites close to or within the mouth of the channel pore and that the Phe/Tyr/Cyc (F385 in Figure 5) residue in the Na channel isoforms is located within the permeation pathway. Thus, the suggested alignment of amino acids sequences is where the glutamate 384 and phenylalanine 385, involved in TTX and STX binding in Na^+ channels are in similar position as the critical residue at position 449 in K channels, a major site in CTX and TEA binding (Figures 4 and 5). Which is to say that the proposed topology of the first domain of the Na channel is similar to that of SS2 in K channels (Figure 5) (Guy and Conti, 1990; Hille, 1992).

Scorpion Toxins Binding Sites

Inactivation of sodium channels is slowed by numerous neurotoxins, including the peptide α-scorpion toxins that act at the neurotoxin receptor site on the extracellular side of the Na channel. Binding of the α-scorpion toxins to this receptor site is strongly voltage dependent, suggesting that the voltage-dependent conformational changes that lead to channel activation and inactivation involve channel segments that also contribute to the formation of the α-scorpion toxin receptor site (Catterall, 1980, 1992). The use of several antibodies (Ab) directed against specific sequence regions in the external linker S5-S6 flanking the SS1-SS2 region in rat brain Na channels (Figure 5) enable the identification of Na channel segments that may associate to form the α-scorpion toxin receptor site. Antibodies that recognized the N-terminal end of SS1 in domain I (antibody 355-371) and IV (antibody 1686-1703, Figure 5) specifically inhibited the binding of α-scorpion toxins to the channel (Thomsen and Catterall, 1989). Similarly, the binding of other scorpion toxins, shown to suppress Na^+ currents in insect axons, was also effectively inhibited by some of these site-directed antibodies, corresponding to the C-terminal end of SS2 in repeats I, III and IV (Gordon et al., 1992) (Figure 5).

Thus, the receptor binding sites of various neurotoxins that modify Na channel function are located at the S5-S6 extracellular linker, in regions flanking the putative pore domain (Figures 3–5). This upholds the channel model that the four repeat domains symmetrically assemble around a central pore and the receptor sites for the scorpion neurotoxins are formed by the close apposition of the S5-S6 extracellular loops in domains I–IV (Catterall, 1992; Gordon et al., 1992) (Figure 2). However, it is not yet understood how binding of α-scorpion toxins to their extracellular site slows the inactivation process. The available results implicate the channel domain in this region to be involved in the coupling of gating processes.

The studies using site-directed mutagenesis and formation of chimeric potassium channels between subtypes with different ion conductances and pharmacological properties have revealed that several amino acid residues located near the extracellular ends of the S5 and S6 segments are required for block of the channel from the extracellular side by the peptide CTX and by TEA. Residues located at equivalent positions in each of the four repeats of Na channels have been shown to be required for block of the channel by TTX and STX. Moreover, residues in the center of the segment containing SS1 and SS2 are important for block of the K channel from the intracellular side by TEA ion. These results argue that the short segments SS1 and SS2 may traverse the membrane in an extended conformation, placing the residues between them on the intracellular side of the channel. These short segments may therefore form the inner walls of the transmembrane pore, and the residues between them may form an intracellular ion binding site (Figure 3). Consistent with this idea, minor changes in the amino acids in this segment have dramatic effects on ion selectivity (see below and Yool and Schwartz, 1991, Kirsch et al., 1992). The S5 and S6 segments may surround the inner walls of the pore and form the wider intracellular and extracellular openings through which ions enter and exit.

Structure of the "Selectivity Filter"

The distinct structure of the ion conducting pore present in each ion channel class is reflected by the high selectivity shown by the Na, Ca, and K channels on the permeable ion allowed to pass.

K channels are highly selective for K^+ over Na^+, and are not efficiently blocked by Ca^{2+} or Mg^{2+}. Site-directed mutagenesis studies have provided evidence that sequence alternations within the P region alter the selectivity and the ion binding properties of the channel. Single amino acid substitutions that replaced the conserved threonine 441, located in the middle of the P region in K channels by serine (T441S) and the phenylalanine 433 by serine (F433S) (Figure 4, g) modify channel selectivity without affecting sensitivity to external TEA or macroscopic kinetic properties of the currents (Yool and Schwartz, 1991). These mutations increased permeability to the larger cations, NH_4^+ and Rb^+, without eliminating the ability of the channel to exclude Na^+. The change in selectivity induced by single amino acid substitutions supports the hypothesis that P forms the pore-lining region of the *Shaker* channel (Figure 3).

All site directed mutations reducing negative charge at the SS2 residues in Na channels (see Figure 5) also caused a marked decrease in single channel conductance, suggesting that they form part of the extracellular mouth and/or pore of the Na channel. Heinemann et al. (1992b) have emphasized a key role for residues which determine the ion selectivity of Na channels as well as the binding of TTX at the putative pore region. The positions corresponding to D384 in SS2 that determine the TTX/STX sensitivity of the Na channel is occupied by glutamate in all four repeats of Ca channels at equivalent positions (Figure 5, boxed residues). Single mutations that replace the lysine at position 1422 in repeat III and/or alanine 1714 in repeat IV of rat brain Na channel by glutamate, which occurs at equivalent positions in calcium channels (Figure 5), altered ion selectivity properties of the Na channel to resemble those of Ca channels (Heinemann et al., 1992b). The mutations K1422E and A1714E in the SS2 segment of repeat III and IV decreased selectivity for Na^+ and increased the sensitivity to block by Ca^{2+} and other divalent cations to micromolar concentrations (wild type Na channel is inhibited by divalent cations in the mMolar range of concentrations). The mutant Na channel K1422E is even permeable to Ca^{2+} and Ba^{2+}, like Ca^{2+} channels. Furthermore, the channel carrying both mutations is not only permeable to Ca^{2+} and Ba^{2+}, but is also selective for Ca^{2+} over Na^+ at their physiological concentrations.

These results indicate that lysine 1422 and alanine 1714 are critical in distinguishing the Na channel from the Ca channel with respect to ion selectivity by causing a dramatic change in the ion selectivity of the sodium channel, from Na^+ selective to Ca^{2+} selective. These also suggest that these sites in the Na and Ca channels form part of the selectivity filter of these channels. In addition, these changes created a high-affinity site for calcium binding and block of monovalent ion conductance through the Na channel, as has been previously described for calcium channels. These results mirror the ion conductance properties of Ca channels and indicate that a key structure determinant of the ion selectivity difference between Ca and Na channels is specified by the negatively charged amino acid residues at the mouth of the putative pore-forming region.

The 19 amino acids of the P region of cloned voltage-gated K channels contain invariably a Gly-Tyr-Gly-Asp (G-Y-G-D) sequence motif towards the C-terminus (see Figure 4). *Shaker* K^+ channels are relatively insensitive to millimolar concentrations of external Ca^{2+} or Mg^{2+} (Heginbotham et al., 1992). Deletion of residues Tyr-Gly from the P region of *Shaker* K channel (YG, 15-16 in P, Figure 4) resulted in channels that produced voltage-activated currents in *Xenopus* oocytes, with some altered gating properties. However, the mutated channels were no longer selective for K^+ and display little selectivity among the alkali metal cations (Li^+, Na^+, K^+, or Cs^+) as the sole monovalent cation in the bath solution. Moreover, the deletion of YG also has created a relatively high affinity Ca^{2+} and Mg^{2+} blocking site in the ion channel pore (Heginbotham et al., 1992). Mutation of the deletion mutant channel by replacing Asp 447 with Glu (D447E) produced functional channels in oocytes with higher affinity to Ca^{2+} by almost ten fold (Figure 4, h; Heginbotham et al., 1992).

Interestingly, the homologous pore-forming region in Ca^{2+} channels reveals a similar motif of the Gly-Glu pair present in three of the four Ca channel domains, which matches the Gly-Asp pair in the K channel deletion mutant (Figure 4, g). Moreover, mutations involving the equivalent residue in Na channels result in Ca channel-like ion conduction properties (Heinemann et al., 1992b) (Figure 5). These results strongly suggest that these positions comprise at least part of the selectivity filter, and, as yet by an unknown mechanism, confer ion selectivity on the K, Na, and Ca channels.

The cyclic nucleotide-gated ion channels (CNG channels) from olfactory and retinal neurons reveal amino acid sequences that have a distant ancestral connection to the voltage activated ion channel. They also contain a pore-forming region that bears a striking resemblance to the equivalent P region of the K channels (Goulding et al., 1992). However, the YG pair is missing and the Asp 447 is replaced by glutamate (Heginbotham et al., 1992), similarly to the Ca channel. The CNG channels do not discriminate between Na^+ and K^+ ions and are blocked by the divalent cations Ca^{2+} and Mg^{2+}, like the mutant K channel described above. The accumulating evidence supports the view that the SS1-SS2 region of voltage gated ionic channels forms parts of the channel lining and comprises the ion selectivity filter, which is responsible for the high and selective ion conductance through the permeation pathway.

To summarize, the model in which P (in the K channel) is tucked into the plane of the membrane and lines the pore is supported by evidence that two amino acids, D431 and T449, strongly influence external TEA and CTX binding, and by the data that mutations in the intervening region alter selectivity but not external TEA binding. Thus, the model that envisages the sequence of 17 amino acids within P

(432-448 in *Shaker* K channels, Figure 4) as a β hairpin that extends about 0.2 nm into the membrane, is supported by these observations. The juxtaposition of β-hairpins from each subunit of the channel could form a central pore as a β barrel that spans the distance between the internal and external vestibules (Figure 3).

VOLTAGE-DEPENDENT ACTIVATION

Voltage-gated ionic channels control the permeability of the cell membrane to specific ions by opening or closing in response to changes in the potential difference across the plasma membrane (Hille, 1992). The voltage-dependent activation of ionic channels, leading to their opening, must involve transmembrane movements of charges that can sense the membrane electric field. Such voltage sensors must be charged or polar structures intrinsic to the channel protein which detect a change in membrane potential and respond to it by moving in the electric field, thereby causing conformational changes that lead to the channel opening. These charged residues are called the gating charges and their movements produce a measurable current called the gating current (for a review see Armstrong, 1992). The localization and identification of these charges within a channel protein are the subject of extensive study.

When the Na channel was cloned in Numa's laboratory (Noda et al., 1984, 1986) it was noticed that the nearly 2000 amino acid residues composing the channel protein contain four homologous internal repeats, each of which has six putative transmembrane segments (S1-S6, see Figure 1). Each of the four Na channel homologous repeats contains a highly conserved transmembrane segment, termed S4, which consists of 20 residues and contains repeats with the motif Lys/Arg-X-X (lysine or arginine followed by two usually nonpolar residues, see Figures 6 and 7), which has been postulated to form a critical element of the voltage sensor (Guy and Conti, 1990).

Present models of sodium channels predict that the S4 segment plays a critical role in the voltage dependent conformational changes that underlie channel gating. In support of this idea, analogous sequences to S4 have been found in Ca and K voltage-activated channels (Figure 7). The unique features of all S4-like sequences is the presence of four to eight positive amino acid residues spaced at three residue intervals along a putative helical segment (Figure 6), although recent models suggest that the α helix may be moving and propagating along the S4 segment upon membrane depolarization (Guy and Conti, 1990). These highly unusual structures are almost certainly part of the voltage sensor, and it seems highly likely that they move outward during activation of the channel (see Figure 6) (Yang et al., 1996; Larsson et al., 1996).

It is assumed that many of the positive charges in the putative voltage sensor segment S4 form ion pairs with negative charges in other transmembrane segments. The breakage of these ion pairs and the displacement of segment S4 within the transmembrane electric field may be the origin of the charge movement that accompanies the opening and closing of the channel pore and confers voltage

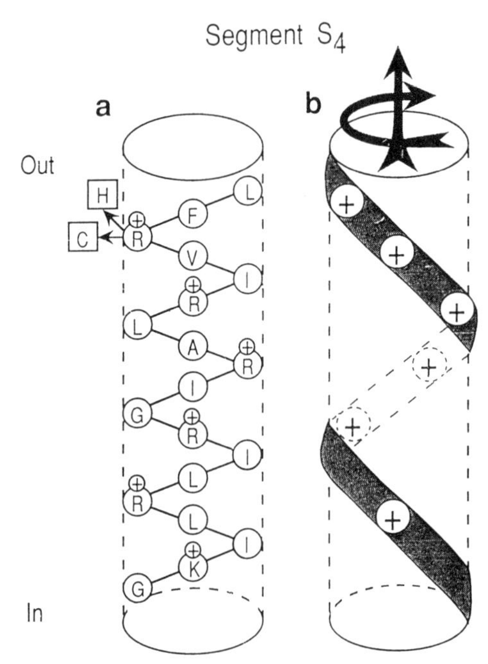

Figure 6. Representation of the putative S4 transmembrane α helix as a voltage sensor. The amino acid sequence of the S4 segment of repeat IV of the Na channel is depicted and presented by a single letter code (*a*). For more details and references see Figure 7. Positively charged amino acids are indicated by a (+) signs to highlight the proposed spiral arrangement of the charge distribution in the S4 helix. These positive charges, illustrated as a spiral ribbon around the cylinder-helix (*b*), are suggested to be paired with fixed negative charges on other transmembrane segments of the channel (see Figure 1). The S4 helix is proposed to move outward following depolarization of the membrane. According to the sliding helix model of voltage-dependent activation, S4 helix is proposed to undergo a spiral motion through a rotation of ~60° and outward displacement of ~0.5 nm (Catterall, 1988, 1992). The two known human mutations of the highly conserved Arg residue, shown to be involved in paramyotonia congenita (PC), are indicated by arrows (a, and see Figure 12).

dependence on this gating process. The negative charges thought to serve as counter-ions to the positive charges on S4 have not been identified with certainty. They are not found on a single helix but are, in some proposals, distributed across several of the other transmembrane segments (Guy and Seetharramulu, 1986). The need for charge pairing provides a strong constraint on models of the folding pattern of the peptide (Armstrong, 1992; Papazian et al., 1995).

One of the most attractive models to explain voltage-dependent activation is the "sliding" helix model (Catterall, 1988). The S4 segments are proposed to adopt a helical conformation, and the arginine residues form a spiral ribbon of positive

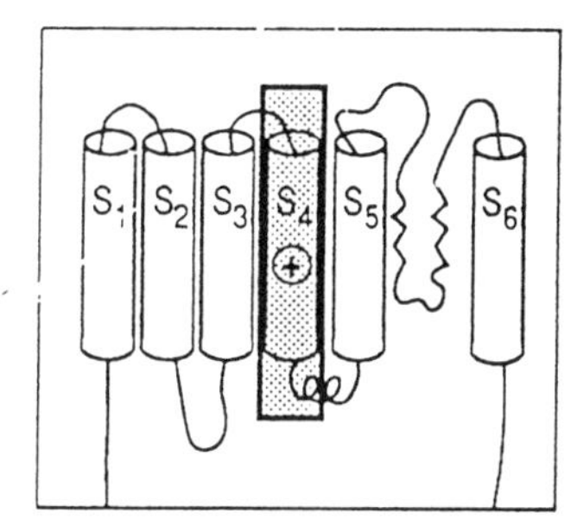

Figure 7. Sequence alignment of S4 segments in voltage-dependent ionic channels and map of site-directed mutagenesis experiments. Sequences of Na, Ca, and K channels are as in Figure 5. The S4 segment is marked according to Guy and Conti (1990). Highly conservative positively charged residues are boxed in heavy line and marked by +1 to +4. Non conserved positive charges are boxed in thin line. The shaded residues correspond to hydrophobic amino acids, substitution of which result in marked alterations in voltage dependent activation of the channel (*rows a-I, a-II*). The stars (*) with numbers indicate the highly conserved Leu residues in voltage gated channels (Leucine heptad motif, that continue in S4-S5 linker and S5 transmembrane segment, see Figure 8). Numbers above or under the mutations correspond to the position of the residue in rat brain Na channel II sequence (*rows a-I, a-II*) or *Shaker* K channel (*row K-a*).

Rows a-I and a-II present amino acid substitution introduced in Na channel sequences in repeats I and II, respectively, and row K-a corresponds to mutations in *Shaker* K channels, that result in alteration of voltage dependent activation properties of the channels. The upper part in row K-a corresponds to substitutions of charged residues and the lower part to mutations of hydrophobic residues in the K channel sequence.

The functional changes that result from the indicated substitution are as follows: (*a-I*) Amino acid substitutions that reduce positive charge by one unit (R220Q, R223Q, K226Q) or mutations K226D/E, that reduce the charge by two units result in a significant shift in activation (Stuhmer et al., 1989). (*K-a*) Neutralization of a basic residue (R368Q) or basic substitution with no charge change (R362K; R371K; R377K) and H378Q, reduce the voltage dependence of activation, suggesting that voltage-dependent activation involves the S4 sequence but is not solely due to electrostatic interactions (Papazian et al., 1991; Choi et al., 1993). Neutralization of the first two positive charges in the S4 sequence of K channel (R362Q; R365Q) change the gating valence of the voltage sensor, producing shifts of the activation curves along the voltage axis. Mutations R365 and K380 showed a linear relation of decreasing total gating valence, consistent with predominantly electrostatic behavior of these charged residues (Liman et al., 1991; Logothetis et al., 1992). Each of the conservative mutations of hydrophobic residues in S4 sequence of *Shaker* K channel (L358A; A359L; L361A; L366A; L375A; L382V) caused large shifts of the voltage dependence curve of activation, specifically alteration of the voltage-dependent properties (shaded residues) (Lopez et al., 1991). Substitutions of L382 with large hydrophobic residues (M,F,I, and V) produced positive shifts in activation gating (McCormack et al., 1993). These mutations emphasize the importance of the precise chemical interactions between S4 residues and the surrounding protein (Liman et al., 1991; Logothetis et al., 1992; McCormack et al., 1993). (*continued*)

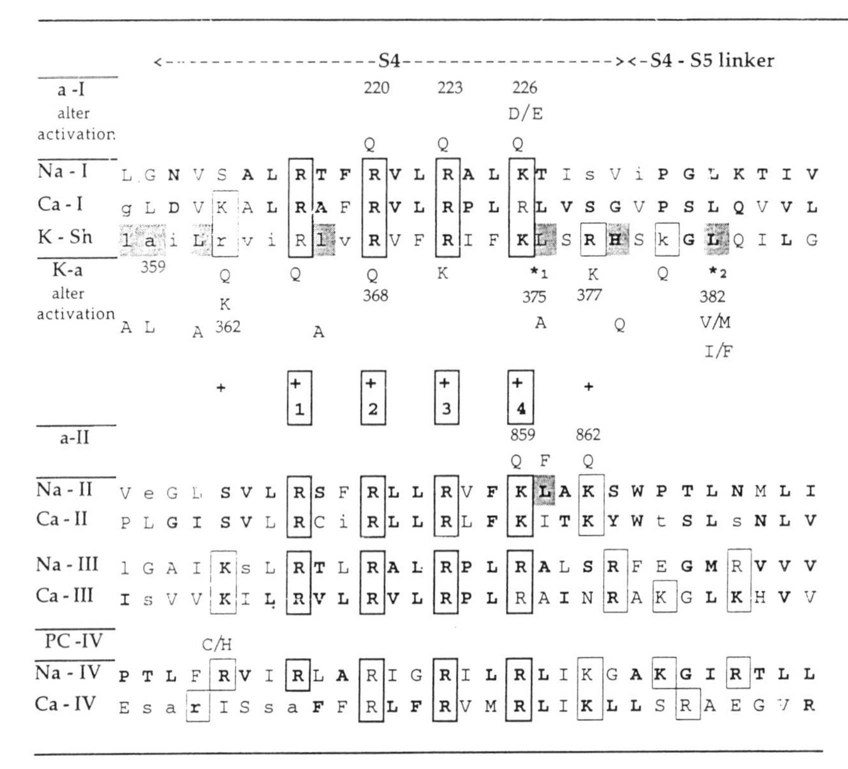

Figure 7. (Continued) (*a-II*) Neutralization of positive charges in the second repeat (K859Q; K862Q) (Stuhmer et al., 1989) or conservative substitution of the Leu (L860F) (Auld et al., 1990) results in altered voltage dependent activation of the Na channel. (*PC-IV*) This row indicates two human mutations in the skeletal muscle Na channel gene that results in paramyotonia congenita (PC) (Ptacek et al., 1992). The Arg residue (R1448 in S4 segment in domain IV of human skeletal muscle Na channel) is absolutely conserved in all known Na channel sequences among highly divergent species. The mutation replaces this Arg with either Cys or His, neutralizes this highly conserved positive charge in PC (see also Figures 6 and 12).

charges around the helix (Figure 6) that are stabilized in their transmembrane position by formation of ion pairs with negative amino acids in the surrounding segments, S1, S2 and S3. The positive charges are drawn into the cell by the force of the electric field at a typical resting potential of the cell, e.g., –80 mV. On depolarization, this force is relieved and the S4 segment slides outward along a spiral path, and then forms a new set of ion pairs. The sequential voltage-driven movement of the S4 helices are proposed to initiate sequential conformational changes in the four repeat domains of the ion channel, resulting in activation of the channel. This idea is supported by gating charge movements that move immediately on depolarization, whereas ionic current begins to flow only after a significant lag time (Armstrong, 1981; Catterall, 1988, 1992). However, the mechanism by which the S4 segment serves as voltage sensor is not yet known.

The opening of voltage-activated ionic channels is a very steep function of voltage and it was shown that the equivalent of about four to eight charges (gating charges) are translocated across the membrane field during activation of Na, Ca, and K channels (Zagotta and Aldrich, 1990a, b; Papazian et al., 1991; Hille, 1992). The magnitude of the charge movement on S4s has been estimated. Recent measurements of gate current fluctuations in sodium channels expressed in *Xenopus* oocytes reveal that a gating charge movement of 2.3 electronic charges cross the membranes (Conti and Stuhmer, 1989). Three such shots are needed to open Na channels. An identical estimate of the single gate charge has been obtained by a totally different method for A-type *Shaker* K channels (Zagotta and Aldrich, 1990b). The charge movement is coupled to the opening or closing of a putative gate; precisely how is still a major question. Analysis of gating currents provides a detailed description of the rates and voltage dependencies of the transitions among the conformational states. For a review see Armstrong (1992). However, to identify functional regions of the channel, the effects of site directed mutations on the channel functional properties had to be investigated.

Structural Elements that Influence Activation

Effects of Charged Substitutions in S4 Segments

Noda et al. (1984) postulated that the S4 segment acts as the voltage-sensing device responsible for sodium channel activation. The direct involvement of S4 in gating has been studied by introducing point mutations into the S4 region so that positively charged amino acid residues are replaced by neutral or negatively charged residues.

The residues in S4 at positions labeled +1, +2, +3, and +4 in Figure 7 appear to be especially important in determining the steepness of the voltage dependency (Stuhmer et al., 1989). Neutralization of one positive charge at positions +2, +3 or +4 by replacement of one arginine (R220Q, or R223Q,) or lysine (K226Q) residue in segment S4 of repeat I (in Na channel, Figure 7) by glutamine residue, causes a significant reduction in apparent single-gate charge for activation. Double mutations or mutations that reduced the net charge at position 226 (the last positive charge in segment S4 of repeat I) by two units (K226D or E), causes stronger reduction in the single-gate charge and yields a positive shift at about 20 mV in the potential dependence of activation. Reduction of two positive charges in segment S4 in repeat II, K859Q and K862Q, also resulted in a positive shift.

Thus, Stuhmer et al. (1989) have shown that reducing the net positive charge in segment S4 causes a decrease in the apparent gating charge, as manifested by the reduction in the steepness of the potential dependence of activation. This finding provides experimental evidence that the positive charges in segment S4 constitute at least part of the gating charge involved in the voltage-dependent activation of the sodium channel. The unique structural features of S4 segments are strikingly well conserved in the Ca channel and the K channels (Figure 7).

Site directed mutagenesis of the seven positively charged residues in the S4 sequence of a *Shaker* K channel and electrophysiological analysis indicate that voltage dependent activation involves the S4 sequence but is not solely due to electrostatic interactions (Papazian et al., 1991) (Figure 7). Basic substitutions that involve no change in charge on the S4 sequence altered the sensitivity of the channel to voltage: Replacing arginine at positions 362, 371 or 377 shifted the midpoint of activation in the depolarized direction: R362K caused a small shift, whereas R371K shifted the midpoint of activation by ~60 mV. In addition, the mutant R377K was substantially less sensitive to changes in voltage than its parent *Shaker* K channel (Figure 7, K-a) (Papazian et al., 1991).

Neutralization of the arginine at position +2 (R368Q, Figure 7) resulted in a large reduction in voltage-dependence and shifted the midpoint of activation by 53 mV in the depolarizing direction (Papazian et al., 1991). Substitution of arginine 362 by glutamine (R362Q, Figure 7, K-a [Liman et al., 1991]) causes an unexpected large decrease in the value of total gating valence, suggesting that R362, in addition to carrying part of the gating charge, is important for stabilizing the higher order structure of the voltage sensor, and that in the absence of R362 (located near the outer membrane surface in S4, see Figure 6) the remaining sensor charges are arranged in a different conformation which results in the greatly reduced value of total gating valence. In contrast, mutation of arginine 365 and lysine 380 (R365Q; K380Q, Figure 7) showed a linear relation of decreasing total gating valence, that can be attributed mainly to changes of the charge at those residues, consistent with predominantly electrostatic behavior of R365 and K380 (Liman et al., 1991; Logothetis et al., 1992).

Thus, the net positive charges in the S4 sequence reduced the sensitivity of the channel to voltage, as predicted by the hypothesis that the S4 sequence functions as a voltage sensor (Stuhmer et al., 1989; Papazian et al., 1991). However, many mutations of basic residues, replaced either with a different basic residue or the neutral amino acid, glutamine, result in a shift of the voltage dependence curves along the voltage axis. These suggest that, in addition to electrostatic interactions between the S4 sequence and the membrane electric field, interactions between the S4 sequence and other parts of the channel, and perhaps cooperativity between subunits, contribute to the sensitivity of the channel to voltage (Papazian et al., 1991; Lopez et al., 1991 and see below). Because the dielectric constant is likely to be small within the hydrophobic interior of a protein and a charge residue may have significant electrostatic interactions with residues over distances as large as 0.1–0.2 nm, mutations of the S4 basic residues have the potential of altering the long range of electrostatic interactions (Lopez et al., 1991).

Effects of Hydrophobic Substitutions in S4 and the Leucine Heptad Repeat

To examine the potential structural involvement of the S4 sequence in channel activation, conservative mutations of hydrophobic residues were made. The hydrophobic residues (mostly leucine) that are highly conserved among the various voltage-activated ionic channels from different species (Figure 7) are more likely to interact with other parts of the channel protein rather than with lipids in the membrane.

The S4 sequence of the *Shaker* K channels has 5 leucine residues. Mutations of each hydrophobic residue in the S4 sequence (Figure 7, shaded hydrophobic residues) gave rise to a large shift along the voltage axis in channel activation (McCormack et al., 1991; Lopez et al., 1991). Two of the six S4 mutants gave a hyperpolarizing shift while the remaining 4 gave a depolarizing shift. Substitution of the leucine 375 (L375A, Figure 7) gave the most dramatic depolarizing shift of 86 mV. There was no correlation between the position of the mutation in the S4 sequence and the direction of the shift. The effects of the hydrophobic mutations on the voltage- independent properties were mild or not significant and the S4 mutations did not affect the time course of macroscopic inactivation, indicating that these mutations did not grossly alter channel structure (Lopez et al., 1991).

Partially overlapping the S4 segment, continuing through the S4-S5 linker and into the S5 segment, is a sequence region in the *Shaker* family K channels, containing leucine at every seventh position. This leucine heptad motif is also present in voltage gated Na and Ca channels (Figures 7 and 8).

The equivalent position to leucine 375 in the S4 segment of the *Shaker* K channel (the first leucine in the heptad motif) is conserved in S4 in domain I and IV of all Ca channels and domain II and IV of all Na channels sequenced to date (Figure 7). Alteration of a single hydrophobic residue in the S4 sequence of the Na channel domain II, leucine 860 to phenylalanine, results in a 20–25 mV shift in the voltage dependence of the probability of channel activation (Auld et al., 1990). This shift is approximately equivalent to the shift in activation observed by Stuhmer et al. (1989) resulting from the combined neutralization of two positive charges, K859Q and K862Q, that are on each side of the L860 (Figure 7). The observation that alteration of leucine 860 causes a dramatic change in gating behavior, similarly to L375 of K channel, suggests that L860 is involved in critical interaction with one or more adjacent transmembrane helixes. Disruption of the possible leucine-heptad motif structure by substituting one of the leucine elements with other hydrophobic residues (see Figure 7) could inhibit the conformational changes or interactions required for channel opening.

The important structural role played by the leucines in the heptad region was further demonstrated by Hurst et al. (1992) who found the second leucine in the leucine-heptad repeat in *Shaker* K channels (leucine 382 [*2], Figure 7), that lies at the C-terminal end of the S4 segment, is well conserved in all voltage-gated K, Na and Ca channel genes isolated thus far. This implies that it plays an important

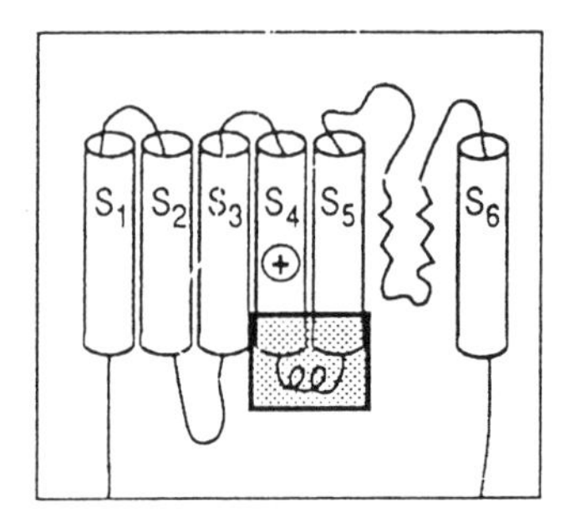

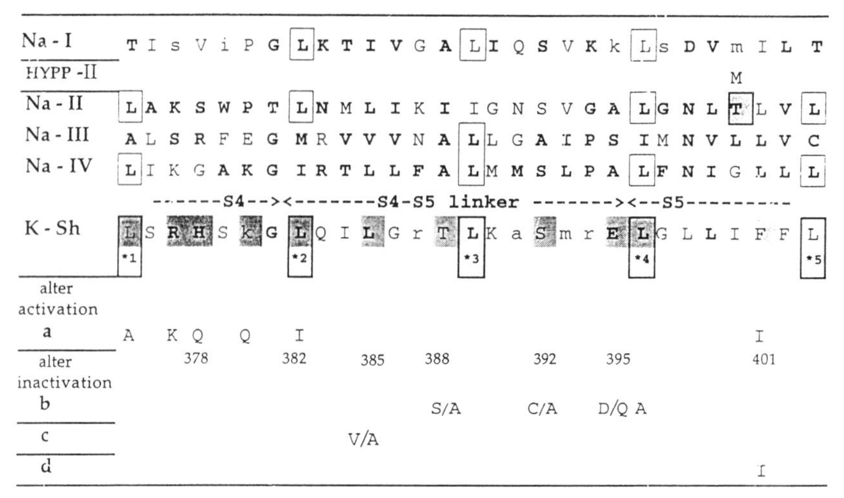

Figure 8. Amino acid alignment of sequences linking segments S4 and S5 (S4-S5 linker). Ending of S4 segment and beginning of segment S5 are indicated by broken lines, limited by > < signs. Boxed Leu residues indicate the highly conserved amino acids in voltage gated channels that consist of the Leu heptad motif, beginning in the S4 segment up to the S5 (see also Figure 7). Shaded residues indicate substitutions resulting in altered inactivation. Darkly shaded residues indicate substitutions that alter activation of the channel. The end and beginning of S4 and S5 segments are marked according to Pongs (1992).

Amino acids that are identical among Na channels in rat brain, skeletal and cardiac muscle subtypes, eel electroplax, and Drosophila (see Figure 5 for references), human brain, skeletal and cardiac muscle Na channel sequences (Ahmed et al., 1992), are indicated in bold-upper case letters; conservative amino acids (only one substitution among all Na channels) are in upper case and non conservative amino acids (two or more substitution) are in lower case letters. See Figure 4 for the Shaker K channel sequence.

Rows a–d present amino acid substitutions that resulted in functional alteration in activation (*a*) or inactivation (b–d) properties of the channels. Row HYPP-II presents the human mutation in skeletal muscle sodium channel, Thr to Met in hyperkalemic periodic paralysis (HYPP) (Ptacek et al., 1993), marked by boxed light shading. (*continued*) structural or functional role. A single point mutation of leucine 382 to isoleucine, (L382I, Figure 7), was made in the equivalent position of 1, 2, 3, or 4 domains of four covalently linked subunit domains of a rat homologue of *Shaker* K channel (Hurst et al., 1992). The depolarization required for channel activation increased approximately linearly with the number of isoleucine residues, indicating that these four subunits interact cooperatively during activation. Expression of this construct channel provides strong evidence that voltage dependent K channels have four subunits positioned symmetrically around a central permeation pathway that act cooperatively to open the channel in response to a voltage change and is stabilized by these leucine residues (Hurst et al., 1992).

Substitution of leucine 382 (*2) with large hydrophobic residues (Figure 7, L382M/F/I/V) always results in a positive shift in the voltage dependence of activation of 15–65 mV (McCormack et al., 1991; Schoppa et al., 1992; McCormack et al., 1993), i.e., toward a reduction in channel open probability at a given potential. Thus, a leucine at this position stabilizes the channel open conformation relative to close conformations, suggesting that this position maximally stabilizes a hydrophobic interaction with other hydrophobic residues in the channel protein, in a conformational state that leads to channel opening. Substitutions of this leucine also affect inactivation properties (see below).

To summarize, several studies have shown that mutations in the positively charged residues of S4 can lead to large changes in the voltage dependence of the channel gating. In Na channels, the substitution of a single residue in one of the

---

Figure 8. (Continued) (*a*) Mutations that alter voltage dependent activation in the C-terminal end of S4 segment are indicated by dark shading (see Figure 7 for references). Substitution of Phe401 by Ile (F401I) alter both activation and inactivation (Zagotta and Aldrich, 1990a; Gautam and Tanouye, 1990). (*b*) Mutations of two polar residues within the linker joining S4 and S5 segments, T388S/A and S392C/A, and of a non polar L396A, all decreased the relative amplitude of the fast component of macroscopic inactivation, as did mutation of the charged Glu residue E395D/Q (Isacoff et al., 1991). (*c*) Mutations of another hydrophobic residue, Leu 385 (L385V/A), increased the relative amplitude of the fast component of inactivation. A Phe substitution at L385 (L385F) did not alter the relative amplitude of the inactivation components but accelerated slow inactivation. Mutations at other positions in the S4–S5 linker had little effect on inactivation (Isacoff et al., 1991). (*d*) Mutation of a Phe to Ile in *Shaker* K channel (F401I) increases the voltage required to activate and inactivate the channel by approximately 20 mV and decreases the steepness of the voltage dependence of steady-state inactivation. The latencies until the channel opens following a voltage step are increased at low voltages (Zagotta and Aldrich, 1990a; Gautam and Tanouye, 1990). (*HYPP-II*) The mutation that alters inactivation in HYPP in domain II of skeletal muscle Na channel (T698M) (Cummins et al., 1993; Cannon and Strittmatter, 1993) and in an equivalent position in *Shaker* K channel (F401I, *row d*) are shaded (see also Figure 12).

four S4 segments resulted in voltage shifts of up to 30 mV and decreases in the voltage sensitivity of activation by as much as 40 mV (Stuhmer et al., 1989). In the case of K channels (Papazian et al., 1991; Liman et al., 1991; Logothetis et al., 1992), the substitution of uncharged or negatively charged residues for individual positively charged residues in the S4 segment of each subunit resulted in channels having voltage shifts up to 70 mV and reduction of voltage sensitivity to as little as half of normal. These charged residues are clearly important determinants in the gating of voltage-gated channels.

Surprisingly, mutations of hydrophobic residues within the leucine heptad motif that do not alter the number of charges in the S4 segment, have effects as large as those resulting from mutations that are presumed to alter the number of the channel voltage sensing charges. Conservative mutations of these leucines, especially the highly conserved second leucine in the leucine-heptad repeat (McCormack et al., 1991; McCormack et al., 1993) result in large voltage shifts and large reductions in voltage sensitivity of gating. Disruption of the possible leucine-heptad motif structure by substitution of one of the leucine elements with other hydrophobic residues (see Figure 7) could inhibit the conformational changes or interactions required for channel opening.

Coupling Between Activation and Inactivation

Single channel analysis has shown that most of the voltage dependence of the *Shaker* K channel resides in the activation process, and that inactivation is voltage independent over a large range of potentials (Zagotta and Aldrich, 1990a,b). Therefore, the voltage sensor of *Shaker* is likely to control primarily activation. Indeed, mutations in the S4 segment alter the voltage dependent properties of the channel and not the global channel structure.

The leucine heptad repeat that overlaps part of the S4 and S5 segments is conserved in most other voltage dependent channels. This suggested that it plays a role in communicating the movement of the S4 helix to the rest of the channel molecule. In support of this notion, substitution of a neutral amino acid for another, a phenylalanine change to isoleucine in *Shaker* K channel (F401I Figure 7) (Zagotta and Aldrich, 1990a; Gautam and Tanouye, 1990), within the region where the leucine repeat motif overlaps the S5 segment, resulted in altered channel voltage dependence for activation and inactivation (Gautam and Tanouye, 1990). The mutation increased the voltage required to activate and inactivate the channel by approximately 20 mV and decreases the steepness of the voltage dependence of steady-state inactivation (Figure 8) (Zagotta and Aldrich, 1990a). The effects of the mutation on both activation and inactivation provide further support that these processes are intimately coupled. Mutations that shift the midpoint of activation shift the midpoint of prepulse inactivation by about the same amount. These confirm that activation and inactivation are strongly coupled in the *Shaker* K channel; that is, most channels open before they inactivate (see also Figure 9). Thus, an alteration

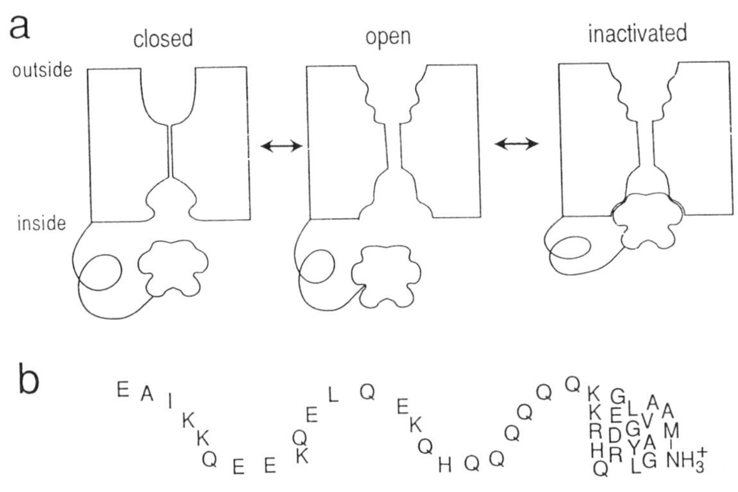

Figure 9. Scheme of mechanism of fast inactivation. Inactivation model of "ball and chain" implies that the inactivation "ball" is a piece of the protein, removable by pronase, that diffuses and binds within a cavity at the inner end of the channels' pore, only once the channel has entered the open state. (*a*) The extra- and intracellular cavities of the channel pore can undergo conformational changes when the channel enters the open state from the close or resting state, following a voltage change. The binding site for the inactivation ball becomes available after the channel has been activated and opened. The inactivation step, namely the binding of the ball to its receptor, is not voltage dependent and its apparent dependence on voltage arises because of the coupling between activation of the channel that exposes the inner receptor site, and the binding of the inactivation ball to the open conformation. Other inactivation mechanisms (slow inactivation and C-type), may result from conformational states (of the inner and/or outer parts) that do not involve the ball particle. See text for more explanation. (*b*) Ball (20 residues) and chain (~60 residues) have been suggested for rapidly inactivating K channels at the N-terminus of the peptide. The sequence presented in Figure 10 is schematically illustrated (see Figure 10 for references).

in the processes leading to opening will affect the inactivation of the channel as well.

Mutations of conserved uncharged residues in the heptad repeat have also been shown to influence the inactivation properties of *Shaker* channels, indicating that residues in this region might form a portion of the receptor site for the inactivation particle (McCormack et al., 1991) (and see below). Correlation between altered inactivation properties and single channel current amplitudes, resulting from substitution of other uncharged residues in the S4-S5 linker, has led to the proposal that this linker determines a portion of both the inactivation particle receptor and the ion permeation pathway (Isacoff et al., 1991).

Several conservative substitutions of highly conserved hydrophobic residues in the S4 but not in the other proposed membrane spanning segments have large and specific effects on the voltage dependence of channel opening. Thus it appears that slight alterations of the hydrophobic interactions between the S4 sequence and its surrounding, are likely to be due to local perturbations of the local interactions between the hydrophobic side chain of an S4 hydrophobic residue (leucine mostly) and its immediate surrounding in the closed and/or open conformation of the channel. These were sufficient to cause a substantial shift in the voltage dependence of channel activation (Schoppa et al., 1992).

These findings have three main implications: (1) functionally, the S4 sequence seems to be intimately involved in voltage dependent activation of the channel, (2) structurally, the S4 sequence seems to be buried within the channel protein so that most of its hydrophobic residues interact closely with other parts of the channel, and (3) mechanistically, the hydrophobic interactions between the S4 sequence and its immediate surrounding are likely to be different in the open and closed states of the channel, so even the most conservative mutations of the S4 hydrophobic residues severely alter the relative stability of the open and closed states (Lopez et al., 1991).

The S4 sequence is present in the second-messenger-gated channel that requires the binding of three or more molecules of cGMP to become activated, unlike voltage-gated channels (Jan and Jan, 1990). It seems likely that some of the hydrophobic residues, and perhaps even some of the basic residues that are highly invariant among the cyclic GMP-gated and voltage gated cation channels, correspond to the core structure of the protein and are important in establishing the basic architecture of the channel (Jan and Jan, 1990). This notion is supported by the finding of Rapaport et al. (1992) that synthetic peptide monomer with sequence corresponding to that of an S4 segment of a Na channel is readily aggregated within the membrane independently of the transmembrane diffusion potential. Thus, the S4 sequence probably arose in an ancestral channel. In addition to its function in the voltage sensing mechanism in some of the descendent voltage-gated channels, it probably forms part of the core structure of both voltage-gated and cyclic GMP-gated channels.

INACTIVATION

Several types of voltage dependent ion channels open upon depolarization (activation) and they spontaneously close, even when the depolarization is maintained (inactivation). These dual effects of depolarization were first described for the voltage gated Na channels that are responsible for the fast depolarizing phase of the action potential. Inactivation of ion channels is important in the control of membrane excitability. For example, the duration of the action potential is at least partly determined by the rate at which Na channels enter the fast inactivated state; delayed-rectifier K channels, which regulate action potential repolarization, are inactivated only slowly, whereas transient K channels encoded by the *Drosophila Shaker* gene (A-type), which affect action potential duration and firing frequency, have both fast and slow inactivation (Hille, 1992).

The *Shaker* K channels are among the most rapid inactivation type of K channels. Like voltage-activated Na channels, the transient *Shaker* K channel enters a long nonconducting state, the inactivated state, within a few milliseconds of opening. Single channel experiments have demonstrated that the inactivation process has little intrinsic voltage dependence and that it is fast and coupled to the activation process. Voltage independent inactivation implies no movement of charged particles through the electric field across the membrane for the inactivation conformational change, thus suggesting that domains that do not span the membrane are involved in inactivation (see Figures 1 and 9).

The Model for Inactivation

Both Na and *Shaker* K channel types inactivate after opening, and the inactivation of both can be removed by mild protease treatment of the intracellular side of the membrane. The similar action of intracellular trypsin on Na and K channels may indicate a conserved mechanism of inactivation among channels with different selectivity (Hoshi et al., 1990; Armstrong, 1981, 1992). The model of a blocking particle tethered to the rest of the channel was proposed many years ago by Armstrong and Benzanilla to explain the properties of inactivation of voltage activated Na channels (see Armstrong, 1992) and was picturesquely termed the "ball and chain" model. This model has three essential structural features: a positively charged inactivation particle (the ball), a polypeptide tether (the chain), and a negatively charged receptor site at the intracellular mouth of the channel that develops a high affinity for the inactivation ball when the channel is activated (see Figure 9).

In the ball and chain model, inactivation is caused by the reversible binding of this particle to the open state. It was proposed that fast inactivation of Na and K channels may result from the blocking of the permeation pathway by a positively charged cytoplasmic gate, an inactivation particle that can swing into position after part or all of the conformational change required for activation has taken place (Hoshi et al., 1990; Zagotta et al., 1990; Armstrong, 1992). The binding site or receptor for the ball is not available until the channel is open, or about to open, producing a coupling of activation and inactivation processes (see Figure 9).

Structural Domains Involved in Inactivation

N-Terminal Domains in K channels—the "Ball and Chain" Region

Genetic manipulation of *Shaker* K channels further supports the ball and chain model (Hoshi et al., 1990; Zagotta et al., 1990; Ruppersberg et al., 1991). The first 83 amino acids of the *Shaker* protein can be divided into two functional domains on the basis of the effects of deletions. Deletion mutation of the first 22 amino acids of *Shaker* K channel disrupts fast inactivation (Hoshi et al., 1990) (Figure 10), whereas deletion of amino acids at the carboxyl side leave inactivation intact. Mutations in the first domain, constituting approximately the first 19 residues, cause a significant slowing or complete removal of the fast inactivation process of the channel.

The first 19 amino acid are characterized by 11 conservative hydrophobic or uncharged residues followed by 8 hydrophilic residues, including four positively charged amino acids. This region is a good candidate for the inactivation ball. The macroscopic inactivation rate is progressively smaller upon decreasing the positive charge (see Figure 10). This is viewed as disrupting the positively charged character

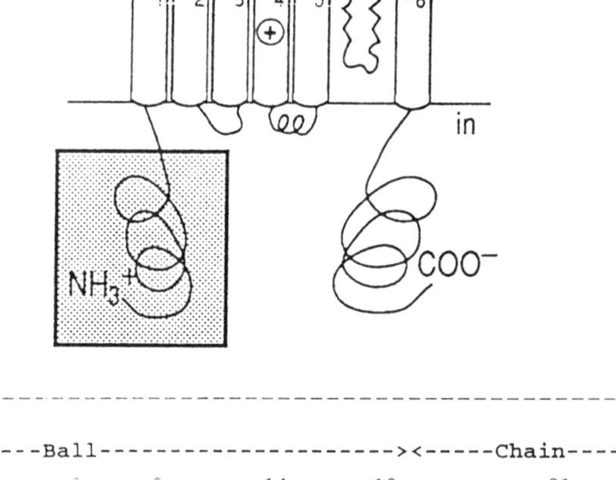

ShB M A A V A G **L** Y G L G E D R Q H R K K QQQQQHQKEQLEQKEEQKKIAE

a <--del---> Q Q Q Q

b R/E

Figure 10. Map of site-directed mutagenesis experiments on *Shaker* K channel N-terminal domain that revealed the "ball and chain" regions. Amino acid sequence of the first ~40 amino acids of *Shaker* (ShB) K channel is shown (Hoshi et al., 1990). The domains suggested to correspond to the "ball" and "chain" region in the model for fast inactivation are indicated above the sequence. Rows a and b present mutations or deletions that disrupt *Shaker* channel fast inactivation in the proposed ball domain (Hoshi et al., 1990). The indicated substitutions resulted in the following effects: (*a*) Neutralization of positive charges K18Q+K19Q and R14Q+R17Q slowed macroscopic inactivation. Deletion of four uncharged residues (6–9) also disrupt inactivation. (*b*) Changing the hydrophobic character of the first amino acid residues (adding a charge in position 7, mutant L7R/E, bold-underlined) also disrupts inactivation.

of the ball (Hoshi et al., 1990). Deletion of uncharged residues at position 6-9 or point mutations that replace the leucine 7 with charged residues, arginine or glutamate, still disrupts inactivation, indicating that conservation of the hydrophobic character in the first ten amino acids of the *Shaker* amino-terminal is important for normal inactivation (Figure 10). Disruption of inactivation by the mutations of hydrophobic residues (Figure 10) could be interpreted as disturbing the structure of the core of the ball (Figure 9).

The first 19 residues are followed by the second extended domain (at least 60 residue), comprising a sequence of hydrophilic residues, including a large number of potential trypsin sites, that make it a good candidate for the chain domain. Large deletions restricted to this region speed up inactivation, as might be expected for a mutation that shortened the chain, while insertions slow inactivation consistent with a longer chain. The effect of trypsin is similar to that of the deletion mutations in the amino terminal variable region that disrupt inactivation (Hoshi et al., 1990).

The basic plan of ten or more hydrophobic or uncharged residues followed by charged residues is repeated in all of the amino variants of alternatively spliced *Shaker* gene K channels. The various amino ends attached onto the core region generate channel variants with diverse inactivation kinetics when expressed in *Xenopus* oocytes (Schwartz et al., 1988). If the amino ends represent ball and chain domains, then the amino acid sequences in these variants form balls with different affinities for their receptor and chains with different lengths (Hoshi et al., 1990).

Although the deletion of 22 first amino acids resulted in no appreciable inactivation in 60 ms, most of the channels entered a long-lived inactivated state after 600 ms. This slow inactivation process may represent a second, fundamentally different, mechanism (Hoshi et al., 1990) (see below).

Restoration of inactivation. Zagotta et al. (1990) tested the ball and chain model by examining the effect of a peptide corresponding to the ball region (first 20 amino acids of *Shaker* K channel, see Figures 9 and 10) on the gating of mutant *Shaker* K channel expressed in *Xenopus* oocyte that contains, near its NH_2-terminus, a large deletion of 41 amino acids that effectively removes fast inactivation. According to this model, the putative ball region should be able to interact with the rest of the channel and produce inactivation even when it is not covalently attached to the rest of the channel protein.

Application of the *Shaker* peptide to the cytoplasmic side of the deletion mutant channels accelerates their inactivation rate. At each voltage, the rate of the macroscopic inactivation increases with increasing concentrations of peptide, as expected for a simple bimolecular reaction. In addition, the macroscopic inactivation rate at a given peptide concentration is dependent on the voltage. At the more positive voltages, where the channels activate more rapidly, peptide-induced inactivation occurs more rapidly. However, the rate of the inactivation transition induced by peptide observed in single channel recording is independent of voltage. These results indicate that the inactivation produced by the peptide is coupled to activation, as is the case for a normal fast inactivation process. Furthermore, a change in only a single amino acid residue, leucine 7 (L7E), can disrupt the ability of the peptide to induce inactivation (see also Figure 10). Thus, alterations that disrupt normal inactivation when made in the NH_2-terminal of *Shaker* also disrupt the peptide induced inactivation when made in the peptide (Zagotta et al., 1990; Toro et al., 1992).

These studies support the model describing the structural domain of the "ball" with the first 20 amino acids in the NH_2-terminus that interacts with parts of the open channel to cause inactivation, and is connected to the rest of the protein by a "chain" sequence of 60 or more amino acids that tether the inactivation ball near its receptor (Figure 9).

Coupling of Inactivation and Channel Blockade

Although the K channel is multimeric, one inactivating particle is probably sufficient to block the channel (Miller, 1992). Furthermore, internal TEA blockade produces a substantial slowing of the fast inactivation, indicating that a simple competitive relationship exists between TEA and the ball peptide (Choi et al., 1991; Toro et al., 1992). These results strongly suggest that the N-terminus behaves like an "inactivation particle" and directly occludes the pore to cause inactivation (Figure 9). However, TEA and the inactivation particle may bind to slightly different sites with some overlap that prohibits both from binding at once, since a site directed mutation of the *Shaker* channel that alters internal TEA affinity by 10-fold has little or no effect on fast inactivation (Yellen et al., 1991; Heginbotham et al., 1992) (Figure 4).

Further support for the similarity between the behavior of the rapid inactivation of *Shaker* K channels and the direct pore blockers (i.e., TEA and CTX) is provided by the observation that recovery from inactivation is speeded by increasing concentrations of permeant K^+ ions on the opposite (external) side of the membrane (Demo and Yellen, 1991), just as blockade can be relieved by transpermeant ions and was described for other pore blockers (see above). A "foot-in-the-door" model, in which the channel cannot close when occupied by a blocking or permeant ion, has been used to explain ion effects on gating in several types of ion channels. This model supposes that the blocker, like TEA, binds within a part of the pore that becomes constricted upon inactivation, thus preventing inactivation (Choi et al., 1991; Lopez-Barneo et al., 1993).

Single channel experiments show that the channel reopens from the inactivated state upon depolarization and that binding of the inactivation particle tends to keep the channel open (Demo and Yellen, 1991). These openings were usually required for recovery, as though the blocking particle must exit the pore before the channel can close (see Figure 9). Channels that are blocked by TEA are protected from inactivation. This finding is analogous to the ability of other blockers to prevent channel closing. Although neither blocked channels nor inactivated channels can carry current, the blocked channels are in rapid equilibrium with the pool of open channels and thus are only temporarily nonconducting. Inactivated channels, on the other hand, do not return rapidly to the open pool.

The nature of the amino terminus determines whether the K channel mediates rapidly inactivating or non inactivating K currents. However, Stocker et al. (1990) have shown that the local change in the amino terminal sequence of the *Shaker* channel may alter the quaternary structure of the K channels and thereby affect its sensitivity to external channel blockers like 4-aminopyridine, TEA and CTX (Stocker et al., 1990). The altered quaternary structure does not, however, affect the single-channel conductance and ion selectivity, which are properties of the pore. These results suggest that the amino terminus of *Shaker* proteins may affect K channel structures on both sides of the membrane (see also Baukrowitz and Yellen, 1995).

The behavior of the mutations in the voltage sensor S4 domain provides further evidence for coupling between blockade and gating. The mutation R377K and H378Q (Figures 7, 8), which affect the voltage dependence of activation, alter internal TEA and alkyl tail-TEA binding in a manner consistent with blockade being directly coupled to the voltage dependent channel opening (Hurst et al., 1992). Furthermore, the mutation R377K and several others in S4 that shift the voltage dependence of activation also shift that of inactivation by about the same amount (Papazian et al., 1991). These results all indicate that TEA blockade and inactivation are strongly coupled to channel activation and that TEA, alkyl-TEA, and the inactivation particle interact with regions of the channel that take part in the charge-moving conformational change required for activation in a similar manner (Hurst et al., 1992).

Measurements of gating currents of the *Shaker* channel indicate that the charge on the voltage sensor of the channel is progressively immobilized by prolonged depolarization. The charge is not immobilized in a mutant of the channel that lacks inactivation—removal of inactivation also removes gating charge immobilization (Benzanilla et al., 1991). These results can be interpreted that the inactivation particle not only blocks the conduction pathway but also prevents the return of the charged segments to their resting, closed positions. Thus, the region of the molecule responsible for inactivation interacts, directly or indirectly, with the voltage sensor to prevent the return of the charge to its original position (Benzanilla et al., 1991). The gating transitions between closed states of the channel appear not to be independent, suggesting that the channel subunits interact during activation.

To summarize, other molecules that interact with the internal mouth of the channel may behave like the inactivation particle (Choi et al., 1991). Perfusion of the internal side of the channel with TEA not only blocks the ionic current but also induces charge immobilization. This implies that the channel must open to be blocked by TEA and for immobilization of the total charge to occur (Figure 9). The amount of charges immobilized by the inactivation particle is about half of the total charge, indicating that at least two of the segments of the tetrameric channel can move back (or all four can move half-way back) while the inactivation particle is still in its blocking position; the rest of the charge can only move back after the ball has left its binding site (Benzanilla et al., 1991).

Receptor for Inactivation Gate: S4–S5 Linker

Several mutations in the linker joining S4 and S5 transmembrane segments in K channels (Figure 8) strongly affect the gating processes. In this linker, mutations of two polar residues (T388S/A and S392C/A, Figure 8) and of a nonpolar leucine (L396A), all decreased the relative amplitude of the fast component of macroscopic inactivation as did mutations of the charged Glu 395 to Gln or Asp (E395 D/Q, Figure 8, b). E395Q is the only mutation that affects inactivation among several glutamine substitutions of highly conserved acidic residues modeled to be on the cytoplasmic side of the membrane. Substitution at E395 greatly reduces the fast component of inactivation. The slower component is also affected (Isacoff et al., 1991; Baukrowitz and Yellen, 1995).

In contrast, substitution at L385 by a valine or alanine (Figure 8) increases the relative amplitude of the fast component of inactivation. L385 in *Shaker* K channels seems to have a special role of destabilizing the fast inactivation state, allowing for multiple openings during a sustained depolarization. Thus, replacing this residue with residues that have smaller side chains stabilized the fast inactivated state, suggesting that the processes of fast and slow inactivation interact (Baukrowitz and Yellen, 1995). The stabilization of the inactivated state in the L385A mutations may be explained by allowing tighter binding of the cytoplasmic inactivation gate to its receptor (see Figure 9). Mutations at other positions in the loop had little effect on inactivation. Other normal channel function were preserved in the S4–S5 mutants, indicating that the effects on inactivation were not due to global disruption of channel structure (Isacoff et al., 1991).

The mutations S392C and L385A (Figure 8) that affected inactivation also reduce the channel conductance for K^+, in contrast to the mutants that did not affect channel inactivation. Thus, those residues that are likely to function as a receptor for the inactivation gate probably lie near the permeation pathway, so that even the most conservative alteration of their side chains impedes the K^+ ion flux through the channel in both directions (Isacoff et al., 1991).

These observations are consistent with the proposal that the S4-S5 linker loop forms part of the inactivation receptor and that this receptor lies at or near the cytoplasmic mouth of the permeation pathway for K^+. Another portion of the permeation pathway, accessible to external and internal TEA is encoded by the P region, where amino acid substitutions affect channel conductance and ion selectivity (Figures 3, 4 and see Baukrowitz and Yellen, 1995).

The residues in the S4-S5 loop that affect channel conductance and activation are distributed in such a way that they would be on the same face of an α-helix whereas residues without effect would face in other directions. Isacoff et al. (1991) suggested that the S4-S5 loop may form an α-helix with one face lying near the cytoplasmic mouth of the pore where it forms part of the inactivation receptor. If the S4 segment moves outward in response to depolarization, then an accompanying conformational change in the attached S4-S5 loop may explain the tight coupling observed between activation and inactivation. Destabilization of the inactive state(s) could account for the increase in steady state open probability and the increased number of reopenings. Perhaps this occurs by decreasing the affinity between an inactivation particle and the vestibule of the pore in both K and Na channels (see Figure 9).

Other Inactivation Mechanisms

The cloning and functional characterization of the *Drosophila Shaker* (Sh) K channel family revealed variant transient A-type K channels with distinct activation and inactivation rates. The K channels are formed by proteins that share a common core region containing six-potential membrane spanning segments (S1-S6) flanked by diverse amino- and carboxyl termini located on the intracellular side of the membrane. All *Shaker* encoded K channels show similar properties of K selectivity, voltage dependence, and sensitivity to 4-aminopyridine, but are markedly different in their kinetics of inactivation and recovery from inactivation. By generating particular combinations of amino and carboxyl domains with the common core channel, both transient and non transient currents can be expressed from *Shaker* RNAs in *Xenopus* oocytes (Schwartz et al., 1988; Iverson and Rudy, 1990). The currents only differ in macroscopic inactivation kinetics. Analysis of currents produced by the various combinations suggest that the divergent amino domains resulted in striking differences in the probability of channel reopening, as observed in single channel recording.

Shaker K channels exhibit at least two types of inactivation mechanism, associated with separate domains of the molecule, referred to as N- and C-type inactivation. These are responsible for the differences in activation time course observed for the various K channels (Timpe et al., 1988; Stuhmer et al., 1989; Iverson and Rudy, 1990; Stocker et al., 1990; Hoshi et al., 1990). N-type inactivation involved the region near the amino terminus, which forms a tethered inactivation particle that can block the internal mouth of the channel, as described above (Hoshi et al., 1990; Zagotta et al., 1990). However, after removal of N-type inactivation by deletions in the amino terminus, another inactivation process remains, that was designated C-type, because it varies among the C-terminal alternatively spliced variants of *Shaker*. The two inactivation mechanisms have been shown to couple (Baukrowitz and Yellen, 1995).

The primary role of the carboxyl domains was suggested to influence the rates of recovery from inactivation. Mutations that abolish the fast inactivation of *Shaker* K channels by deletion of 20 first amino acids, produce a channel that remains open much longer, but is still able to inactivate during a very long pulse lasting seconds (Hoshi et al., 1990). This slow inactivation process may represent a second, fundamentally different mechanism which resembles the slow inactivation of delayed rectifier channels.

VanDongen et al. (1990) who made a series of deletions in the N- and C-termini of the delayed rectifier K channel with an unusually long C-terminus found that deletions in either terminus affected both activation and inactivation and that the kinetics were largely restored in truncated channels that had most of both termini removed. It is therefore concluded that the N- and C-termini are not essential for voltage dependent activation and inactivation of the slowly inactivated K channels, although they can have profound modulatory effects on these parameters. The C-terminus is suggested to be directly responsible for the functional changes caused by the N-terminus deletion. Since the delayed rectifier inactivates much more slowly than the Na channels and A type *Shaker* K channels, these results do not exclude the ball and chain model as a possible mechanism for fast inactivation.

Inactivation properties of many types of K channels have been shown to be altered by external application of TEA. The residue located at position 19 in the pore region of the K channels (Figures 3, 4) (Jan and Jan, 1990; Busch et al., 1991 and see above) determines the affinity to external TEA blockade. This residue is histidine in the equivalent position in K channel types that slowly inactivated (over hundreds of milliseconds) in several species. However, external TEA not only reduces the current through these K channels, but also markedly slows inactivation, suggesting that a channel which is open and blocked by TEA cannot inactivate. When histidine was replaced by tyrosine in the equivalent position of the homologous non-inactivating channel, the amount of inactivation was dramatically decreased, suggesting that this residue takes place in both TEA binding and channel inactivation (see Figures 3, 4) (Busch et al., 1991). The results suggest a structural basis for another type of mechanism for inactivation, involving an extracellular blockade of the channel pore. They also support the notion of coupling between gating and blockade of the channels (Baukrowitz and Yellen, 1995).

Choi et al. (1991) have demonstrated that *Shaker* K channels have two distinct inactivation processes that can be differentiated by their sensitivity to internal and external TEA. They tested the effects of both external and internal TEA on the remnant slow inactivation (C-type) of a *Shaker* deletion mutant in 6-46 N-terminal amino acids. In these channels where N-type inactivation is abolished, external TEA has been found to reduce the current and slows down the inactivation, while internal TEA simply reduces the current equally at all times. The remnant inactivation is, therefore, a separate process and involves a conformational change near the external mouth of the channel (Choi et al., 1991; Lopez-Barneo et al., 1993) (and see Figure 9). Replacement of threonine at position 449, near the outer mouth of the K channel (Figure 4, h) by glutamate or lysine leads to a 100-fold acceleration of inactivation. In contrast, substitution by valine results in channels that do not inactivate. The activation kinetics were not affected. Thus, slow (C-type) inactivation is a process which strongly depends on the amino acid at the outer position in the pore region (Lopez-Barneo et al., 1993; Baukrowitz and Yellen, 1995).

To summarize, the fast inactivation process appears to be due to the ball and chain mechanism, and internal TEA appears to compete with the binding of the ball. On the other hand, the slower inactivation revealed by disruption of the fast process is insensitive to internal TEA but slowed by external TEA (Hoshi et al., 1990; Choi et al., 1991). This is clearly different from the fast process but similar to the slow inactivation of delayed rectifiers K channels. It thus seems likely that slow *Shaker* inactivation and delayed rectifier K channel inactivation occur by a common mechanism, which is different from the specialized mechanism that produces the fast inactivation of *Shaker* K channels and Na channels (see below).

Na Channel Inactivation Domain—the III–IV Linker

The transient inward Na current that results from membrane depolarization is terminated within a few milliseconds by the process of inactivation. The duration of the action potential is at least partly determined by the rate at which Na channels enter the fast inactivated state (Hille, 1992). The inactivation of Na channels have been attributed to the short cytoplasmic segment that links homologous repeats III and IV of the Na channel (III–IV linker, see Figure 1) and is highly conserved in most neuronal and muscle sodium channels (Figure 11) (Gordon et al., 1988; 1990).

The critical role played by this short cytoplasmic linker III–IV in the fast inactivation process and Na channel gating is based on the findings that: (a) antibodies directed against a peptide sequence within this region delay the Na current inactivation when applied from the cytoplasmic surface and completely block fast inactivation of single sodium channels (see Figure 11) (Vassilev et al., 1988, 1989), (b) expression of the sodium channels as two polypeptides with a molecular cut between domains III and IV by co-injection of two cRNAs coding for repeats I–III and III–IV, which is analogous to removal of the linker region between III and IV, produces slowly inactivating Na currents in *Xenopus* oocytes (Stuhmer et al., 1989). Further, small insertions in this linker also slow inactivation (Patton and Goldin, 1991), and (c) similarly, inactivation can be removed by proteolytic treatment of native sodium channels, and the probable site of cleavage has been mapped to the linker between domains III and IV (Agnew et al., 1991), indicating together that this region, located on the cytoplasmic side of the membrane, is involved in the inactivation of the sodium channel. Site-directed mutagenesis techniques provides further evidence for the critical role of the III–IV linker in the inactivation process of Na channels.

The region corresponding to the synthetic peptide SP19 (see Figure 11, Ab SP19), which is known to inhibit channel inactivation (Vassilev et al., 1988, 1989), represents a portion of the intracellular loop between domains III and IV that contains the consensus site for protein kinase C viz., Lys-Lys-Leu-Gly-Ser-Lys-Lys. This highly conserved motif in Na channels (see Figure 11), suggests a common modulatory mechanism by PKC in the various tissues and species. The modulation of Na channels by stimulation of PKC in cells expressing Na channels

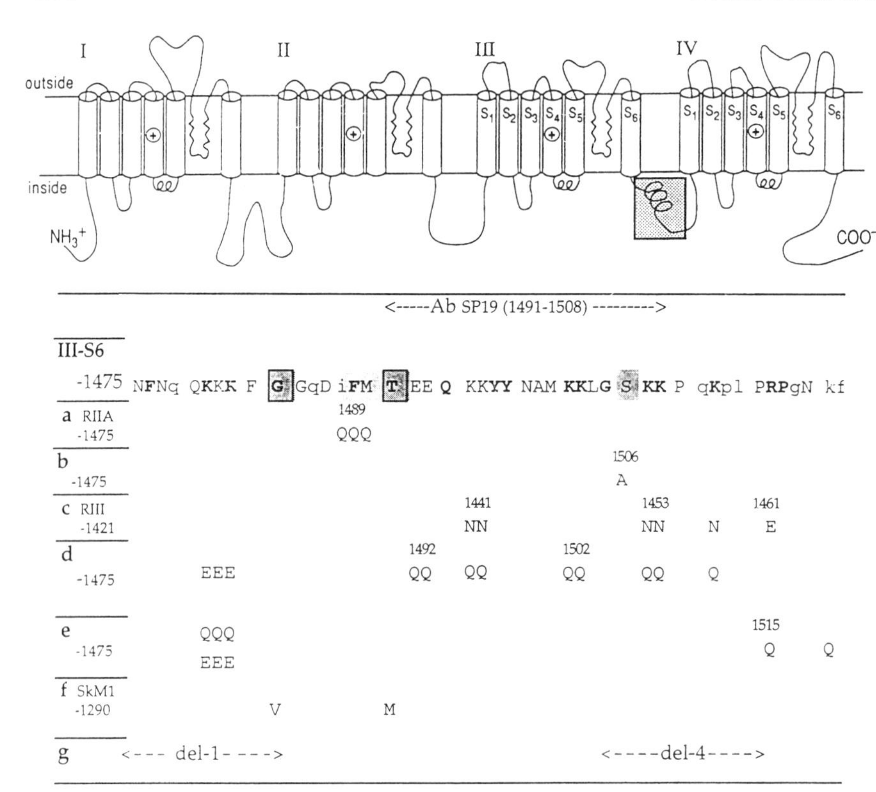

Figure 11. Map of site-directed mutagenesis on Na channel intracellular sequence linking repeats III and IV, and their functional effects. Amino acids residues, presented in a single-letter code, that are identical among Na channels in rat brain (RI, RII, RIII), rat skeletal and cardiac muscle subtypes, eel electroplax, Drosophila (see Figure 5 for references) and squid (Sato and Matsumoto, 1992), are indicated in bold-upper case letters; conservative amino acids are in upper case and non-conservative amino acids (two or more substitution among all channels) are in lower case letters.

Shaded residues indicate amino acid substitutions that result in marked slowing of the inactivation process. The numbers in the beginning of each row and above or under the substitutions correspond to amino acid position within the respective Na channel subtype. Sequence directed antibody (Ab SP19) that inhibits fast inactivation of Na channels (Vassilev et al., 1988, 1989) is indicated. Shaded boxed residues correspond to human mutations causing paramyotonia congenita (PC) (McClatchey et al., 1992) (see also Figure 12).

The substitutions presented in rows a–g result in the following functional changes in Na channel activity: (*a*) Inhibition of the fast inactivation. Substitution of three hydrophobic amino acid residues (IFM 1488-1490QQQ in rat brain Na channel IIA-RIIA) to glutamines and especially that of the Phe, dramatically inhibit the fast inactivation (West et al., 1992). (*b*) Mutation of the PKC phosphorylation site (in RIIA, S1506A) blocks the reduction of Na^+ current and the slowing of inactivation caused by activation of PKC (West et al., 1991). (*continued*) induces a rapid decrease in peak current and slowing of inactivation, without a change in the voltage dependence of activation. The reduction of Na^+ current and the substantial slowing of Na channel inactivation results from an increased lifetime of single Na channel openings and an increased probability of reopening of Na channels during prolonged depolarization (Numann et al., 1991). Site-directed mutagenesis in which serine 1506 was replaced with alanine (S1506A, Figure 11, b) blocks the reduction of Na current and the slowing of inactivation caused by PKC (West et al., 1991). Thus phosphorylation of a single serine residue in linker III–IV by protein kinase C is required for the effects of PKC, suggesting that the slowing of inactivation may be caused by a direct effect of phosphorylation on the function of this protein segment in inactivation gating.

In contrast to the ball and chain model, neutralization of all the charged residues in this III–IV linker, that contains at least 12 positively charged and 3 negatively charged residues, results in mutant channels that inactivate with kinetics similar to the wild type channel (see Figures 11, c, d, e). Thus, none of the charge residues in the linker III–IV are essential for fast inactivation of the Na channel (Moorman et al., 1990a; Patton et al., 1992). The kinetics of the single Na channel currents are consistent with a model in which Na channels enter a long lived inactivated state in a mainly voltage independent manner, and activation gating is strongly voltage dependent. Slowly inactivating Na channels do so because of reopenings. Point mutations that convert two or three positively charged residues (lysine) in the III–IV linker of a slowly inactivating Na channel (RIII, Figure 11, c) to asparagines (non-charged), produce faster inactivation of the Na channel (contrary to the

---

Figure 11. (*continued*) (c) Mutations that cause more rapid decay of current in the slowly inactivating rat brain Na channel III (K1441N + K1442N; K1453N + K1454N + K1457N; R1461E). Activation gating was notably slower for the R1461E currents (Moorman et al., 1990a). (*d*) Mutations that affect the voltage dependence of peak conductance. Mutant that neutralizes five net charges (in RIIA, E1492Q + E1493Q + K1495Q + K1496Q + K1502 + K1503Q + K1507Q + K1508Q + K1511Q) displayed a shift of about +9 mV. Mutations K1480E + K1481E + K1482E displayed a +12 mV shift in the voltage dependence of peak conductance (Patton et al., 1992). (*e*) Mutations that affect the voltage dependence of inactivation. The mutant RIIA channel (K1480Q + K1481Q + K1482Q) displayed a shift of about +6 mV. Mutations in which the Lys residues were replaced by Glu (E) residues, displayed a +11 mV shift in the voltage dependence of inactivation. Mutations R1515Q + K1519Q displayed a shift of about –7mV (Patton et al., 1992). (*f*) Mutations in human skeletal muscle Na channel 1 (G1299V; T1306M), substitution of which causes PC (McClatchey et al., 1992), see Figure 12. (*g*) Deletions that dramatically slow inactivation. Deletion 1 (deletion of 10 amino acid residues (N1475 - G1485 in RIIA) completely eliminated the fast inactivation. Deletion 4 (amino acid residues G1505 - P1515) displayed markedly slower kinetics of inactivation (Patton et al., 1992).

expectation from the ball and chain model) (Moorman et al., 1990a, b; Patton et al., 1992) (see above).

Although the charged residues are not critical for fast inactivation, mutations altering them do have significant effects on the voltage-dependence properties of the channel (see Figures 11, d, e). In particular, replacement of the three lysine residues in the amino end of the linker shifts the voltage dependence of peak conductance and fast inactivation in the positive direction. These results are consistent with the hypothesis that these charged residues interact electrostatically from the cytoplasmic side with a voltage sensor of the channel, presumably one of the S4 segments, as suggested for K channels.

Whereas the mutations affecting charged residues in linker III–IV did not have substantial effects on the kinetics of inactivation, deletions of 10 amino acids in the linker did markedly slow Na channel inactivation (Patton et al., 1992) (see Figure 11, g). Deletion of the first 10 amino acids completely eliminated fast inactivation, yet the channel still slowly inactivated with a time constant of several seconds, indicating that a much slower inactivation process is still present. Deletion of the fourth group of 10 amino acids slowed inactivation by about 70-fold. The effect was not simply shortening the linker, since deletion of a different group of 10 amino acids has no substantial effect on the kinetics of inactivation. The results with the deletion mutants demonstrate that the linker III–IV is critical for fast inactivation and are consistent with the idea that the III–IV linker is the inactivation gate of the Na channel. Point mutations in this region support this notion.

In contrast to the nonessential role of positively charged amino acids in linker III–IV, replacement of three contiguous hydrophobic amino acids in the amino terminal region of the linker with glutamine (I1488Q, F1489Q, M1490Q, Figure 11, a) removes fast inactivation nearly completely (West et al., 1992). In contrast to the mild effects of replacing Ile and Met, mutant F1489Q displayed greatly slowed, biphasic inactivation. The time course of decay of the Na current was almost as slow as for mutant IFM/Q3, identifying Phe-1489 as the critical amino acid residue within the hydrophobic cluster IFM. Single F1489Q channels demonstrated increased probability of reopening, evidently responsible for the noninactivating component of the Na current observed at the macroscopic level. This residue is proposed to be an essential part of the fast inactivation gate of the Na channel (West et al., 1992).

The results obtained by the hydrophobic mutations and the slow and incomplete inactivation seen with these mutations is consistent with the hypothesis that reducing their hydrophobicity destabilizes the interaction of the inactivation gate with its receptor (see Figure 9). This may cause open channels to inactivate more slowly and allow them to return to the open state after inactivation.

Two single mutations in this region of one glycine residues to valine and threonine residue to methionine (Figure 11, f) in the skeletal muscle Na channel cause a human muscular disorder, paramyotonia congenita (PC) (McClatchey et al., 1992) (see also Figure 12 and below).

West et al. (1992) suggested that the loop structure of the Na channel inactivation gate closely resembles the hinged-lid structures of allosteric enzymes that consist of structured loop of 10–20 residues between two hinge points and serve as rigid lids that fold over the active sites of allosteric enzymes to control substrate access (Joseph et al., 1990). By analogy, the linker III–IV may function as a rigid lid to control Na entry to and exit from the intracellular mouth of the pore of the Na channel. This hinged lid may be held in the closed position during inactivation by a hydrophobic latch formed by the hydrophobic cluster IFM. As in the ball and chain model, the hydrophobic cluster IFM serves as an essential component of the inactivation particle, which occludes the intracellular mouth of the activated Na channel. However, positively charged residues are not required. The effects of some peptide neurotoxins (e.g. α scorpion toxins, *Conus striatus* toxin, *Conus textile* δTxVIA [Gonoi et al., 1987; Catterall, 1992; Fainzilber et al., 1994]) that induce the slowing or inhibition of the Na channel inactivation process by binding to the external side of the channel, may resemble, by analogy, binding of allosteric ligands that produce a conformational change in the channel causing the lid to open the active site, namely the pore (West et al., 1992; Catterall, 1992; Fainzilber et al., 1994). α-scorpion toxin increased reopening in both fast and slowly inactivating Na channel subtypes (Catterall, 1992).

There are some analogies between the inactivation gating structures of the K and Na channels, although the inactivation gate of the Na channel is not at its amino terminus and their is no significant similarity in amino acid sequence. The linker III–IV links domains III and IV of sodium channel α subunit, and therefore is the amino terminus of domain IV. Its position between domain III and IV may be similar to the position of the much larger *Shaker* amino terminus in a homotetrameric *Shaker* K channel. Both positively charged and hydrophobic amino acid residues are essential components of the inactivation gate in K channels, although mutations of a single hydrophobic residue, Leu-7 (Figure 10), have the most dramatic effects on inactivation as was found for mutation of Phe-1489 (Figure 11) on Na channels (West et al., 1992).

The primary structure of the linker III–IV, which positions the IFM hydrophobic cluster only 14-amino acid residues from transmembrane segment IIIS6 (see Figures 11, 1, 9) in a loop that is predicted to have an ordered, partially α-helical structure (Guy, 1988) and is tethered at both ends, is most consistent with an ordered conformational transition that folds linker III–IV into the intracellular mouth of the transmembrane pore, similarly to the ball structure suggested for the K channels (Figure 9). A conformational change of this kind is consistent with the results showing that antibodies against linker III–IV that block inactivation can do so by binding to the noninactivated state of the channel, as if their binding site is made inaccessible by the process of inactivation (Catterall, 1992).

Mutations in Na Channel in Muscular Disorders Define Additional Structures Related to Inactivation

Genetic linkage studies have led to evidence indicating that SCN4A, the locus encoding the human skeletal muscle sodium channel α subunit, is the site of the primary defect in both paramyotonia congenita (PC) and hyperkalemic periodic paralysis (HYPP). These are two autosomal dominant disorders characterized by episodes of abnormal muscle membrane excitability (Ptacek et al., 1993). The evidence that at least two different muscular disorders may involve molecular defects in the Na channel gene presents a unique opportunity to correlate a disturbance in channel function with a structural change in a specific domain of the sodium channel protein.

Clinically, the two disorders are distinct. PC involves a mutation that leads to cold-induced myotonia (stiffness due to muscle membrane hyperexcitability). By contrast, during HYPP periodic attacks of muscle paralysis (associated with elevated serum K^+), the resting potential of a muscle fiber is markedly depolarized, which renders the sarcolemma electrically inexcitable and causes flaccid paralysis. In both disorders, electrophysiological studies have revealed an abnormal, TTX-blockable Na current in diseased muscle. In HYPP, depolarization of the muscle membrane appears to result from incomplete channel inactivation during which single channels open repetitively and the open time is prolonged (Cannon et al., 1991). By contrast, in PC an increased Na^+ flux is produced by cooling the muscle below 27°C (McClatchey et al., 1992).

Several mutations have been identified in the skeletal muscle Na channel gene. Two PC mutations in SCN4A are located in the III–IV cytoplasmic loop of the sodium channel (Figure 12). The Gly-Val alternation and, most likely, the Thr-Met substitution (Figure 11, f) are associated with a temperature-sensitive clinical phenotype. The Gly affected is present in this position in all sodium channels sequenced to date, including species as widely separated as man, *Drosophila* and squid (see Figure 11 for references). The extraordinary conservation of this Gly implies that it is of critical functional importance, probably giving flexibility to this domain of the sodium channel protein. McClatchey et al. (1992) suggest that the substitution of the more rigid valine for Gly stiffens this normally flexible domain of the molecule, restricting its movements in response to changes in transmembrane potential. At normal temperatures, this mutation has minimal clinical significance. However, even a minor drop in temperature may impede movement of the loop sufficiently to allow an abnormal Na^+ flux. The Thr residue affected in other PC mutations is also remarkably conserved (Figure 11), though the effect of substituting the nonpolar Met for the polar Thr is less predictable. The two mutations in the cytoplasmic loop may alter the relative stability of different conformational states (McClatchey et al., 1992).

A third PC mutation in domain IV occurs in the S3 segment, resulting in the replacement of a neutral leucine with a positively charged arginine (Ptacek et al., 1993) (see Figure 12). This will introduce not only a change in formal charge, but

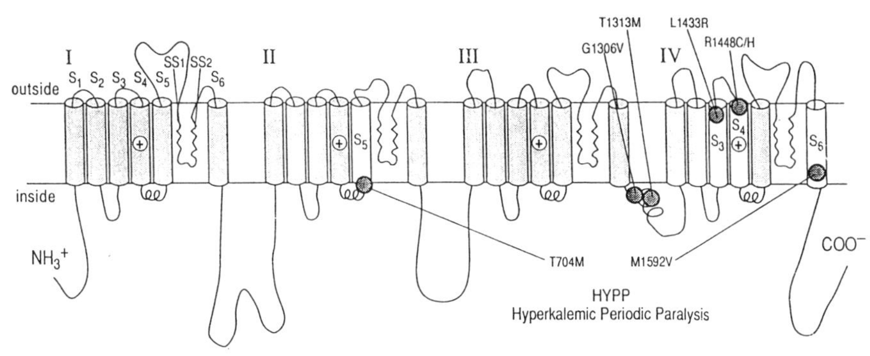

Figure 12. Sodium channel model with known PC and HYPP mutations. The location of each of the 7 mutations are shown within the channel's primary structure (indicated by the numbers) (Ptacek et al., 1993). Single letter codes for amino acids are used. The mutation position within the channel model is denoted by dark circles. Note that all identified mutations are localized near the extra- or intracellular face of the channel segments. For additional information, see Canon (1996).

also an increase in side chain size. These factors will affect charge distribution as well as packing density in this region of the molecule. Although this change does not occur in the S4 segment, it is located near the outer membrane surface and might be in close proximity to the S4. Change in local packing or in electrostatic or hydrophobic interactions conferred by this single substitution might explain alternations in the kinetics of processes such as channel activation and inactivation (Ptacek et al., 1993) (see Figure 12).

Two of the reported mutations in PC occur in an S4 segment of domain IV and neutralize one of the positively charged arginine residues (Figures 7, 12) (Ptacek et al., 1992, 1993). These mutations fit with the model suggesting that changing the net charge in a positive voltage sensor segment (S4) might alter the response to membrane depolarization. This model received support from studies in which positively charged residues in the S4 segment of Na or K channels were replaced by neutral or negatively charged amino acids, as discussed above. These mutations decreased steepness of the voltage dependence of activation (see Figure 7) (Stuhmer et al., 1989; Papazian et al., 1991).

In HYPP muscle, a small fraction of Na channels demonstrate abnormal inactivation kinetics in the presence of elevated extracellular K^+. These channels alternate between a normal kinetic mode and an abnormal one characterized by intermitted failure of inactivation and persistent openings during prolonged depolarization. The resultant noninactivating Na^+ current leads to persistent membrane depolarization, inactivation of normal Na channels, failure of action potential propagation and eventual muscle paralysis (Rojas et al., 1991; Ptacek et al., 1993).

The two naturally occurring mutations in the Na channel α subunit that have been reported in patients with HYPP (Rojas et al., 1991; Ptacek et al., 1993) define new functional domains of the sodium channel that are involved in the process of inactivation. Both HYPP mutations affect transmembrane regions far from the

III–IV cytoplasmic linker (see above). Both mutations are due to substitutions of a single nucleotide and cause rather conservative changes of one uncharged residue for another: threonine 698 to methionine residue in transmembrane segment S5 in domain II (T/M, see Figures 8, 12) (Ptacek et al., 1993), and methionine to valine at residue 1585 in S6 in domain IV (Figure 12). The amino acid substitutions occur at positions that are completely conserved in virtually all Na channel genes sequenced to date.

Structural models of Na channel suggest that both mutations involving HYPP lie near the cytoplasmic face of transmembrane segments (Figure 12). When either mutant was expressed in a mammalian cell line, macroscopic Na currents failed to inactivate completely and single channels showed repetitive openings with prolonged open times. Both mutations disturbed inactivation without affecting the time course of the onset of the Na current or the single channel conductance (Cannon and Strittmatter, 1993). Together, these portions of the protein may serve as the docking site for the inactivation gate. In this model the channel should activate normally. Inactivation would begin, but the conformation of the inactivated channel might not be as stable in the mutants where the interaction of the gate with the docking site has been altered, thereby leading to persistent sodium ion flux.

The T698M mutation substitutes a nonpolar neutral residue (Met) for a polar neutral residue (Thr) in the cytoplasmic end of S5 in domain II (Figures 8, 12). Analogous substitution of a neutral amino acid for another (F401I, Figure 8) in the homologous place in the *Shaker* K channel also influences inactivation (Figure 8), resulting in decreased steepness of the voltage dependence of steady-state inactivation (Zagotta and Aldrich, 1990a). Moreover, site directed mutations of polar (Ser 392) and nonpolar (Leu 396) neutral residues to alanine in the homologous S4-S5 cytoplasmic linker of the *Shaker* K channel (Figure 8) decreased the proportion of the rapidly inactivating current and hastened the recovery from inactivation (Isacoff et al., 1991). These effects were interpreted as destabilizing the inactive state by possibly altering the structure of a receptor for an inactivation particle.

Structural models of the Na channel place the S6 segment adjacent to the putative pore-forming segment between S5 and S6 (SS1 and SS2, see Figures 2, 3). Thus it is reasonable to imagine that the mutation near the cytoplasmic end of S6 in the Na channel, M1585V (Figure 12), also lies near the vestibule of the pore and interacts with an inactivation gate formed by the cytoplasmic linker between domains III and IV. The HYPP mutations, located near the inner end of the channel, could interact with this putative inactivation gate when it is closed. Its effect on Na channel inactivation may arise from a destabilization of this interaction. This may be the basis for the similarity of the functional defect in inactivation for mutation T698M in S5 of domain II and M1585V in S6 of domain IV (Cannon and Strittmatter, 1993). Additional ion-channel defects are described in Cannon (1996).

These neuromuscular disorders are interesting particularly since they provide clues to the relationship between the structure of the sodium channel and its function at the molecular, tissue and organismal levels.

Are Functional Properties Represented by Distinct Structural Domains?

The classical concept of the voltage-activated ion channel, formulated by Hodgkin and Huxley in the early '50s, postulated that the gating factors or gates, responsible for activation and inactivation of the channel were completely independent of each other: activation and inactivation were distinct parts of a single structure. Gating current measurements designed to test this idea provided clear evidence that this is not the case: no gating current was detected that had the time course of inactivation, predicted to exist by that model. Inactivation was shown to hinder the closing of the activation gate, thus reducing the flow of gating charge by producing charge immobilization, indicating coupling between activation and inactivation (see Armstrong, 1992 for a review and see above). Later studies provided more insight into the complex mechanisms of activation and inactivation.

Single channel recordings (Moorman et al., 1990a, b) suggested that individual channels can switch between two modes of gating, which might underlie the two phases of inactivation, namely the fast (normal) and slow one. The same channels actually spend part of their time in a "normal" gating mode that has rapid inactivation. A slowly inactivating mode is not absent but probably is a less common occurrence (Zhou et al., 1991). This idea, that the general gating behavior of Na channels might be modulated according to the equilibrium among gating modes, may shed some light on slowing of inactivation caused by peptide toxins (see above). This could be due to channels being preferentially found in a slowly inactivation mode when the toxin is bound. Zhou et al. (1991) who considered the effect of site directed mutations on channel inactivation, suggest the possibility that they might instead bias the distribution of gating modes and arise from a change in the equilibrium among gating modes instead of assuming that these would work directly through interference with the inactivation mechanism. The functional equilibrium is determined by interactions between all amino acid residues constituting the channel.

The process of channel activation involves a large rearrangement of charges, as expected from the steep voltage dependence of activation and also as observed in gating current measurements (Schoppa et al., 1992). In a single *Shaker* K channel, the rearrangement corresponds to the movement of about 12 elementary charges across the entire membrane field, or three charges per subunit. It is not clear at present if the mobile charges are carried by the S4 region in the form of a helix (Catterall, 1988; Durell and Guy, 1992) or, alternatively, by a large secondary structural change, e.g. the unwinding of a helical S4 segment (Guy and Conti, 1990). Whatever the case, if gating charge is carried by the movement of the S4 segment, it is reasonable to expect that residues that participate in this movement are likely to encounter different environments in the conformation corresponding to closed and open states of the channel. Indeed, modifications of uncharged residues in S4 can also cause shifts in the potential of half activation. This suggests that in addition to the overall charge of S4, the interactions between the residues, their properties, and the overall conformation are also important. This is to be expected because the equilibrium is determined by interactions between all residues (Stuhmer and Parekh, 1992).

In summary, mutational analysis of the S4 domain has shown that the S4 residues influence both channel activation and inactivation properties. Auld et al. (1988) observed that a mutation in the S4 cytoplasmic end in the second domain of the Na channel (F860L, Figure 7), near the S4–S5 linker produced roughly 5-fold lower currents. Heptad residues outside S4 (in S4–S5 linker) also produce voltage shifts when conservatively substituted; many of these shifts are as large as those from nonconservative substitutions or neutralizations of S4 charges (Figure 8). Thus, it appears that the S4 and heptad repeat regions may encompass one functional domain involving determination of the stability of open, closed and inactivated conformational states (see Figure 9).

Studies focusing on the III–IV loop of the Na channel and the analogous amino terminus of the K channel indicate that the III–IV linker is not restricted in its effects to inactivation. These studies indicate also that the S4 forms only part of the voltage sensor: other parts of the protein have to interact with the S4 segment. Indeed, conversion of one positive charge by replacing Arg by Glu in the linker region between domains III and IV in the Na channel delays activation (Moorman et al., 1990a) (R1461E, Figure 11, c), suggesting that activation is a property of several different regions of the channel and a structural interaction between inactivating domains and part of the channel involved in activation (Moorman et al., 1990a; Benzanilla et al., 1991; Stuhmer and Parekh, 1992). These findings support the concept of highly conserved elements that are involved in both activation and inactivation gating.

Finally, this chapter has stressed the similarities in structure and function among the voltage gated ion channels. However, it should be borne in mind that the Na, K and Ca channel types are in fact different in terms of structure and function; they differ in ion selectivity, sensitivity to toxins and drugs, and gating kinetics. Certain mutations in S4 in the Na channel indicate that activation and inactivation can be influenced independently. However, other mutations in S4 of the Na and especially the K channel, indicate that activation and inactivation are always modified in parallel. These results may suggest that at least some aspects of gating are coupled. Some may not be coupled, and may vary among the different K and Na channels, as may be implied by their different structures and functions.

SUMMARY

Molecular cloning of the cDNAs of different ion channels combined with altering channel structure using molecular biology techniques, coupled with detailed electrophysiological studies of heterologously expressed native and mutated ion channels has greatly expanded our knowledge of channel function. These enable the elucidation of structure-function relationships in ion channels. Although high resolution atomic structural information is not yet available, the accumulating evidence, including amino acid sequence hydropathy analysis, antibody and toxin binding mapping and various pharmacological and electrophysiological data, suggest that the channel is organized as shown in Figures 1 and 2.

Mutagenesis experiments have identified several functional domains of voltage-sensitive *Shaker*-like K channels and Na channels. Many different mutations in the region between S5 and S6 have suggested that this pore region (P) spans the membrane twice in a β sheet conformation and is involve in actual ion conduction (Figures 3–5). Mutations in the S4 segment, the S4–S5 linker and the S5 segment affect the voltage sensitivity of the channels (Figures 6–8). The fast inactivation of sodium and potassium channels has been shown to involve intracellular regions, the N-terminus in K channels and the III–IV linker in Na channels (Figures 9–11). This has been suggested to act as a "ball" to occlude the pore (Figure 9). Another type of inactivation has been clearly detected (slow inactivation) and, in K channels, is linked to the ion permeation process. Other domains related to inactivation have been discovered in studies carried out on human mutations involved in muscular disorders (Figure 12).

Although we have learned a great deal about the structural parts involved in key functions of voltage sensitive ion channels (such as activation, inactivation, single-channel conductance and selectivity, toxin block and modulation), our understanding of these channels is limited mostly to phenomenology. The understanding of channel function will require knowledge of the molecular and biophysical mechanisms.

REFERENCES

Agnew, W.S., Cooper, E.C., Shenkel, S., Correa, A.M., James, W.M., Ukomad, U.C., & Tomiko, S.A. (1991). Voltage-sensitive sodium channels: Agents that perturb inactivation gating. Ann. NY Acad. Sci. 625, 200–223.

Ahmed, C.M.I., Ware, D.H., Lee, S.C., Patten, C.D., Ferrer-Montiel, A.V., Schinder, A.F., McPherson, J.D., Wagner-McPerson, C.B., Wasmuth, J.J., Evans, G.A., & Montal, M. (1992). Primary structure, chromosomal localization, and functional expression of a voltage-gated sodium channel from human brain. Proc. Natl. Acad. Sci. USA 89, 8220–8224.

Armstrong, C.M. (1981). Sodium channels and gating currents. Physiol. Rev. 61, 644–682.

Armstrong, C.M. (1992). Voltage dependent ion channels and their gating. Physiol. Rev. 72(4), 5–13.

Armstrong, C.M. & Benzanilla, F. (1977). Inactivation of sodium channel. II. Gating current experiments. J. Gen. Physiol. 70, 567–590.

Auld, V.J., Goldin, A.L., Krafte, D.S., Catterall, W.A., Lester, H.A., Davidson, N., & Dunn, R.J. (1990). A neutral amino acid change in segment IIS4 dramatically alters the gating properties of the voltage-dependent sodium channel. Proc. Natl. Acad. Sci. USA 87, 323–327.

Auld, V.J., Goldin, A.L., Krafte, D.S., Marshall, J., Dunn, J.M., Catterall, W.A., Lester, H.A., Davidson, N., & Dunn, R.J. (1988). A rat brain Na+ channel α subunit with novel gating properties. Neuron 1, 449–461.

Backx, P.H., Yue, D.T., Lawrence, J.H., Marban, E., & Tomaselli, G.F. (1992). Molecular localization of an ion-binding site within the pore of mammalian sodium channels. Science 257, 248–251.

Barchi, R.L. (1988). Probing the molecular structure of the voltage-dependent sodium channel. Ann. Rev. Neurosci. 11, 455–495.

Baukrowitz, T. & Yellen, G. (1995). Modulation of K^+ current by frequency and external $[K^+]$: A tale of two inactivation mechanisms. Neuron 15, 951–960.

Beam, B.P. (1989). Classes of calcium channels in vertebrate cells. Ann. Rev. Physiol. 51, 367–384.

Benzanilla, F., Perozo, E., Papazian, D.M., & Stefani, E. (1991). Molecular basis of gating charge immobilization in Shaker potassium channels. Science 254, 679–683.

Bontems, S., Roumestand, C., Gilquin, B., Menez, A., & Toma, F. (1991). Refined structure of charybdotoxin. Common motifs in scorpion and insect defensins. Science 254, 1521–1523.

Busch, A.E., Hurst, R.S., North, R.A., Adelman, J.P., & Kavanaugh, M.P. (1991). Current inactivation involves a histidine residue in the pore of the rat lymphocyte potassium channel RGK5. Biochem. Biophys. Res. Commun. 179, 1384–1390.

Cannon, S.C. (1996). Ion-channel defects and aberrant excitability in myotonia and periodic paralysis. TINS 19(1), 3–10.

Catterall, W.A. (1986). Molecular properties of voltage-sensitive sodium channels. Ann. Rev. Biochem. 55, 953–986.

Catterall, W.A. (1988). Structure and function of voltage sensitive ionic channels. Science 242, 50–61.

Catterall, W.A. (1991). Structure and function of voltage-gated sodium and calcium channels. Curr. Opin. Neurobiol. 1, 5–13.

Catterall, W.A. (1992). Cellular and molecular biology of voltage-gated sodium channels. Am. J. Physiol. 72(4) 15–48.

Choi, K.L., Aldrich, R.W., & Yellen, G. (1991). Tetraethylammonium blockade distinguishes two inactivation mechanisms in voltage-activated K channels. Proc. Natl. Acad. Sci. USA 88, 5092–5095.

Choi, K.L., Mossman, C., Aube, J., & Yellen, G. (1993). The internal quaternary ammonium receptor site of Shaker potassium channels. Neuron 10, 533–541.

Christie, M.M., North, R.A., Osborne, P.B., Douglass, J., & Adelman, J.P. (1990). Heteropolymeric potassium channels expressed in Xenopus Oocytes from cloned subunits. Neuron 2, 405–411.

Cohen, S.A. & Barchi, R.L. (1993). Voltage-dependent sodium channels. Intern. Rev. Cytol. 137C, 55–102.

Conti, F. & Stuhmer, W. (1989). Quantal charge redistributions accompanying the structural transitions of sodium channels. Eur. Biophys. J. 17, 53–59.

Cummins, T.R., Zhou, J., Sigworth, F.J., Ptacek, L.J., & Agnew, W.S. (1993). Functional consequences of a Na channel mutation causing hyperkalemic periodic paralysis. Neuron 10, 667–678.

Demo, S.D. & Yellen, G. (1991). The inactivation gate and the Shaker K channel behaves like an open channel blocker. Neuron 7, 743–753.

Durell, H. & Guy, H.R. (1992). Atomic scale structure and functional models of voltage-gated potassium channels. Biophys. J. 62, 238–250.

Fainzilber, M., Kofman, O., Zlotkin, E., & Gordon, D. (1994). A new neurotoxin receptor site on sodium channels is identified by a conotoxin that affects sodium channel inactivation in molluscs, and acts as an antagonist in rat brain. J. Biol. Chem. 269, 2574–2580.

Fujita, Y., Mynlieff, M., Dirksen, R.T., Kim, M.S., Niidome, T., Nakai, J., Friedrich, T., Iwabe, M., Myiata, T., Furuichi, T., Furutama, D., Mikoshiba, K., Mori, Y., & Beam, K.G. (1993). Primary structure and functional expression of the w-conotoxin sensitive N-type calcium channel from rabbit brain. Neuron 10, 585–598.

Gautam, M. & Tanouye, M.A. (1990). Alteration of potassium channel gating: Molecular analysis of the Drosophila Sh 5 Mutation. Neuron 5, 67–73.

Goldstein, S.A.N. & Miller, C. (1992). A point mutation in a Shaker K channel changes its charybdotoxin binding site from low to high affinity. Biophys. J. 62, 5–7.

Gonoi, T., Ohizumi, Y., Kobayashi, J., Nakamura, H., & Catterall, W.A. (1987). Action of polypeptide toxin from the marine snail *Conus striatus* on voltage-sensitive sodium channels. Mol. Pharmacol. 32, 691–698.

Gordon, D. (1996). Sodium channels as targets for neurotoxins. In: Cellular and molecular mechanisms of toxin action (Lazarowici, P. and Gutman, Y., eds.). Harwood Press, Amsterdam. In press.

Gordon, D., Merrick, D., Wollner, D.A., & Catterall, W.A. (1988). Biochemical properties of sodium channels in a wide range of excitable tissues studied with site-directed antibodies. Biochemistry 27, 7032–7038.

Gordon, D., Moskowitz, H., Eitan, M., Warner, C., Catterall, W.A., & Zlotkin, E. (1992). Localization of receptor sites for insect-selective toxins on sodium channels by site-directed antibodies. Biochemistry 31, 7622–7628.

Gordon, D., Moskowitz, H., & Zlotkin, E. (1990). Sodium channel polypeptides in central nervous systems of various insects identified with site directed antibodies. Biochim. Biophys. Acta 1026, 80–86.

Goulding, E.H., Nagi, J., Kramer, R.H., Colicos, S., Axel, R., Siegelbaum, S.A., & Chess, A. (1992). Molecular cloning and single channel properties of the cyclic nucleotide-gated channel from catfish olfactory neurons. Neuron 8, 45–58.

Guy, H.R. (1988). A model relating the structure of the sodium channel to its function. In: Current Topics in Membrane and Transport (Agnew, W., ed.), vol. 33, pp. 289–308, Academic Press, San Diego.

Guy, H.R. & Conti, F. (1990). Pursuing the structure and function of voltage-gated channels. Trends Neurosci. 13, 201–206.

Guy, H.R. & Seetharramulu, P. (1986). Molecular model of the action potential sodium channel. Proc. Natl. Acad. Sci. USA 83, 508–512.

Hartmann, H.A., Kirsch, G.E., Drewe, J.A., Taglialatela, M., Joho, R.H., & Brown, A.M. (1991). Exchange of conduction pathways between two related K channels. Science 251, 942–944.

Heginbotham, L., Abramson, T., & MacKinnon, R. (1992). A functional connection between the pores of distantly related ion channels as revealed by mutant K channels. Science 258, 1152–1155.

Heginbotham, L. & MacKinnon, R. (1992). The aromatic binding site for tetraethylammonium ion on potassiumm channels. Neuron 8, 483–491.

Heinemann, S.H., Terlau, H., & Imoto, K. (1992a). Molecular basis for pharmacological differences between brain and cardiac sodium channels. Pfluegers Arch. Eur. J. Physiol. 4222, 90–92.

Heinemann, S.H., Terlau, H., Stuhmer, W., Imoto, K., & Numa, S. (1992b). Calcium channel characteristics conferred on the sodium channel by single mutations. Nature 356, 441–443.

Hille, B. (1992). Ionic Channels of Excitable Membranes. Sinauer, Sunderland, MA.

Hoshi, T., Zagotta, W.N., & Aldrich, R.W. (1990). Biophysical and molecular mechanisms of Shaker potassium channel inactivation. Science 250, 533–538.

Hurst, R.S., Kavanaugh, M.P., Yakel, J., Adelman, J.P., & North, A. (1992). Cooperative interactions among subunits of a voltage-dependent potassium channel: Evidence from expression of concatenated cDNAs. J. Biol. Chem. 267, 23742–23745.

Isacoff, E.Y., Jan, Y.N., & Jan, L.Y. (1990). Evidence for the formation of heteromultimeric potassium channels in Xenopus oocytes. Nature 345, 530–534.

Isacoff, E.Y., Jan, Y.N., & Jan, L.Y. (1991). Putative receptor for the cytoplasmic inactivation gate in the Shaker potassium channel. Nature 353, 86–90.

Isom, L.L., De-Jongh, K.S., Patton, D.E., Reber, B.F.X., Offord, J., Charbonneau, H., Walsh, K., Goldin, A.L., & Catterall, W.A. (1992). Primary structure and unctional expression of the beta-1 subunit of the rat brain sodium channel. Science 256, 839–842.

Isom, L.L., De-Jongh, K.S., & Catterall, W.A. (1994). Auxiliary subunits of voltage-gated ion channels. Neuron 12, 1183–1194.

Iverson, L.E. & Rudy, B. (1990). The role of the divergent amino and carboxxyl domains on the inactivation properties of potassium channels derived from the *Shaker* gene of Drosophila. J. Neurosci. 10, 2903–2916.

Jan, L.Y. & Jan, Y.N. (1990). A superfamily of ion channels. Nature 345, 672.

Joseph, D., Petsko, G.A., & Karplus, M. (1990). Anatomy of a conformational change: Hinged "Lid" motion of the Triosephosphate isomerase loop. Science 249, 1425–1428.

Kavanaugh, M.P., Hurst, R.S., Yakel, J., Varnum, M.D., Adelman, J.P., & North, R.A. (1992). Multiple subunits of a voltage-dependent potassium channel contribute to the binding site for tetraethylammonium. Neuron 8, 493–497.

Kavanaugh, M.P., Varnum, M.D., Osborne, P.B., Christie, M.J., Busch, A.E., Adelman, J.P., & North, A. (1991). Interaction between tetraethylammonium and amino acid residues in the pore of cloned voltage-dependent potassium channels. J. Biol. Chem. 266, 7583–7587.

Kayano, T., Noda, M., Flockerzi, V., Takhashi, H., & Numa, S. (1988). Primary structure of rat brain sodium channel III deduced from the cDNA sequence. FEBS Lett. 288, 187–194.

Kirsch, G.E., Drewe, J.A., Hartmann, H.A., Taglialatela, M., de Biasi, M., Brown, A.M., & Joho, R.H. (1992). Differences between the deep pors of K channels determined by an interacting pair of nonpolar amino acids. Neuron 8, 499–505.

Larsson, H.P., Baker, O.S., Dhillon, D.S., & Isacoff, E.Y. (1996). Transmembrane movement of the *Shaker* K^+ channel 54. Neuron 18, 387–397.

Liman, E.R., Hess, P., Weaver, F., & Koren, G. (1991). Voltage-sensing residues in the S4 region of a mammalian potassium channel. Nature 353, 752–756.

Liman, E.R., Tytgat, J., & Hess, P. (1992). Subunit stochiometry of a mammalian K channel determined by construction of multimeric cDNAs. Neuron 9, 861–871.

Logothetis, D.E., Movahed, I.S., Satler, C., Lindpaintner, K., & Nadal-Ginard, B. (1992). Incremental reductions of positive charge within the S4 region of a voltage-gated K channel result in corresponding decrease in gating charge. Neuron 8, 531–540.

Lopez, G.A., Jan, Y.N., & Jan, L.Y. (1991). Hydrophobic substitution mutations in the S4 sequence alter voltage-dependent gating in Shaker K channels. Neuron 7, 327–336.

Lopez-Barneo, J., Hoshi, T., Heinemann, S.H., & Aldrich, R.W. (1993). Effects of external cations and mutations in the pore region on C-type inactivation of Shaker potassium channels. Receptors and channels 1, 61–71.

Loughney, K., Kreber, R., & Ganetzky, B. (1989). Molecular analysis of the pare locus, a sodium channel gene in Drosophila. Cell 58, 1143–1154.

MacKinnon, R. (1991). Determination of the subunit stochiometry of a voltage-activated potassium channel. Nature 350, 232–235.

MacKinnon, R. & Miller, C. (1989). Mutant potassium channels with altered binding of charybdotoxin, a pore-blocking peptide inhibitor. Science 245, 1382–1385.

MacKinnon, R. & Yellen, G. (1990). Mutations affecting TEA blockade and ion permeation in voltage-activated potassium channels. Science 250, 276–279.

MacKinnon, R., Heginbotham, L., & Abramson, T. (1990). Mapping the receptor site for charybtotoxin, a pore-blocking potassium channel inhibitor. Neuron 5, 767–771.

McClatchey, A.I., Van den Bergh, P., Paricak-Vance, M.A., Raskind, W., Verellen, C., McKenna-Yasek, D., Rao, K., Haines, J.L., Bird, T., Brown, R.H., Jr., & Gusella, J.F. (1992). Temperature-sensitive mutations in the III–IV cytoplasmic loop region of the skeletal muscle sodium channel gene in Paramyotonia Congenitia. Cell 68, 769–774.

McCormack, K., Tanouye, M.A., Iverson, L.E., Lin, W., Ramaswami, M., McCormac, T., Campanelli, J., Mathew, M.K., & Rudy, B. (1991). A role of hydrophobic residues in the voltage-dependent gating of Skaker K channels. Proc. Natl. Acad. Sci. USA 88, 2931–2935.

McCormack, K., Lin, L., Iverson, L.E., Tanouye, M.A., & Sigworth, F.J. (1992). Tandem linkage of Shaker K channel subunits does not ensure the stochiometry of expressed channels. Biophys. J. 63, 1406–1411.

McCormack, K., Lin, L., & Sigworth, F.J. (1993). Substitution of a hydrophobic residue alters the conformational stability of Shaker K^+ channels during gating and assembly. Biophys. J. 65, 1740–1748.

Miller, R.J. (1992). Voltage-sensitive Ca^{2+} channels. J. Biol. Chem. 267, 1403–1406.

Miller, C. (1992). Ion channel structure and function. Science 258, 240–241. Curr. Biology. 2(11), 573–575.

Moorman, J.R., Kirsch, G.E., Brown, A.M., & Joho, R.H. (1990a). Changes in sodium channel gating produced by point mutations in a cytoplasmic linker. Science 250, 688–691.

Moorman, J.R., Kirsch, G.E., Vandongen, A.M.J., Joho, R.H., & Brown, A.M. (1990b). Fast and slow gating of sodium channels encoded by a single mRNA. Neuron 4, 243–252.

Newland, C.F., Adelman, J.P., & Tempel, B.L. (1992). Repulsion between Tetraethylammonium ions in cloned voltage-gated potassium channels. Neuron 8, 975–982.

Noda, M., Shimuzu, S., Tanabe, T., Takai, T., Kayano, T., Ikeda, T., Takahashi, H., Nakayama, H., Kanaoka, Y., Minamino, N., Kangawa, K., Matsuo, H., Raftery, M., Hirose, T., Inayama, S., Hayashida, H., Miyata, T., & Numa, S. (1984). Primary structure of Electrophorous electricus sodium channel deduced from cDNA sequence. Nature 312, 121–127.

Noda, M., Ikeda, T., Kayano, T., Suzuki, H., Takeshima, H., Kurasaki, M., Takahashi, H., & Numa, S. (1986). Existence of distinct sodium channel messenger RNAs in rat brain. Nature 320, 188–192.

Noda, M., Suzuki, H., Numa, S., & Stuhmer, W. (1989). A single point mutatuion confers tetrodotoxin and saxitoxin insensitivity on the sodium channel II. FEBS Lett. 259, 213–216.

Numann, R., Catterall, W.A., & Scheuer, T. (1991). Functional modulation of brain sodium channels by protein kinase C phosphorylation. Science 254, 115–118.

Papazian, D.M., Timpe, L.C., Jan, Y.N., & Jan, L.Y. (1991). Alteration of voltage-dependence of Shaker potassium channel by mutations in the S4 sequence. Nature 349, 305–310.

Papazian, D.M., Shao, X.M., Seoh, S.A., Mock, A.F., Huang, Y., & Wainstock, D.H. (1995). Electrostatic interactions of S4 voltage sensor in Shaker K^+ channel. Neuron 14, 1293–1301.

Park, C.S., Hausdorff, S.F., & Miller, C. (1991). Design, synthesis, and functional expression of a gene for charybdotoxin, a peptide blocker of K channels. Proc. Natl. Acad. Sci. USA 88, 2046–2050.

Park, C.S. & Miller, C. (1992). Interaction of charybdotoxin with permenant ions inside the pore of a K channel. Neuron 9, 307–313.

Patton, D.E. & Goldin, A.L. (1991). A voltage-dependent gating transition induces use-dependent block by totrodotoxin of rat IIA sodium channels expressed in Xenopus oocytes. Neuron 7, 637–647.

Patton, D.E., West, J.W., Catterall, W.A., & Goldin, A.L. (1992). Amino acid residues required for fast Na channel inactivation: Charge neutralizations and deletions in the III–IV linker. Proc. Natl. Acad. Sci. USA 89, 10905–10909.

Pongs, O. (1992). Molecular biology of voltage-dependent potassium channels. Physiol. Rev. 72(4) 69–88.

Ptacek, L.J., George, A.L., Barchi, R., Robertson, M., & Leppert, M.F. (1992). Mutations in an S4 segment of the adult skeletal muscle sodium channel cause paramyotonia congenita. Neuron 8, 891–897.

Ptacek, L.J., Gouw, L., Kwiecinski, H., McManis, P., Mendell, J.R., Barohn, R.J., George, A.L., Barchi, R.L., Robertson, M., & Leppert, M.F. (1993). Sodium channel mutations in paramyotonia congenita and hyperkalemic periodic paralysis. Ann. Neurol. 33, 300–307.

Rapaport, D., Danin, M., Gazit, E., & Shai, Y. (1992). Membrane interactions of the sodium channel S4 segment and its fluorescently-labeled analogues. Biochemistry 31, 8868–8875.

Rogart, R.B., Cribbs, L.L., Muglia, L.K., Kephart, D.D., & Kaiser, M.W. (1989). Molecular cloning of a putative tetrodotoxin-resistant reat heart sodium channel isoform. Proc. Natl. Acad. Sci. USA 86, 8170–8174.

Rojas, C.V., Wang, J., Schwartz, L.S., Hoffman, E.P., Powell, B.R., & Brown, R.H., Jr. (1991). A Met-to-Val mutation in the skeletal muscle sodium channel alpha-subunit in hyperkalemic periodic paralysis. Nature 354, 387–389.

Rudy, B. (1988). Diversity and ubiquity of K channels. Neuroscience 25, 729–749.

Ruppersberg, J.P., Schroter, K.H., Sakmann, B., Stocker, M., Sewing, S., & Pongs, O. (1990). Heteromultimeric channels formed by rat brain potassium channel proteins. Nature 345, 535–537.

Ruppersberg, J.P., Stocker, M., Pongs, O., Heinemann, S.H., Frank, R., & Koenen, M. (1991). Regulation of fast inactivation of cloned mammalian I-K(A) channels by cysteine oxidation. Nature 352, 711–714.

Satin, J., Kyle, J.W., Chen, M., Bell, P., Cribbs, L.L., Fozzard, H.A., & Rogart, R.B. (1992). A mutant of TTX-resistant cardiac sodium channels with TTX-sensitive properties. Science 256, 1202–1205.

Sato, C. & Matsumoto, G. (1992). Primary structure of squid sodium channel deduced from the complementary DNA sequence. Biochem. Biophy. Res. Comm. 186, 61–68.

Schoppa, N.E., McCormack, K., Tanouye, M.A., & Sigworth, F.J. (1992). The size of gating charge in wild-type and mutant shaker potassium channels. Science 255, 1712–1715.

Schwartz, T.L., Tempel, B.L., Papazian, D.M., Jan, Y.N., & Jan, L.Y. (1988). Multiple potassium channel components are produced by alternative splicing at the Shaker locus in Drosophila. Nature 331, 137–142.

Stocker, M., Pongs, O., Hoth, M., Heinemann, S.H., Stuehmer, W., Schroter, K.H., & Ruppersberg, J.P. (1991). Swapping of functional domains in voltage-gated K channels. Proc. Royal Soc. London Series B. Biological Sciences 245, 101–107.

Stocker, M., Stuhmer, W., Wittka, R., Wang, X., Muller, R., Ferrus, A., & Pongs, O. (1990). Alternative Shaker transcripts express either rapidly inactivating or noninactivating K channels. Proc. Natl. Acad. Sci. USA 87, 8903–8907.

Striessing, J., Glossmann, H., & Catterall, W.A. (1990). Identification of a phenylalkylanine binding region within the a1 subunit of a skeletal muscle Ca^{2+} channels. Proc. Natl. Acad. Sci. USA 87, 9108–9112.

Stuhmer, W. (1991). Structure-function studies of voltage-gated ion channels. Ann. Rev. Biophys. Chem. 20, 65–78.

Stuhmer, W., Conti, F., Suzuki, H., Wang, X., Noda, M., Yahagi, N., Kubo, H., & Numa, S. (1989). Structural parts involved in activation and inactivation of the sodium channel. Nature 339, 597–603.

Stuhmer, W. & Parekh, A.B. (1992). The structure and function of Na channels. Curr. Opin. Neurobiol. 2, 243–246.

Tempel, B.L., Papazian, D.M., Schwartz, T.L., Jan, Y.N., & Jan, L.Y. (1987). Sequence of probable potassium channel component encoded at Shaker locus of Drosophila. Science 237, 770–775.

Terlau, H., Heinemann, S.H., Stuhmer, W., Pusch, M., Conti, F., Imoto, K., & Numa, S. (1991). Mapping the site of block by tetrodotoxin and saxitoxin of sodium channel II. FEBS Lett. 293, 93–96.

Thomsen, W.J. & Catterall, W.A. (1989). Localization of the receptor site for a-scorpion toxins by antibody mapping. Implications for sodium channel topology. Proc. Natl. Acad. Sci. USA 86, 10161–10165.

Timpe, L.C., Schwartz, T.L., Tempel, B.L., Papazian, D.L., Jan, N.Y., & Jan, L.Y. (1988). Expression of functional potassium channels from Shaker cDNA in Xenopus oocytes. Nature 331, 143–145.

Toro, L., Stefani, E., & Latorre, R. (1992). Internal blockade of a Ca^{2+} activated K channel by Shaker B inactivating "Ball" peptide. Neuron 9, 237–245.

Trimmer, J.S. & Agnew, W.S. (1989). Molecular diversity of voltage-sensitive Na channels. Ann. Rev. Physiol. 51, 401–418.

Trimmer, J.S., Cooperman, S.S., Tomiko, S.A., Zhou, J., Crean, S.M., Boyle, M.B., Kallen, R.G., Sheng, Z., Barchi, R.L., Sigworth, F.J., Goodman, R.H., Agnew, W.S., & Mandel, G. (1989). Primary structure and functional expression of a mammalian skeletal muscle sodium channel. Neuron 3, 33–49.

VanDongen, A.M.J., Frech, G.C., Drewe, J.A., Joho, R.H., & Brown, A.M. (1990). Alternation and restoration of K channel function by deletion at the N- and C-termini. Neuron 5, 433–443.

Vassilev, P., Scheuer, T., & Catterall, W.A. (1989). Inhibition of inactivation of single sodium channels by a site-directed antibody. Proc. Natl. Acad. Sci. USA 86, 8147–8151.

Vassilev, P., Scheuer, T., & Catterall, W.A. (1988). Identification of as intracellular peptide segment involved in sodium channel inactivation. Science 241, 1658–1661.

West, J.W., Numann, R., Murphy, B.J., Scheuer, T., & Catterall, W.A. (1991). A phosphorylation site in the sodium channel required for modulation by protein kinase C. Science 254, 866–868.

West, J.W., Patton, D.E., Scheuer, T., Wang, W., Goldin, A.L., & Catterall, W.A. (1992a). A cluster of hydrophobic amino acid residues required fast sodium channel inactivation. Proc. Natl. Acad. Sci. USA 89, 10910–10914.

Yang, N., George, A.L., & Horn, R. (1996). Molecular basis of charge movement in voltage-gated sodium channels. Neuron 16, 113–122.

Yellen, G., Jurman, M.E., Abramson, T., & Mackinnon, R. (1991). Mutations affecting internal TEA blockade identify the probable pore-forming region of a potassium ion channel. Science 251, 939–942.

Yool, A.J. & Schwarz, T.L. (1991). Alteration of ionic selectivity of a K channel by mutation of the H5 region. Nature 349, 700–704.

Zagotta, W.N. & Aldrich, R.W. (1990a). Alterations in activation gating of single shaker A-type potassium channels by the Sh-5 mutation. J. Neurosci. 10, 1799–1810.

Zagotta, W.N. & Aldrich, R.W. (1990b). Voltage-dependent gating of Shaker A-type channels in Drosophila muscle. J. Gen. Physiol. 95, 29–60.

Zagotta, W.N., Hoshi, T., Aldrich, R.W. (1990). Restoration of inactivation in mutant of Shaker potassium channels by a peptide derived from ShB. Science 250, 568–571.

Zhou, J., Potts, J.F., Trimmer, J.S., Agnew, W.S., & Sigworth, F.J. (1991). Multiple gating models and the effect of modulating factors on the mI sodium channel. Neuron 7, 775–785.

RECOMMENDED READINGS

Armstrong, C.M. (1992). Voltage dependent ion channels and their gating. Physiol. Rev. 72(4) 5–13.

Barchi, R.L. (1988). Probing the molecular structure of the voltage-dependent sodium channel. Ann. Rev. Neurosci. 11, 455–495.

Beam, B.P. (1989). Classes of calcium channels in vertebrate cells. Ann. Rev. Physiol. 51, 367–384.

Cannon, S.C. (1996). Ion-channel defects and aberrant excitability in myotonia and periodic paralysis. TINS 19(1), 3–10.

Catterall, W.A. (1992). Cellular and molecular biology of voltage-gated sodium channels. Am. J. Physiol. 72(4) 15–48.

Catterall, W.A. (1988). Structure and function of voltage sensitive ionic channels. Science 242, 50–61.

Catterall, W.A. (1991). Structure and function of voltage-gated sodium and calcium channels. Curr. Opin. Neurobiol. 1, 5–13.

Cohen, S.A. & Barchi, R.L. (1993). Voltage-dependent sodium channels. Intern. Rev. Cytol. 137C, 55–102.

Hille, B. (1992). Ionic Channels of Excitable Membranes. Sinauer, Sunderland, MA.

Miller, R.J. (1992). Voltage-sensitive Ca^{2+} channels. J. Biol. Chem. 267, 1403–1406.

Miller, C. (1992). Ion channel structure and function. Science 258, 240–241; Curr. Biology. 2(11), 573–575.

Pongs, O. (1992). Molecular biology of voltage-dependent potassium channels. Physiol. Rev. 72(4) 69–88.

Snutch, T.P. & Reinez, P.B. (1992). Ca^{2+} channels: Diversity of form and function. Curr. Opin. Neurobiol. 2, 247–253.

Stuhmer, W. (1991). Structure-function studies of voltage-gated ion channels. Ann. Rev. Biophys. Chem. 2065–2078.

Stuhmer, W. & Parekh, A.B. (1992). The structure and function of Na channels. Curr. Opin. Neurobiol. 2, 243–246.

Chapter 13

Ligand-Gated Ion Channels in Vertebrates

NOMI ESHHAR

Principles of Medical Biology, Volume 7A
Membranes and Cell Signaling, pages 307–340.

ISBN: 1-55938-812-9

INTRODUCTION

Nerve cells interact and communicate with each other and with muscle cells primarily through electrical signals. Electrophysiological and pharmacological studies have shown that rapid neurotransduction events (less than a millisecond) at neuronal chemical synapses are mediated by specialized receptors containing integral ion-channels that are selectively permeable to either cations (sodium, potassium and calcium ions) or anions (chloride ions). These receptors are able to detect and convert neurotransmitter binding (and various ligands) into electric signals, i.e., induce alterations in membrane potential. Partial isolation and molecular cloning led to the structural characterization of the ligand-gated ion channel receptors. Biochemically, ligand-gated ion channels are multimeric proteins composed of homologous membrane spanning subunits. The subunits are assembled (most likely) into pentameric complexes arranged around a central ion-permeating pore. Inspection of primary protein sequences reveals several structural motifs common to all subunits cloned thus far. Receptor subunits are typified by a large extracellular NH_2-terminal region and four putative hydrophobic membrane embedded domains designated M1, M2, M3, and M4 segments.

The best known members of ligand-gated ion channels include the glutamate, glycine and GABA (γ-aminobutyric acid) receptors, and the muscle and brain nicotinic acetylcholine receptors (nAChRs). These receptors are widely distributed in vertebrates, play key roles in diverse physiological processes, and are implicated in the pathophysiology of various neurodegenerative disorders.

This chapter covers the diversity, distribution and molecular characteristics of ligand-gated ion channels in vertebrates, and focuses on their participation in integrative brain function.

TERMINOLOGY: LIGAND-GATED ION CHANNELS

In the nervous system, the term ligand-gated ion channels designates a group of pore-forming proteins, tightly embedded in the membrane, that allow ion transport across membranes in response to neurotransmitter binding. Hence, these proteins have been broadly classified as ionotropic receptors.

LIGAND-GATED ION CHANNELS

Ligand-gated ion channels serve as receptors for fast transmembrane signaling at postsynaptic membranes of nerve-nerve and of nerve-muscle chemical synapses. Once neurotransmitter molecules have been released from vesicles present at presynaptic nerve terminals, they diffuse across the synaptic cleft and interact with membranous receptors located postsynaptically. Neurotransmitter binding triggers conformational changes in the receptor protein which in turn results in channel opening and increased ion fluxes across the membrane (Figure 1). Specialized

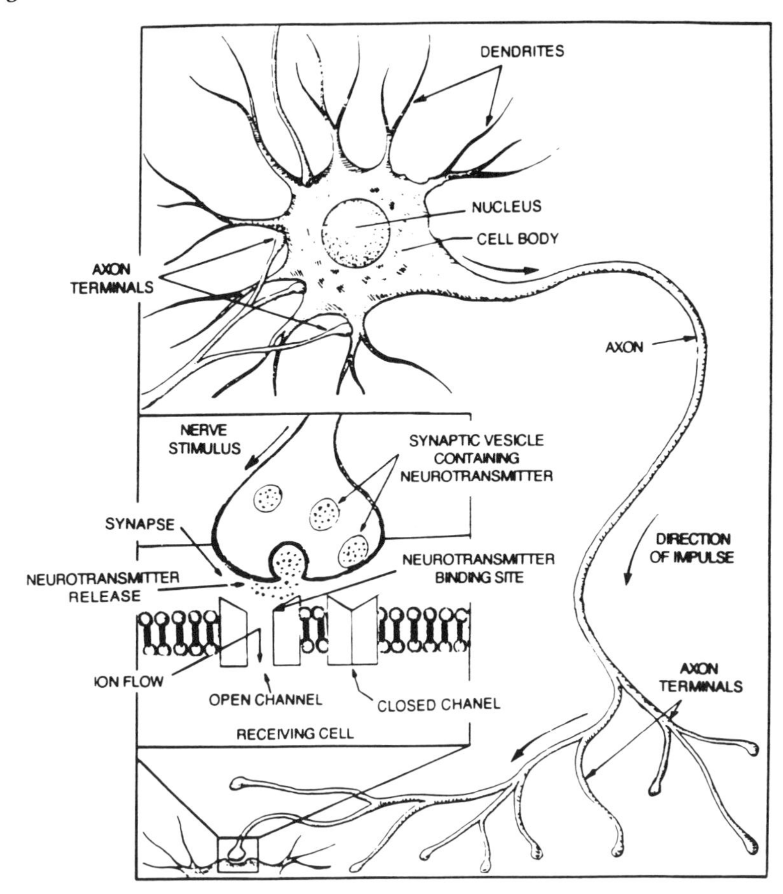

Figure 1. Schematic representation of signal transmission at chemical synapses. Transmission of electric impulses requires release of a neurotransmitter by the presynaptic nerve cell. Receptors containing integral ion channels located at post-synaptic terminals open in response to neurotransmitter binding and mediate ion flow across the post-synaptic membrane.

transmitter and ligand recognition sites are thought to be coupled directly or indirectly with the ionic channel. The passage of ions causes transient alterations in the electric properties of the postsynaptic cell membrane. Receptor activation produces either rapid excitatory or inhibitory postsynaptic potentials. At the resting membrane potential, fast excitatory neurotransmission is produced by membrane depolarizing responses resulting from a receptor-mediated increase in membrane permeability to certain cations, mainly Na^+, K^+, and Ca^{2+} ions. Rapid inhibition of neuronal firing, by contrast, is produced by membrane hyperpolarization responses resulting in activation of receptors forming chloride ion channels. The operation of ligand-gated ion channels can be monitored using electrophysiological methods.

NEUROTRANSMITTERS FOR RECEPTOR CHANNELS

Distinct amino acids are the prevalent transducers of most excitatory and inhibitory synaptic signaling in the mammalian central nervous system (CNS). At central synapses, ligand-gated Na^+ channels are activated principally by the dicarboxylic acid L-glutamate. L-glutamate is after glutamine, the most abundant amino acid (about 10 mM) in the CNS of vertebrates where it plays a key role in cerebral metabolism and is the likely neurotransmitter at most excitatory synapses. Glutamate was the first excitatory amino acid (EAA) to be recognized following the discoveries made by Hayashi (1954) and Curtis (1959) who showed that glutamate exerts convulsive effects and is able to excite single central neurons. Along with glutamate, several endogenous EAAs such as aspartate, quinolinate, and homocysteate are now recognized as candidate excitatory neurotransmitters. EAA pathways are abundant in the vertebrate CNS and include cortical association, corticifugal and subcortical pathways. Activation of ligand-gated ion channels is primarily induced by GABA and glycine. Both amino acids gate Cl^- channels of similar conductance properties but distinct pharmacologies. Glycine constitutes the major inhibitory neurotransmitter in the spinal cord and in the lower brainstem, and is less abundant in higher regions of the brain. It participates in regulation of both motor and sensory functions, thereby inducing skeletal muscle relaxation. GABA is the predominant inhibitory neurotransmitter in higher regions of the neuraxis, and is widely distributed in synaptic terminals throughout the vertebrate brain (at approximately 20% of CNS synapses). GABAergic neurotransmission includes cerebellar corticonuclear neuronal connections, and striatonigral, nigrothalamic, and nigrotectal pathways. Acetylcholine (ACh) is the transmitter used by motor neurons of the spinal cord and is the main excitatory neurotransmitter at the vertebrate neuromuscular junction where it initiates striated muscle contraction. ACh is abundant in neurons of the basal ganglia, but not a widespread transmitter in the brain.

COMMON CHARACTERISTICS OF LIGAND-GATED ION CHANNELS

The physiology, pharmacology, and anatomical organization of ligand gated ion channel receptors have been studied extensively over the past 30 years using a combination of different experimental techniques. These include electrophysiology, receptor binding studies, and immunocytochemistry. Very recent advances in protein purification and molecular cloning and expression technologies made molecular and genetic studies of these receptors possible. As evidenced by *in vitro* electrophysiological (patch-and voltage-clamp experiments) and receptor binding measurements, excitatory and inhibitory neurotransmitters in the CNS elicit their physiological activity via interaction with numerous receptor subtypes. Ligand gated-ion channel receptor subtype diversity was initially established by com-

Table 1. Ligand-Gated Ion Channel Subtypes of Vertebrate Neurons

Transmitter	*Receptor Type*	*Antagonists*	*Modulators*	*Ions Induced in Ionotropic Function*
GABA	GABA → BZ_1, BZ_2	Bicuculline Picrotoxin	Barbiturates Benzodiazepines Alcohol, Zn^{2+} Steroids Glutamate	Cl^-
Glycine	Glycine	Strychnine	—	Cl^-
Acetylcholine	Neuronal-nACh	Neuronal-bungarotoxin (at some sites)	—	Ca^{2+}, Na^+, K^+
Glutamate	NMDA	ADV, APH CPP, Ketamine MK-801, PCP	Glycine, Mg^{++} Polyamines, Zn^{++} Phencyclidine-like drugs	Ca^{2+}, Na^+, K^+
Glutamate	AMPA/Quis	CNQX, γDGG	—	Na^+, K^+
Glutamate	Kainate	CNQX, γDGG	—	Na^+, K^+

pounds that either mimic (agonize) or block (antagonize) the activity of the natural neurotransmitter. The effects of selective neurotransmitter analogues (agonists and/or antagonists) on neuronal membranes have distinguished distinct subclasses of receptors, named according to the prototype neurotransmitter-like agonist with which they interact preferentially (see examples in Table 1).

Ligands of the channel receptors that display high affinity and selectivity were indispensable tools for isolation, pharmacological characterization and localization of these receptor proteins. Anatomical distribution of the ligand gated ion-channel receptor binding sites in the CNS of vertebrates was examined in brain slices by radioligand binding studies and by quantitative autoradiography. Autoradiography demonstrated a differential subtype distribution throughout the brain, i.e., each receptor subtype has a unique pattern of neuroanatomical distribution. Nevertheless, auoradiographic observations were in good agreement with known target areas of excitatory and inhibitory pathways as predicted by electrophysiological measurements. Changes in receptor densities during ontogeny and in the diseased brain, provided strong evidence that they are involved in various physiological processes during normal brain development and play key roles in the etiology of several neurological disorders (examples are given later on). Recently, ligand-gated ion channels have been suggested to represent "receptive sites" in brain for ethanol, and have been examined in brains of chronic alcoholics with and/or with no associated disease.

Ligand gated-ion channel receptors contain recognition sites for a variety of modulatory drugs (some of which have broad therapeutic use, e.g., tranquilizers).

Neurotransmitter-mediated channel activity was found to be allosterically modulated by compounds that bind to the receptor at specialized regulatory sites, topographically *distinct* from the neurotransmitter recognition site. Allosteric interactions, taking place within receptor complexes, were ascribed mainly for $GABA_A$ (a GABA receptor subtype) and NMDA (a glutamate receptor subtype) receptors (see Table 1 for receptor subtypes, and for the muscle nicotinic acetylcholine receptor; allosteric modulations are described in more detail further on). The modulatory actions of drugs, i.e., potentiation and/or attenuation of neurotransmitter-elicited channel activity, were determined primarily electrophysiologically-complemented by binding of radioligands for modulatory sites.

Biochemical analyses of affinity purified receptors have shown that ligand-gated ion channels are composed of a number of homologous polypeptide subunits. Subunit composition of partially purified receptor preparations was analyzed by dissociation of the receptor complexes in a denaturing detergent and subsequent gel electrophoresis separation, and with subunit-specific monoclonal antibodies. Subunits responsible for ligand recognition were identified by photoaffinity labeling procedures, based on ultraviolet illumination-induced incorporation of radiolabeled ligands into membrane-bound receptors (receptor subunit composition and subunits carrying neurotransmitter binding sites are described in Table 2). Use of reversible cross-linking agents and monoclonal antibodies made against specific glycine receptor subunits indicated a pentameric structure for the glycine receptor channel. These observations resembled the subunit organization of the muscle nAChR (which was known to contain five membrane-spanning subunits), and were the first indication that different ionotropic receptor families are members of a receptor superfamily. Immunological studies performed with subunit specific antibodies (antibodies that do not recognize the ligand binding recognition site) revealed common antigenic determinants shared between subunits assembled within a given receptor complex, and have provided the first evidence for amino acid sequence homologies between subunits. For example, monoclonal antibody cross-reactivity indicated homology between glycine receptor α and β subunits and between subunits of the GABA ionotropic receptor. Such studies have suggested that receptor subunits are evolutionarily related.

Biochemical isolation and purification of ion channels facilitated the first steps in molecular cloning of the individual receptor subunits. The primary structure of channel receptor subunits has been deduced from cloned complementary DNAs encoding ligand-gated ion channels. Molecular cloning strategies based on partial peptide sequence of a purified protein and/or on *in vitro* functional expression of cloned cDNAs in a *Xenopus* oocyte system have been adopted by several research groups attempting to clone ionotropic receptors. Tryptic cleavage of affinity-purified receptor protein subunits, followed by HPLC separation of the peptide fragments, generated several amino acid sequences used for the design of synthetic oligodeoxyribonucleotide probes. Brain cDNA libraries were screened with the synthetic probes and positive hybridizing clones have been isolated and sequenced.

Table 2. Subunits of Ligand-Gated Ion Channels

Receptor	*Subunit*	*Binding Site*	*Stoichiometry*	*Molecular Weight*	*Variant*
nAChR Torpedo-electric-organ	α	ACh, α-bungarotoxin	$\alpha_2\beta\gamma\delta$ in the embryo, $\alpha_2\beta\delta\varepsilon$ in the adult	40 kD	α_2–α_5
	β	—		48 kD	β_2–β_4
	γ	—		58 kD	
	δ	—		64 kD	
				~250–290 kD	
Neuronal-nACh	α	ACh	$\alpha_2\beta_3$		α_2–α_7
	β	—			β_2–β_4
				~300 kD	
Glycine	α	Glycine, Strychnine	?$\alpha_2\beta_3$	48 kD	α_1–α_4
	β	—		53 kD	
	93 kD			~225 kD	
$GABA_A$	α	Benzodiazepines, GABA, Flunitrazepam	?	50 kD	α_1–α_7
	β	GABA Muscimol required for high affinity		55 kD	β_2–β_3
	γ	benzodiazepine binding		50 kD	γ_1–γ_3
	δ	—		~220–355 kD	
	ρ_1	—			
NMDA	NMDAR1 NMDAR2A NMDAR2B NMDAR2C NMDAR2D		?	~112 kD	NMDAR1-1a, 2a, 3a, 4a, 1b, 2b, 3b, 4b
AMPA/ Kainate	GluR-1(A) GluR-2(B) GluR-3(C) GluR-4(D)	all subunits bind AMPA (low affinity-binding for kainate)	?	~110 kD ~500 kD	flip & flop spliced variants for each subunit
Kainate	KA-1 KA-2 GluR-5 GluR-6 GluR-7	all subunits bind kainate	?		

While technical difficulties have impeded receptor purification, cDNA cloning was performed through functional expression techniques. mRNAs transcribed from cDNA clones were injected into *Xenopus* oocytes and protein expression was tested by the measurement of electrophysiological responses to ligand application.

Deduced amino-acid composition demonstrates a high degree of sequence homology among subunits of a particular receptor type, and a significant overall amino acid identity between subunits of different ligand gated-ion channel members. For example, 30–40% sequence homology is shared between $GABA_A$ receptor subunits (α, β, γ, and δ subunits), 47% identity between glycine receptor subunits, and 35% between the α subunits of the inhibitory receptors $GABA_A$ and glycine. $GABA_A$ and glycine receptor subunits have 15–24% sequence homology with the nAChR α subunits. By contrast, there is a poor overall homology between glutamate receptor subunits and the AChR family. Therefore, it is thought that the three receptor types, the glycine, GABA and nACh receptors constitute a protein superfamily evolved by gene duplication from a common ancestor early in phylogeny and that the glutamatergic receptors are a separate family and/or more distantly related to this superfamily.

Neurotransmitter-gated ion channel function is most likely modulated by phosphorylation, a major mechanism regulating synaptic plasticity. As evidenced over the past years, protein phosphorylation is one of the primary mechanisms for posttranslational regulation of protein function. Complementary DNA cloning has shown consensus potential phosphorylation sites for a variety of protein kinases in most receptor subunits cloned so far (Figure 2, panel C and D). The effects of protein phosphorylation on receptor function have been studied electrophysiologically, biochemically and by mutation analysis. Nearly all members of the superfamily (including the glutamate receptor family) of ligand-gated ion channels display desensitization, a process by which receptors become inactive due to prolonged agonist stimulation. The receptor phosphorylation appears to influence the rate of receptor desensitization, and regulates channel activity, receptor aggregation (e.g., nAChR aggregation induced by agirin), and assembly of receptor subunits.

Molecular cloning and expression studies revealed a far more complex picture of ligand gated-ion channel pharmacology as judged by electrophysiology and ligand binding criteria. Isolation of cDNAs encoding several closely related subunit variants (see Table 2) indicated that ligand gated-ion channel subunits display great heterogeneity; multiple genes encode different isoforms of the individual subunits. The pattern of receptor subunit assembly within a receptor complex *in vivo* plays a crucial role in the pattern of elicited electric postsynaptic potentials, and in determining receptor pharmacological properties. In order to understand which subunit combinations form naturally occurring receptor subtypes, co-expression studies in concert with immunological analysis (co-localization of receptor subunits) and electrophysiological measurements were employed. Expression of different receptor subunit combinations in heterologous systems, i.e., expression in *Xenopus* oocytes and/or in cell culture systems, generated receptors which display much of the pharmacology of their neuronal counterparts. However, the ability of some subunits to form functional homomeric ion channel (e.g., glycine receptor α subunits), albeit with a lower efficiency, raises the possibility that native homomeric receptors exist in brain. Ligand-gated ion channels have a hetero-oligomeric structure comprising of several polypeptide subunits assembled (most probably) into a pentameric complex (Figure 2A). Nevertheless, the exact stoichiometry of subunit assembly in "true *in vivo*" (native) ligand-gated ion channels involved in

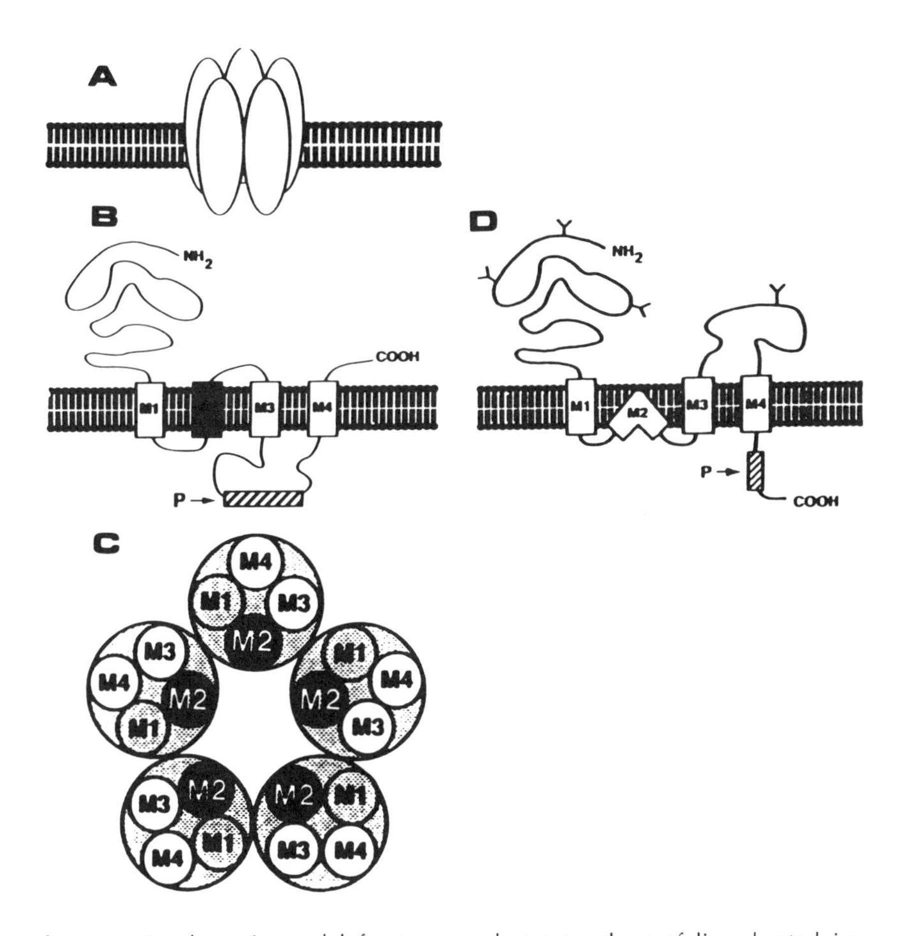

Figure 2. A schematic model for transmembrane topology of ligand-gated ion channels. **A.** Pentameric subunit structure viewed through the plane of the membrane. **B.** Predicted membrane topology of glycine, GABA and nACh receptor subunits, viewed through the plane of the membrane of an individual subunit composed of four transmembrane domains designated M1 through M4. The carboxy terminus of glutamate receptor subunits is predicted to reside intracellularly rather than extracellularly. **C.** Schematic model representing the arrangement of the transmembrane α-helix within each subunit around the central ion channel viewed by cross-section through the transmembrane core. The M2 segment of each subunit lines the wall of the ionic pore. **D.** The most recent proposed transmembrane topology model for the ionotropic glutamate receptors subunits. P indicates the region of the subunit that contains the phosphorylation sites for the various protein kinases. Panels A, B, and C: From S. L. Swope, S. J. Moss, C. D. Blackstone and R. L. Huganir. (1992). FASEB J. 6, 2514–2523. Reprinted with permission from The FASEB Journal.

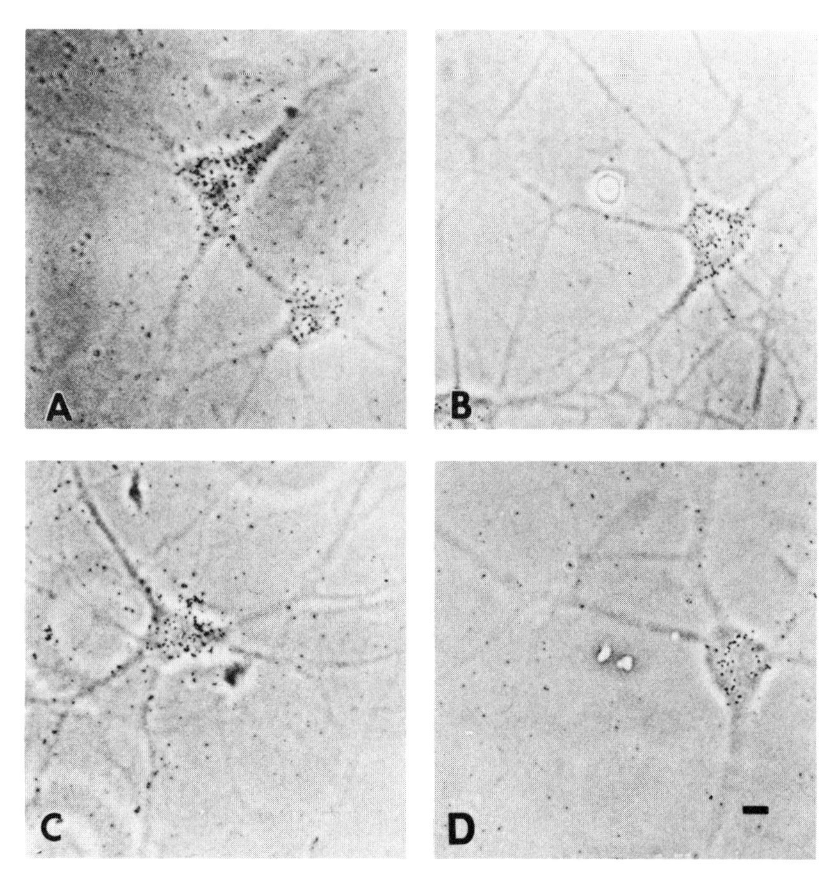

Figure 3. The distribution of GluR1-4 mRNAs in mature, cultured hippocampal neurons by *in situ* hybridization. Phase contrast micrographs (A-D) illustrate hippocampal cells that were hybridized with pan oligonucleotide probes (45-mers). designed to detect specific GluR1 (A), GluR2 (B), GluR3 (C) and GluR4 (D) receptor mRNAs. Location of autoradiographic silver grains over individual neurons shows identical patterns of distribution for all GluR subunits. Intense distribution of grains is observed over the cell body with lower abundance for GluR4. The absence of GluR1-4 mRNAs from dendrites and axons indicates that the synthesis of GluR1-4 proteins in neurons is confined to the cell body. Line scale, 10 μM. From N. Eshhar, R. S. Petralia, C. A. Winters, A. S. Niedzielski, and R. J. Wenthold. (1993). Reprinted with permission from Pergamon Press, Ltd., Oxford.

single-neuron responses is unknown at present (except for Torpedo and muscle nACh-R, see Table 2).

The pattern of receptor subunit gene expression throughout the brain was investigated by *in situ* hybridization histochemistry with oligonucleotides and/or

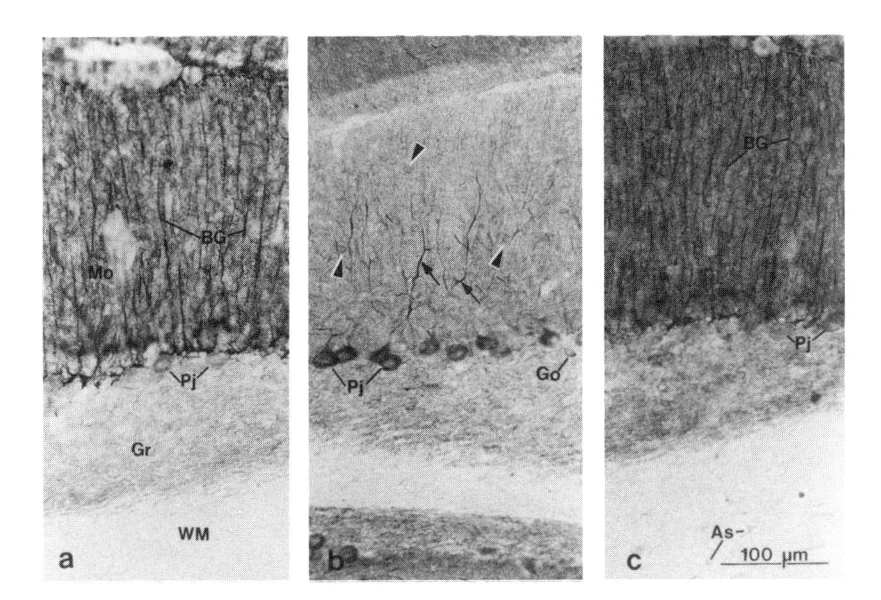

Figure 4. Individual AMPA receptor subunits demonstrate a differential distribution in the rat cerebellum. Sagittal sections of the cerebellar cortex immunolabeled with antibodies to GluR1 (a), GluR2/3 (b,e). and GluR4 (c). Bergmann glia of the molecular layer stained densely when immunolabeled with antibodies to GluR1 and GluR4 and not with antibodies to GluR2/3. Purkinje cell bodies and dendrites are stained densely with antibody to GluR2/3 only. AS, astrocyte-like cells; BG, Bergmann glial processes; Go, Golgi cell; Gr, granular layer; Mo, molecular layer; Pj, Purkinje cell body; WM, white matter; (*small arrow*), Purkinje cell dendrite; (*arrowhead*), molecular cell layer. From R. S. Petralia and R. J. Wenthold. (1992). J. Comp. Neurol. 318, 329–354. Photographs (a)–(c) courtesy of R. S. Petralia and R. J. Wenthold. Reprinted with permission from Wiley-Liss, a division of John Wiley and Sons, New York.

cRNA probes corresponding to unique subunit sequences. Genes coding for subunit variants of a particular receptor subtype are expressed in a cell specific manner in different regions of the brain, e.g., subunit mRNAs exist in a largely distinct neuronal subpopulation. It is likely therefore that subunit gene expression is differentially regulated. *In situ* hybridization detection of receptor subunit mRNAs (e.g., mRNAs for $GABA_A$ and AMPA receptors) was confined to the cell body and largely excluded from dendrites and axons (Figure 3). These observations suggest that mRNAs encoding ligand gated-ion channel proteins are not transported out of the cell body. Rather, receptor proteins are transported to their sites of expression via selective mechanisms following synthesis in the cell body.

Receptors with different properties may be expressed at different times in development. The composition and probably the functional properties of embryonic

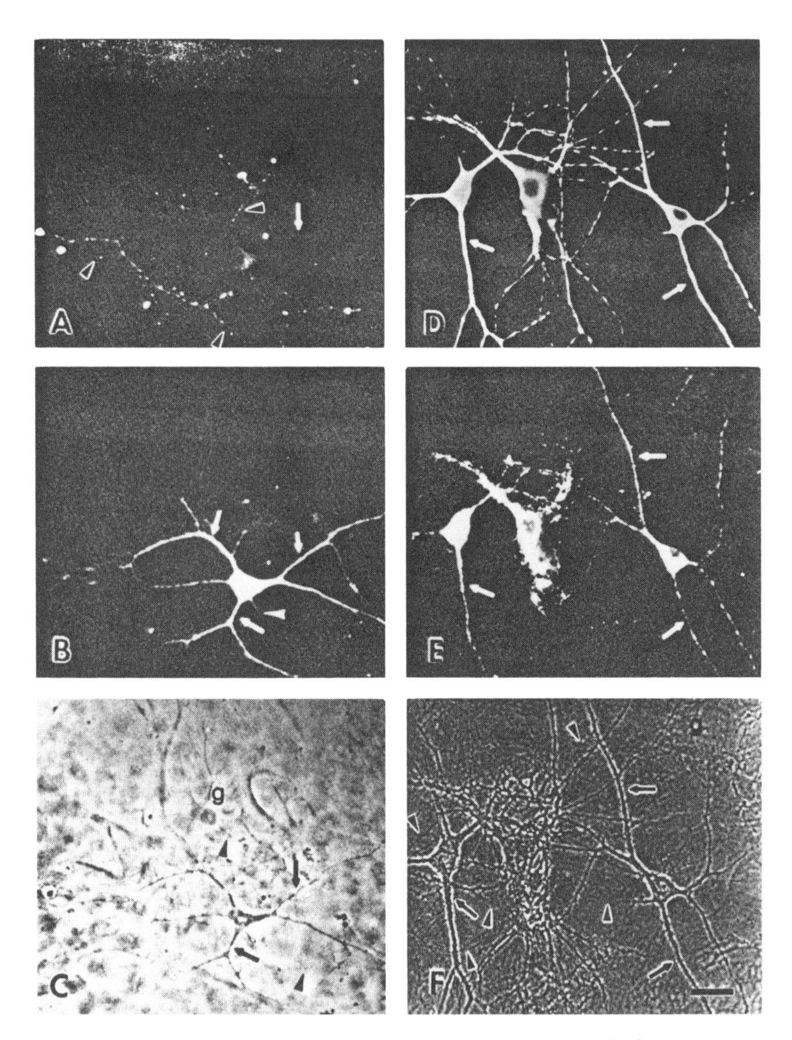

Figure 5. Immunocytochemistry substantiates the distribution of GluR1 AMPA receptor subunit within hippocampal neurons. Receptor localization is compared to the expression of neuronal markers in mature hippocampal neurons (A,B,D,E). Double labeling of cells with antibodies to Tau-1 (**A**) and GluR1 (**B**) illustrates the compartmentalization of Tau-1 to axons and of GluR1 to the cell body and dendrites (**B**). Little or no labeling of axons is seen with antibodies to GluR1. Double labeling with antibodies to MAP-2 (**D**) and to GluR1 (**E**) illustrates their colocalization in the somatodendritic compartment. Cell culture organization is illustrated in the corresponding phase contrast views (**C, F**). Axons, black arrowheads; dendrites, arrow; glial cells with light immunolabeling, g. Commonly, axons run along dendrites (e.g., dendrites labeled with white arrowhead in B). Line scale, 30 μM. From N. Eshhar, R. S. Petralia, C. A. Winters, A. S. Niedzielski, and R. J. Wenthold. (1993). Reprinted with permission from Pergamon Press, Ltd., Oxford.

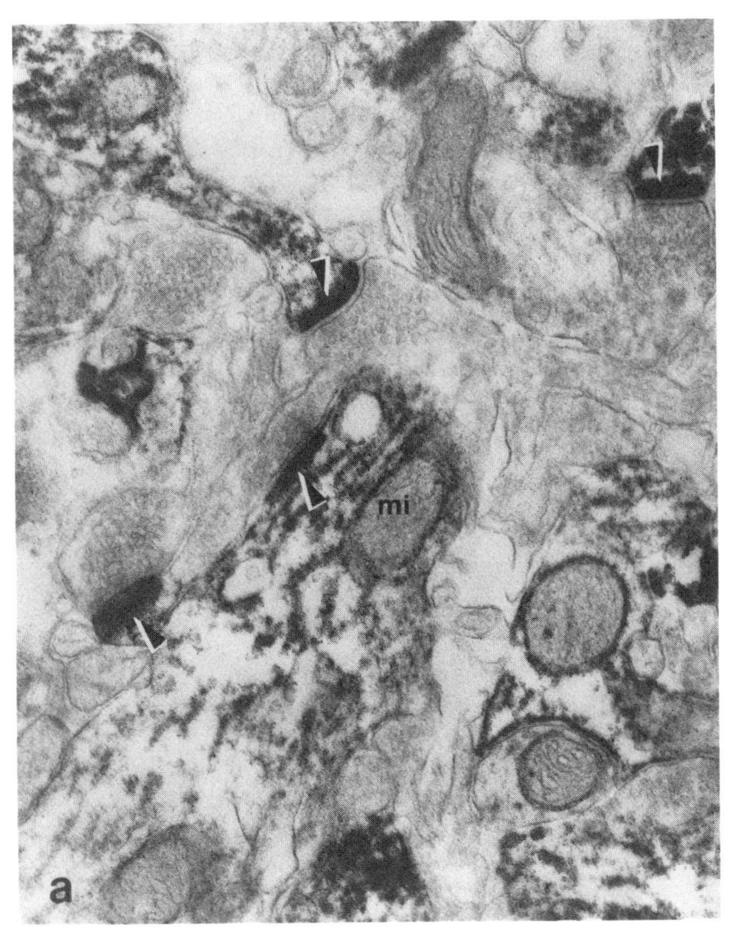

Figure 6. Electron micrograph of cerebral cortex immunolabeled with antibodies to GluR1. GluR1 receptor subunit is mainly associated with postsynaptic densities and adjacent dendritoplasm. mi, mitochondrion in labeled dendrite; (arrowhead), stained synaptic density. From R. S. Petralia and R. J. Wenthold. (1992). J. Comp. Neurol. 318, 329–354. Photograph (a) courtesy of R. S. Petralia and R. J. Wenthold. Reprinted with permission from Wiley-Liss, a division of John Wiley and Sons, New York.

and early postnatal ligand-gated ion channels differs markedly from those expressed in the adult brain. A switch-on, or switch-off in mRNA expression of subsets of ligand gated-ion channel receptor subunits may occur during brain development. For example, a single subunit exchange distinguishes the embryonic muscle nACh and glycine receptors from the adult receptor forms (Table 2). These findings indicate that receptor subunit gene expression is developmentally regulated. One can assume that receptor subunits intensely expressed in the neonatal brain might play an important role in neuronal development, whereas receptor subunits highly expressed postnatally are involved in neurotransmission in the mature brain.

The localization of receptors and their individual subunits on particular cell membranes and at particular synaptic junctions was demonstrated for the first time through immunocytochemistry with polyclonal antibodies generated either against the ion channel receptor complex or against receptor individual subunits by light and electron microscopy (e.g., the ultrastructure localization of the glycine and GABA receptors). Immunocytochemical studies have confirmed the regional distribution of the receptors as described by ligand-binding autoradiographic studies. Moreover, they provided the first evidence that channel receptors are localized within neuronal somatodendritic compartments (immunoreactivity was absent from axon nerve terminals), and are localized not only at synaptic junctions but also at some nonjunctional sites (extrasynaptic localization was described for GABA-gated channels). It is therefore possible that channel receptors at sites distant from the neurotransmitter-releasing terminals may be affected, depending on the level of neuronal activity.

The regional cellular and subcellular distribution of individual receptor subunits in the brain investigated using antibodies against synthetic peptides corresponding to sequences of the carboxy termini portion of several cloned receptor subunits (e.g., antipeptide antibodies that recognize the C-terminal domains of the individual subunits of the AMPA preferring glutamate receptors and $GABA_A$ receptors). As found by immunocytochemistry using light and electron microscopy, there is individual receptor subunit overlapping and differential distribution in the brain (Figure 4). Subunit immunoreactivity was localized within cell bodies and dendritic spines of specific subsets of neurons and not in axons (Figure 5). Electron microscopy revealed receptor subunit association mainly with postsynaptic densities and adjacent dendritoplasm, and with neuronal cell body cytoplasm (Figure 6).

LIGAND-GATED ION CHANNELS: SUBUNIT PRIMARY STRUCTURE

Molecular characterization of ligand gated ion channel subunits has greatly advanced our understanding of interactions taking place between small molecules and channel receptors, and the transport of ions across a biological membrane in response to these interactions. At first (1989), analysis of primary protein sequences (deduced from cDNA sequences) disclosed a similar structure organization for all ligand-gated ion channel receptor subunits. At that time, it had been generally accepted that all ligand gated-ion channel receptors are composed of the same number of subunits and have the same distribution of membrane-spanning helices, resulting in an identical transmembrane topology. Each receptor subunit is composed of an amino-terminal hydrophobic cleavable signal sequence, a large extracellular NH_2-terminal region followed by four hydrophobic α-helical membrane-spanning domains designated M1 through M4; to each subunit spans the membrane four times with the M3-M4 loop in the cytoplasm, and a relatively short carboxy terminus, predicted to reside on the extracellular side of the lipid bilayer membrane (Figure 2B). In recent years (1993–1995), the topology of the glutamate receptor subunit has undergone radical revision. In the new model, the glutamate receptor crosses the cell membrane three times and not four, thus placing the big protein loop outside rather inside the cell (Figure 2D). Contrary to an original model (1989), the M2 segment of the putative channel lining the hydrophobic domain does not span the membrane, but lies in close proximity to the intracellular face of the plasma membrane or loops into the membrane without traversing it. Moreover, while the C-terminus is extracellular in the subunits of the nAChR superfamily (Figure 2B), the C-terminus of the glutamate receptor subunit is suggested to be intracellular (Figure 2D).

The N-terminal domain of the glutamate receptor subunits is longer than that of other ligand-gated ion channel members; it contains 400–500 amino acids instead of 200. Among NMDA receptor (a glutamate receptor subtype) subunits, some possess a longer carboxy terminus as compared to other members of ligand-gated ion channels. The membrane domains M1, M2 and M3 are closely aligned in the middle of the protein and an intracellular stretch of hydrophilic residues (81–125 amino acids long) separates the M3 and M4 segments. Structural alignment of selected cloned subunits is illustrated in Figure 7.

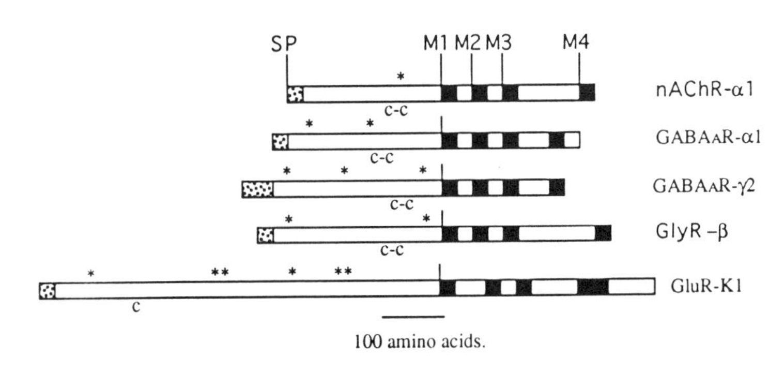

Figure 7. Structure of selected amino acid receptor subunits and the nAChR α subunit. All subunits are aligned relative to the M1 transmembrane segment. SP, putative signal peptide is indicated by spotted boxes; transmembrane domains M1-M4 are represented by black boxes. Hydrophilic regions are designated by open bars. Asterisks (*) denote potential glycosylation sites in the extracellular domains. C-C refers to the Cys-Cys loop common to all subunits of $GABA_A$, glycine and nACh receptors. The GluR-K1 subunit contains only one cysteine residue indicated by C in the homologous region.

Deduced protein sequences enabled researchers to correlate structural motifs with receptor functional properties. The amino-terminal extracellular domain contains multiple glycosylation sites and a conserved 15-amino-acid region flanked by cysteines (Figure 7). In the muscle nAChR, the Cys-Cys domain has been shown to form a disulfide loop configuration which is probably essential for stabilization of tertiary structure and/or for subunit assembly. Mutagenesis experiments have revealed the importance of the Cyc-Cys loop in maintaining the proper receptor topology essential for ligand binding and channel activation, e.g., Cys-Cys loop is needed for the maintenance of α-bungarotoxin (an nAChR antagonist) binding to the nAChR α subunit. The Cys-Cys loop has been proposed as a hallmark of all ligand gated-ion channel receptors. However, recent cloning of glutamate receptor subunits has not identified an identical Cys-Cys loop in their structure.

A conserved proline residue is present within segment M1 of glycine, $GABA_A$, and nACh receptors. Proline residues producing a flexure in the α-helix may contribute to receptor structural flexibility, which is required for reversible conformational transitions of transmembrane regions responsible for opening of the ion channel. In contrast, the M1 segments of glutamatergic receptor subunits do not contain a proline residue. It is possible that glutamate receptors transform external ligand binding into electric signals by a different mechanism(s).

Domains implicated in ligand binding are not known precisely. Most likely they are located in the N-terminal extracellular domains. Deglycosylation of rat $GABA_A$ receptor subunits modified the binding of benzodiazepine agonists and antagonists (GABA receptor subtypes are described in Table 1). Based on structural considerations, sequence comparison and mutagenesis analysis, agonist binding recognition sites involve multiple interaction sites present within a stretch of charged residues around the two neighboring cysteines in front of the M1 segment. For example, sequences contributing to the recognition of cholinergic ligands by the α_7 neuronal nicotinic receptor subunits were established by mutation of three aromatic amino acids at three different peptide loops (region α 180–200) : the apparent binding affinities of cholinergic ligands significantly decreased in all mutated sequences.

Identical structural organization, shared by all ligand gated-ion channel receptors, suggests that ion permeation across membranes in response to neurotransmitter binding is mediated by a common mechanism. The M2 segment is highly conserved in most of the clones characterized so far and appears to be a major determinant of channel function (e.g., M2 segment displays a very high degree of similarity between $GABA_A$ and glycine α subunits). Recent studies by Wood et al. (1995) have shown a high degree of homology between the voltage gated K channel pore-forming region and the M2 membrane domain of the glutamate receptor. These sequences suggested the hairpin model for the M2 segment. The striking homology in the pore regions of glutamate receptor and voltage-gated K channels suggests a common origin for these ion channels. As inferred from physiological experiments involving determination of pore diameter (diameter is 6.6 Å for muscle type nAChR, and 5.2 Å for glycine receptor), only a single α-helix from each

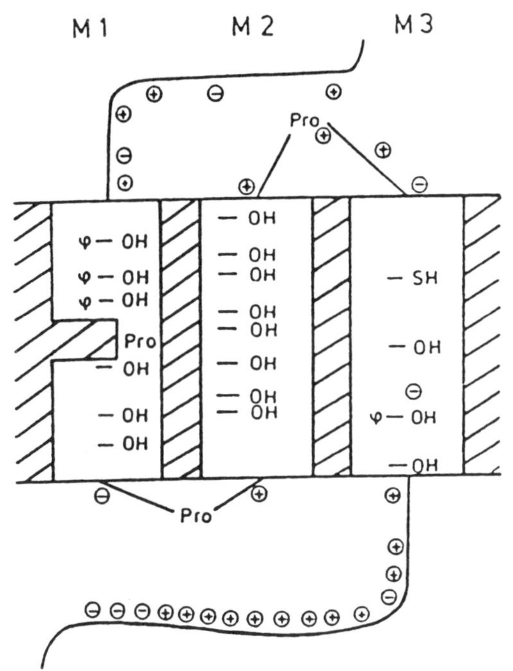

Figure 8. Enlargement of segments M1 through M3 of the α GlyR subunit. These segments are composed of charged and hydroxylated amino acids. Positively charged residues flanking the M2 segment confer anion selectivity. ι-OH, serine or threonine residues; ϕ-OH, tyrosine; + and –, charged amino residues. From G. Grenningloh, A. Rienitz, B. Schmitt, C. Methfessel, M. Zensen, K. Beyreuther, E. D. Gundelfinger and H. Betz. (1987). Nature 328, 251–220. Reprinted with permission from Macmillan Magazines Limited, London.

subunit is needed to surround the ionic pore. Each of the five subunits may contribute its M2 segment to line the inner wall of the ion channel (Figure 2C). The M2 helix of $GABA_A$, glycine and nAChR subunits is rich in uncharged polar amino acid residues (especially concentrated at the cytoplasmic end) such as threonine and serine whose hydroxy side chains (eight conserved hydroxy side chains) are thought to contribute to the formation of a hydrophilic pore allowing ion flow across the membrane (Figure 8).

Mutagenesis studies combined with covalent modification were used to identify determinants within the M2 region that control conductance, gating and ion selectivity. Radioactively labeled channel blockers were found to interact covalently with amino acids within the M2 segment following irradiation. Exchange of three serine residues, present in the M2 transmembrane segment of different nAChR subunits, by alanine, modified both the single channel conductances and the equilibrium binding affinity of an open channel blocker, i.e., single channel outward currents decreased and dissociation of channel blockers was increased. Moreover, mutation of negatively charged residues flanking the M2 region, such as glutamic and aspartic acid residues, to neutral or positive residues (e.g., residues were mutated to lysine residues) affected markedly the rate of ion movement (the number of ions that can cross the channel during its open state was reduced) and ion selectivity. It was shown that the positively charged amino acid residue arginine present within the M2 segment of the GluR-2 subunit (see Table 2 for AMPA subunits) determines ion flow properties in AMPA gated channels. Calcium conductance through AMPA channels was observed *only* in GluR-1,3 and 4 subunit assemblages, and was absent in receptor complexes containing GluR-2 subunits. AMPA receptors conductance properties are most likely determined by a glutamine residue present in the M2 segment of GluR1, 3, and 4 subunits at a homologous position. This functionally critical site is referred to as the Q/R site (see Wisden and Seeburg, 1993). Replacement of asparagine residues within the M2 segment of NMDAR1 by glutamine (see Table 2 for NMDA receptor subunits), demonstrated that calcium permeability and magnesium blockade are controlled by the former residue.

Amino acids flanking the M2 segment that have a net positive charge are thought to confer anion selectivity, e.g., clusters of arginine and lysines at the channel mouth of $GABA_A$ and glycine receptors increase the driving force for anion passage upon channel opening (Figure 8). Excesses of negatively charged residues bordering the M2 segment (as shown for nAChR) are found in cationic selective channels.

The amino acid sequence of the intracellular loop connecting M3-M4 domains of glycine, GABA and nAChR varies between subunits within a family of proteins as well as between different subunit classes. Site-specific mutagenesis technology combined with electrophysiological or biochemical methods has identified in this major intracellular loop multiple potential sites for phosphorylation (Figure 2B), i.e., different protein kinase recognition motifs. These include recognition sites for serine/threonine kinases such as, cAMP-dependent protein kinase (PKA), calcium/calmodulin-dependent protein kinase II, protein kinase C (PKC) and protein tyrosine kinases (PTK). Phosphorylation of subunits of a given receptor, and/or phosphorylation of an individual subunit by different kinases, is most likely responsible for different functional properties elicited by the receptor (Swope et al., 1992; Raymond et al., 1993). Biochemical and electrophysiological studies performed by Roche et al. (1996) have not revealed phosphorylation sites in the loop between M3 and M4 segments of kainate and AMPA receptor subunits but have detected multiple phosphorylation sites located on the C-terminus of the AMPA receptor GluR1 subunit. These data support the new transmembrane topology model for glutamate receptors and indicate the presence of the loop between the M3 and M4 segments in an extracellular position.

PHARMACOLOGY AND FUNCTIONAL DIVERSITY OF LIGAND-GATED ION CHANNELS

Excitatory Receptors

The Muscle and Torpedo Electric Organ Nicotinic Acetylcholine-Receptor (nAChR)

The nicotinic AChR is one of the most extensively studied cellular receptors. It was the first ligand-regulated ion channel to be purified and cloned and has thus served as a prototype for the entire group of ligand-gated ion channels. It is so far the only neurotransmitter channel receptor for which three dimensional contour maps have been obtained by crystallographic analysis. Electron microscopy and X-ray analysis have shown that much of the AChR receptor mass is on the extracellular side of the membrane whereas a smaller receptor extension resides within the cytoplasmic side. The five subunits are arranged symmetrically around a central axis perpendicular to the membrane.

The nAChR was isolated from the electric organs of the marine ray *Torpedo* and the eel *Electrophorus* where it is extremely abundant in cholinergic synapses. Receptor purification and biochemical characterization was facilitated by α-bungarotoxin, a toxin extracted from *Bungarus multicinctus* (snake) venum which binds in a high affinity manner to the receptor. At the adult mammalian neuromuscular junction, nAChRs are clustered on the muscle surface in the endplate region. Clustering of AChR takes place early during neuromuscular synapse formation, and increases as synapse matures. AChR channel opening is triggered by neurotransmitter and/or ligand binding to sites within the extracellular NH_2 terminal domain of each α-subunit (see Table 2 for nAChR subunits). Expression of incomplete combinations of *Torpedo* nAChR subunits and mouse muscle nAChR subunits in a heterologous system has shown that the α subunit alone is unable to bind neither acetylcholine nor *d*-tubocurarine (an acetylcholine antagonist). However, expression of the α subunit in combination with the γ subunit produced a site that recognizes *d*-tubocurarine with high affinity, and acetylcholine with a low affinity. A combination of α and δ subunits generated a protein recognizing *d*-tubocurarine with low affinity and acetylcholine with high affinity.

The cationic channel formed upon activation of the receptor by acetylcholine is selectively permeable to Na^+ and K^+ ions and to some extent to Ca^{2+} ions. Channel activation can be blocked by specific receptor antagonists such as *d*-tubocurarine and α-bungarotoxin which prevent the formation of the ligand-receptor complex. Because these toxins block neuromuscular transmission they are potent paralytic agents when administered to intact animals. On long exposure to acetylcholine the receptor undergoes transitions between small numbers of discrete conformations with distinct binding properties and states of ion channel opening. At least four interconvertible states including a resting, active and two desensitized states, have been described (Changeux, 1990). While agonists stabilize the active and desensitized states, antagonists stabilize the resting state. When the AChR is persistently exposed to ACh and/or to AChR agonists, it finally becomes unresponsive to ligand binding, i.e., desensitized and is unable to initiate muscle contraction. Upon removal of the agonist, the receptor reaches its original resting state, and neuromuscular transmission is renewed. The state of desensitization can be caused by a prolonged accumulation of ACh in the synaptic cleft due to inhibition of acetylcholinesterase which is the prime enzyme responsible for hydrolysis of ACh. Nerve gases such as organophosphates bind irreversibly to the enzyme active site, thereby causing muscle paralysis.

As revealed by functional expression studies, not all AChR subunits are required for channel gating. It was shown that the presence of the α subunit is essential for ligand gated channel activity. In the absence of either β, or γ, or δ subunits, channel activity is still achieved.

Multiple allosteric sites for a variety of regulatory ligands and pharmacological effectors are distributed throughout the extracellular and intracellular parts of the AChR, and in its ionic channel and the protein-lipid interface. These include sites for divalent cations, neuropeptides, local and general anesthetics and protein kinases which differentially influence the equilibrium and kinetics of receptor desensitization and activation, in a positive or negative fashion.

Autoantibodies, which reduce the number of functional receptors or/and impede the interaction of ACh with the receptor, are involved in myasthenia gravis, a muscular dystrophy autoimmune disease.

Neuronal Nicotinic Acetylcholine Receptors

Neuronal nAChRs are receptive sites for nicotine in the brain, and are associated with nicotine's addictive action. The first demonstration (in the mid-1950s) of functional nicotinic receptors in neurons and their participation in neurotransmission was made in cat Renshaw cell recurrent collateral synapses. As shown by Eccles and coworkers, synaptic stimulation of these cells could be blocked by nicotinic antagonists. Later, functional nicotinic receptors were detected in the cerebral cortex, hippocampus, hypothalamus and other brain areas. Although neuronal nAChR subunits are very similar in their amino acid sequences, and are highly homologous to the subunits of the muscle AChR, they differ markedly in their pharmacological and functional properties. Neuronal nAChRs are more permeable to calcium ions as compared to the muscle nicotinic receptor, and are modulated by calcium ions acting externally. The nervous system expresses two distinct nicotinic receptor subtypes: the α-bungarotoxin-insensitive, and α-bungarotoxin-sensitive subtypes. However, α-bungarotoxin fails to block activation of mammalian neuronal nAChRs by nicotinic agonists. The distribution of α-bungarotoxin-binding sites in the brain is distinct from that of the high-affinity nicotine-binding site. Possibly, these receptors represent a pharmacologically distinct neuronal nicotinic receptor subtype. As revealed by molecular cloning, the neuronal-type seems to be made up of only two types of subunits rather than four, a neuronal α-subunit which binds ligand, and a neuronal β-subunit. These subunits are most likely assembled in an $\alpha_2\beta_3$ subunits stoichiometry (see Table 2). It is possible that in the future, additional subunits of the neuronal nAChR will be isolated. Several subtypes of each neuronal subunit have been cloned (see Table 2) and their elicited functional characteristics were determined by expression in *Xenopus* oocytes. Functional and pharmacological diversity of neuronal nicotinic receptors results from different variants of α and β subunits and variable subunit combinations. Homoligomeric-channels formed with the α_7 subunit have a Ca^{2+} permeability higher than that of other nAChRs and even higher than the NMDA subtype of glutamate receptors. For example, individual expression of the α_4 variant generated weakly functional AChR, while α_7 expression formed robust channels. α-Bungarotoxin acts as a potent blocker of the rat $\alpha_3\beta_2$ receptor expressed in oocytes, and blocks to a lesser extent, the $\alpha_4\beta_2$ subunit combination. In addition, acetylcholine gated currents via the α_7 channel are completely blocked by α-bungarotoxin. The α_4-containing receptors show a higher sensitivity to acetylcholine as compared to that of receptor containing the α_3 subunits. High sensitivity to nicotine is exhibited by channels assembled in oocytes from the avian α_7 subunit.

In situ hybridization demonstrated that nAChR subunits are differentially colocalized throughout the brain, suggesting that multiple forms of the receptor exist, each with a different subunit composition. While the β_2 subunit was found in most regions of the brain, the α subunits revealed unique (although there are some areas of overlap) patterns of expression.

Functional diversity, compositional heterogeneity and differential neuroanatomical distribution of neuronal nAChRs suggest that nicotinic addiction may involve a particular subset of nicotinic binding sites in the brain.

Glutamate Receptors

Chemistry has played a crucial role in the rapid increase in knowledge of EAA receptors. Due to newly developed drugs, the identification of multiple receptors for EAAs is now well established. The postsynaptic depolarizing responses elicited by glutamate are mediated through disparate receptors that are either ionotropic (coupled to an ion channel) or metabotropic. The metabotropic receptors are activated by quisqualate and are coupled to effector molecules such as phospholipase C by a G protein, thereby modulating the generation of second messengers. Based on pharmacological and electrophysiological responses to selective agonists, ionotropic glutamate receptors have been divided into three distinct subtypes (see Table 1): those that are sensitive to NMDA (N-methyl-D-aspartate), and those which are preferentially activated by kainate (kainic acid) and AMPA (alpha-amino-3-hydroxy-5-methylisoxazole-propionic acid), also referred to collectively as non-

NMDA receptors. AMPA receptors are also activated by quisqualate and are thus referred to as the quisqualate receptors of the ionotropic type.

The native NMDA receptor-channel complex is a complicated ionophore. It possesses several characteristic features, and can be distinguished from the non-NMDA channels by gating kinetics, ion permeation, modulation, and sensitivity to antagonists. Intracellular recording of membrane potential, especially single channel recording, e.g., voltage-clamp experiments in cultured neurons, have revealed that kainate and AMPA receptors are associated with rapid neuronal depolarizing responses while NMDA receptors are associated with slower depolarizing responses. NMDA-induced currents are of relatively long duration as compared to those elicited by non-NMDA receptors. While non-NMDA receptors activate channels permeable to Na^+ and K^+ ions, selective NMDA receptor agonists permit also the passage of Ca^{2+} ions into the neuron. The NMDA receptor has an unusual form of voltage regulation, e.g., NMDA elicited activity is controlled by a voltage dependent Mg^{2+} blockade, and by several dissociative anesthetics such as PCP, ketamine and MK-801 (Table 1). Radioligand binding studies and mutagenesis analyses have shown that these anesthetics bind to sites within the NMDA channel, thereby blocking channel opening. Furthermore, the NMDA-mediated responses are regulated (altered) at several modulatory sites. Receptor activation is positively controlled by glycine (e.g., glycine increases the frequency of NMDA receptor-induced channel opening) via an action at a recognition site related to the NMDA receptor complex which is distinct from that of the well known strychnine-sensitive sites of the glycine receptor. Two molecules of glutamate and two molecules of glycine are needed to fully activate an NMDA receptor. NMDA induced channel activity is negatively modulated by Zn^{2+} ions, and is influenced by certain polyamines. In addition, zinc ions potentiate rather than inhibit agonist-induced currents at some homomeric NMDAR1 receptors assembled from different NMDAR1 splice variants (see Table 2 for NMDA receptor subunits). AMPA receptors on the other hand appear to be regulated by a modulatory protein and chaotropic agents such as potassium thiocyanate (KSCN).

By the use of selective glutamate antagonists (Watkins and Olverman pioneered in the development of selective EAA receptor antagonists), it was established that some of these receptors participate in excitatory neurotransmission in the CNS and are therefore of physiological relevance. It has been reported that EAA receptors are intimately involved in mechanisms of neuronal plasticity in both the developing and mature brain and in neuronal cell degeneration. Glutamatergic receptors have been implicated in processes associated with memory acquisition and learning, motor control, cognitive processes, neuritogenesis and synaptogenesis. Within hippocampal neurons (the CA1 region of the hippocampus is highly enriched with NMDA binding sites), NMDA receptors were found to be involved in the induction of long term potentiation (LTP), a phenomenon widely regarded as a physiological mechanism of memory. LTP is a long-lasting potentiation of the synaptic response that can last for days, and occurs following a brief high-frequency stimulation. LTP induced by NMDA could be blocked by competitive and non-competitive NMDA receptor antagonists such as MK-801, AP5 and Mg^{2+} ions. Activation of quisqualate receptors was found to lead to NMDA-induced LTP in hippocampal neurons. This suggests that modulation of several glutamatergic receptor subtypes affect processes of learning and memory. The first behavioral evidence linking NMDA receptors to learning was obtained using the water maze test. In this test, spatial learning of rats is determined by the ability of rats to swim in a water tank and reach the platform situated within the tank. By the early 1980s, NMDA antagonists were known to block an ever increasing number of excitatory pathways in the brain (Watkins and Evans, 1981) and were further found to display significant anticonvulsant activity and induce muscle relaxation in certain experimental models of epilepsy (Meldrum, 1985; Kemp et al., 1987), and displayed remarkable neuroprotective properties in models of ischemia or hypoglycemia (other examples are described below).

The pharmacology of glutamate responses in the vertebrate brain is complex and diverse. The diversity arises from the presence of several genes for each receptor subunit, from alternative splicing of transcribed messages and from various subunit combinations assembled within a given receptor (see Table 2 for glutamate receptor subunits). For example, NMDAR1 subunit can form homomeric ion-channels, i.e., expression of NMDAR1 subunit produced a functional channel complex possessing properties characteristic of the NMDA receptor. However, the current amplitude recorded is much smaller than that observed with the native receptor. In contrast, functional homomeric or heteromeric channels are not formed from NMDAR2 subunits alone. The NMDAR1 subunit is widespread through the mature brain, while NMDAR2 mRNAs exhibit overlapping and differential distribution. Expression and *in situ* hybridization studies clearly indicated that NMDAR1 subunit serves as a fundamental subunit required for the NMDA receptor channel complex, and forms a heteromeric configuration with different members of the NMDAR2 subunits. As deduced from pharmacological and electrophysiological measurements, complexes of AMPA receptor subunits (GluR1-4 subunits) exhibit high-affinity AMPA binding and low-affinity kainate binding, form channels gated by AMPA as well as by kainate and L-glutamate (with AMPA acting as a partial agonist), and evoke various electrical responses depending on subunit composition. Furthermore, it was shown that GluR1-4 currents evoked by kainate do not desensitize, while currents evoked by L-glutamate and AMPA produced initial fast desensitization responses. It is thus possible that in some brain areas different responses to kainate and AMPA occur at the same receptors. However, the isolation and expression of additional subunits representing high-affinity kainate receptors (see Table 2) suggests that definitive native kainate receptor subtypes may be composed of these subunits in heteromeric configurations. One can conclude that non-NMDA receptors *in vivo* play a variety of functions according to their subunit composition.

Inhibitory Amino Acid Receptors

Glycine Receptors

Glycine receptors (GlyRs) of spinal cord neurons can be activated by several α and β amino acids such as glycine, β alanine, taurine, L-alanine, L-serine and proline. On the other hand, glycine inhibitory neurotransmission is selectively antagonized by only a few compounds. These include the convulsive plant alkaloid strychnine and its steroid derivatives. The consequence of strychnine poisoning is overexcitation of the motor system resulting in muscular convulsions. Strychnine binding can be displaced by the above mentioned amino acids in a competitive manner. However, the fashion by which strychnine binding is displaced by glycine indicates that they do not bind to an identical site, but rather to two closely related sites of the α subunit (see Table 2 for GlyR subunit composition). The activation of the GlyR channel demands the binding of at least three glycine molecules. The GlyR was the first ligand gated ion channel protein to be isolated from mammalian nervous tissue. Highly purified glycine receptor preparation was obtained from spinal cord membranes followed by detergent extraction and purification on a strychnine affinity matrix. Compositional analysis of purified receptor demonstrated the presence of two glycosylated proteins representing integral membrane-spanning subunits, i.e., the α and β receptor subunits (see Table 2), and a larger nonglycosylated peripheral membrane protein known as gephyrin localized at the cytoplamic face of the postsynaptic glycine receptor complex. Since gephyrin binds with high affinity to polymerized tubulin, this protein is most probably responsible for anchoring the receptor to postsynaptic cytoskeletal elements, e.g., to subsynaptic microtubules. Pharmacological, immunological and biochemical properties distinguish between neonatal GlyR and adult receptor forms. The neonatal form is characterized by a lower strychnine binding affinity, and possesses a different molecular mass and different antigenic epitopes from those carried by the α subunit of the adult form. Glycine receptor purified from cultured neurons derived from fetal rodent spinal cord displays only a major polypeptide band of 49 kD, suggesting a homo-oligomeric structure of the neonatal form. Functional and pharmacological diversities displayed by GlyRs in the vertebrate CNS are now explained by the existence of several homologous GlyR α-subunit genes (see Table 2). Expression studies in *Xenopus* oocytes have shown that agonist-gated chloride channels with rather low efficiency were formed by β subunits, whereas strychnine-sensitive chloride channels were formed by α_3 subunits. The α_2^* subunit is significantly expressed in neonatal spinal cord and its abundance was found to be strongly decreased along postnatal spinal cord development. Functional expression has shown that the α_2^* subunit generates chloride channels of low strychnine sensitivity suggesting that the α_2^* subunit represents a ligand binding subunit of the neonatal GlyR. Glycine receptor α_1, α_2, and β subunit transcripts show increased levels during spinal cord development. Glycine receptor β subunit transcripts are highly expressed in the mature cortex and cerebellum. Low levels of α_3 transcripts are found in regions involved in sensory or motor processing such as the olfactory bulb, cerebellum and spinal cord.

Elucidation of glycinergic inhibition in higher brain regions requires further physiological analysis.

GABA$_A$ Receptors

Biochemical and electrophysiological studies have shown that GABA mediates neuronal inhibition by binding to at least two types of GABA receptors, the GABA-gated anion channel (the GABA$_A$ receptor) and the G-protein-coupled GABA$_B$ receptor. The GABA$_A$ receptor-Cl^- ionophore complex is of particular interest, for it is known to be a site of action for a variety of pharmacologically and clinically important drugs which interact with distinct binding sites (at least five types of binding sites) and allosterically modulates GABA-induced Cl^- ion flux. These include: tranquilizers (benzodiazepines), anxiogenic (b-carbolines), convulsant (bicuculline or picrotoxin), and depressant (barbiturates) agents, and some other anesthetics and anticonvulsant steroids. Benzodiazepines (BZ) such as diazepam or flunitrazepam exert their anticonvulsant, anxiolytic and hypnotic activity by enhancing GABA binding at GABA$_A$ receptors, thereby potentiating GABA activity by increasing the frequency of Cl^- channel opening. The BZ receptor appears to be part of an oligomeric complex possessing binding sites for barbiturates and convulsants such as picrotoxin. Benzodiazepine receptors were studied and visualized in brain with tritiated benzodiazepine analogues using quantitative autoradiography, and with monoclonal antibodies raised against a purified GABA/BZ receptor complex using immunocytochemistry. Diversity in BZ receptors is generated by at least two different GABA$_A$ receptors containing distinct benzodiazepine binding sites BZ$_1$ and BZ$_2$ receptors (see Table 1). The BZ$_1$ subtype is predominantly distributed in the cerebellum, whereas BZ$_2$ receptors are mainly in the cortex, hippocampus and spinal cord. Differential BZ subtype organization suggests that each subtype mediates different physiological properties in the brain. The GABA/BZ receptor complex has been affinity purified from bovine cerebral cortex. As revealed by biochemical analysis, the native GABA$_A$ receptor complex is composed of two major glycoproteins, the α and β subunits (see Table 2). Photoaffinity labeling with tritiated flunitrazepam has assigned the BZ binding site to the α subunit, while the GABA agonist muscimol is incorporated preferentially into the β subunit (see Table 2).

Characterization of GABA$_A$ receptors assembled from different subunits has shed some light on the structural requirements needed for the action of GABA$_A$ receptor modulators. Expression studies indicate that each subunit can form a functional channel and thus must have a binding site for GABA. GABA, bicuculline and barbiturates act on receptors formed from only α or β subunits. Additional subunits, the γ, δ, and p_1 subunits and subunit variants, have been identified by molecular cloning (see Table 2). Their role in binding GABAergic ligands is unclear. However, a functional role has been assigned to the β subunit. Co-expression of the α, β and γ subunits confers high affinity binding of BZ, e.g., the γ subunit potentiates BZ modulatory activity. Potentiation of GABA by BZ is not observed in receptors lacking the γ_2 subunit. Moreover, the γ_2 subunit determines $GABA_A$ receptor sensitivity to zinc ions. Channels containing only and/or combination of α_1 or γ_2 subunits are blocked by zinc ions, while channels containing the γ_2 subunit alone or in combination with α and/or β subunits are insensitive to zinc ions.

In contrast to muscle nAChR, and GlyR, GABA "adult-type" and "embryo-type" receptors have not been identified. However, pharmacological diversities of $GABA_A$ receptors during cortical development are explained by the presence of different compositions of $GABA_A$ receptors in embryonic or neonatal neurons and in the adult brain. Changes in the expression of each $GABA_A$ receptor subunit gene during development appear to correlate with the alteration of GABA's role during brain ontogeny. For example, the immature brain, in contrast to the adult brain, was found to be poorly protected against seizure disorders by the GABA system. Developmental studies have shown that α_2, α_3, α_5, and β_3 mRNAs are highly expressed in embryonic stages and are suppressed in the adult by α_1, α_4, β_2, and δ subunit mRNAs. Similarly, γ_1 and γ_3 gene expression drops markedly during development. A different prenatal role was assigned to the $GABA_A$ receptors. While in the adult, GABA mediates inhibitory neurotransmission, in early developmental stages, GABA depolarizes membranes of neonatal cerebellar and cortical neurons; the resulting currents reveal little desensitization. Moreover, GABA together with glutamate elicit neurotropic activity, and influence neurite outgrowth, differentiation and synaptogenesis in early stages of brain development. Functional and pharmacological differences between GABA receptors present in distinct parts of the adult brain are most likely associated with the particular α subunits isoforms present in the receptor oligomer. High concentrations of the α_1 subunit are found in the cerebellum, while high concentrations of α_4 and α_2 subunits have been detected in the hippocampus. Different types of hippocampal neurons contain different proportions of α subunits. As confirmed by coexpression studies, α subunits most likely define recognition characteristics of the BZ site. Coexpression of different human α isoforms in association with β_1 and γ_2 subunits have shown that receptors containing the α_1 subunit display the BZ_1 phenotype, whereas receptors containing the α_2 or α_3 subunits display a BZ_2 phenotype. As revealed by site directed mutagenesis, the BZ phenotype is determined by a glycine residue located at position 201 of the α_1 subunit. The $BZ\alpha_6$ subunit is probably associated with receptor sensitivity to ethanol. The functional importance of the δ and γ_1 subunits is uncertain. The δ mRNA was found in areas showing high-affinity muscimol binding without accompanying BZ binding; high levels were especially observed in the cerebellum granular layer. Binding sites for picrotoxin, barbiturates and probably steroids seem to be formed by assembly of most $GABA_A$ receptor subunits. Potentiation of GABA responses by certain steroids and barbiturates, and inhibition by picrotoxin was observed with all subunit combinations examined. Thus, it can be concluded that expression of $GABA_A$ receptor subunits and subunit variants in a tissue-specific manner determines the pharmacological properties of GABA synapses in a given brain region.

THE ROLE OF LIGAND-GATED ION CHANNEL RECEPTORS IN THE PATHOGENESIS OF NEUROLOGICAL DISORDERS

The involvement of ligand-gated ion channels in the pathophysiology of the nervous system has been studied in various *in vivo* and *in vitro* models. These include animal models of diseases characterized by receptor alterations and receptor dysfunction, and *in vitro* preparations of brain slices and primary neuronal tissue culture systems.

Motor Disorders are Associated with GlyR Deficit. Severe impairment of glycinergic synaptic function was studied in two distinct animal models of inherited motor disorders. These include the inherited myoclonus in poll Herford calves, a disease characterized by involuntary muscle contraction, and the mutant *spastic* mouse which develops tremor and episodic spasms. In both models, a specific and marked reduction of spinal strychnine-sensitive GlyRs (about 90% reduction) was determined and visualized using autoradiography. Biochemical analysis has shown that receptor polypeptide composition remained normal. However, the molecular mechanism causing a substantial loss of GlyRs is unclear and remains to be elucidated.

Excessive Activation of Glutamatergic Receptors in Brain is Associated with a Wide Range of Neurodegenerative Processes. As shown by various laboratories (Choi, 1988), receptor activation by glutamate contributes to neuronal cell death in hypoxic-ischemic brain injury, physical trauma and stroke, and possibly participates in the pathogenesis of some neurological illnesses such as epilepsy and schizophrenia as well as chronic neurodegenerative disorders such as Huntington's and Alzheimer's diseases, amyotrophic lateral sclerosis (ALS) and Parkinsonism. The linkage between EAA and neurodegenerative conditions remains more speculative in other cases. Evidence to support such speculation has emerged over recent years. Selective glutamatergic agonists provided the means to examine the role played by EAA receptors in neuronal dysfunction and in processes of neuronal cell death. Glutamate mediates both rapid and delayed forms of neurotoxicity. Rapid excitotoxic neuronal cell death is associated with cell swelling, probably related to increased sodium ion permeability, thereby leading to entry of sodium, chloride and water into the cell. Delayed excitotoxicity, on the other hand, depends on the intracellular transport of calcium ions, which in turn most likely activate a variety of enzymes responsible for cellular damage. The NMDA receptors appear to be the main mediators of neurotoxicity. In the hippocampal slice preparation, functional failure in the vulnerable hilar neurons of the dentate gyrus as a result of brief periods of stimulation is prevented by EAA antagonists. In cultured neurons, MK-801 produces a long lasting blockade of NMDA currents. Exposure of cultured neurons to NMDA induced cell mortality which was completely prevented by the presence of MK-801. MK-801 neuroprotective activity is illustrated in Figure 9.

Elegant neuronal culture studies have substantiated the role of excitotoxicity in hypoxic-ischemic neuronal cell death. Cultured hippocampal or cerebrocortical neurons subjected to anoxic conditions *in vitro* were protected from acute neuronal necrosis by the addition of NMDA antagonists. The hippocampus and the dorsolateral striatum are especially vulnerable to ischemic lesions. *In vivo* evidence of NMDA antagonism of ischemic brain damage was demonstrated in a model of 30 min of severe forebrain ischemia by bilateral carotid occlusions and hypotension (global ischemic injury). Unilateral injection of NMDA antagonists into ischemic hippocampus protected neurons from acute necrosis. Administration of APH protected against development of early cytopathology in hippocampal layers CA1, CA3, CA4 and the dentate gyrus. Systemic administration of MK-801 to ischemic gerbils, induced via bilateral carotid clamping, prevented or reduced damage to CA1 hippocampal pyramidal neurons. In a model of focal cerebral ischemia, extensive reduction of cortical infarct volume has been achieved both with competitive and noncompetitive NMDA antagonists. In the retinal ischemic model, combined use of NMDA and non-NMDA antagonists exhibited dramatic cytoprotection. Neuronal damage in rat striatum produced by insulin-induced hypoglycemia coma could be prevented (90% neuroprotection) by NMDA antagonists such as APH. Glutamate toxicity is as well mediated by non-NMDA receptors. In brain slice preparations or in cell culture systems, exposure to AMPA produced neuronal degeneration which could be inhibited by the presence of the antagonist CNQX or DNQX.

Kainate Receptors Play a Significant Role in CNS Pathology. The effects of kainic acid on vertebrate CNS are of interest because kainic acid exerts potent neuroexcitatory, neurotoxic and convulsive effects (Nadler, 1979; Mayer and Westbrook, 1987). The neurotoxic effects of kainic acid are known to be mediated via its specific interactions with kainate receptors and kainate-induced lesions resemble these found in the brain of patients suffering from Huntington's chorea and temporal lobe epilepsy.

NMDA Receptors Participate in the Initiation and/or Propagation of Epileptiform Discharge in Epileptic and Non-epileptic Animals. The role played by EAA receptors in epilepsy was studied in several animal models (Dingledine et al., 1990). Audiogenic seizures in DBA/2 inbred mice were abolished when competitive NMDA antagonists like APH and AP5 were administered intracerebroventricularly 45 minutes prior to sound stimulation. Focal seizures induced by implantation of pure metal cobalt into rat cerebral cortex decreased significantly following the

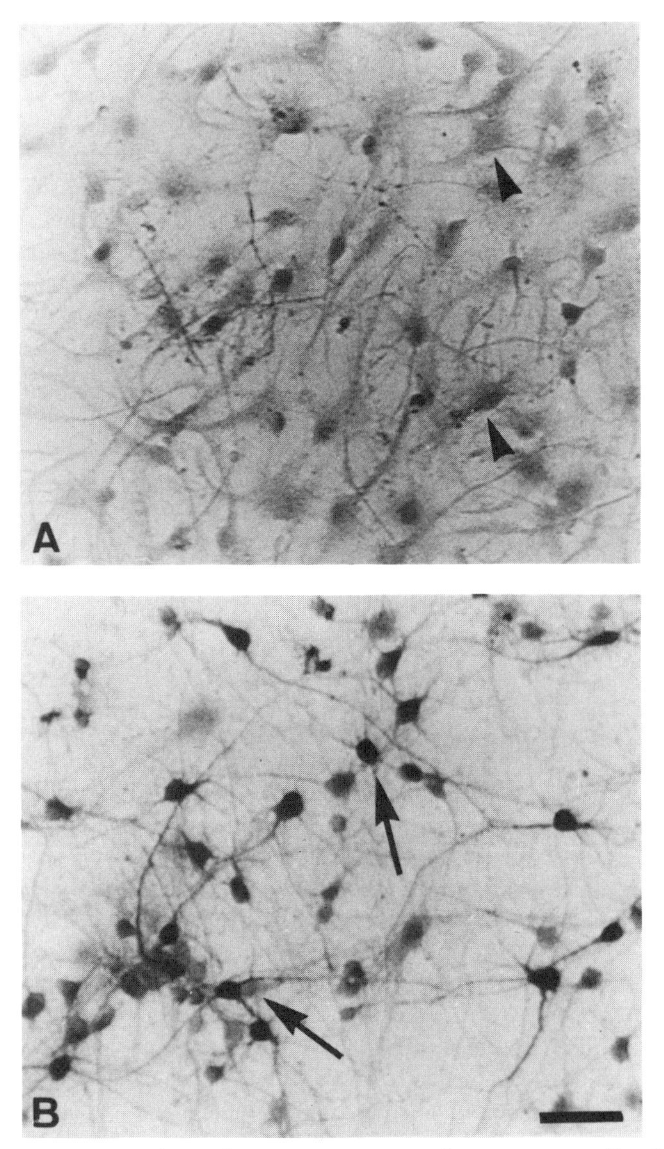

Figure 9. Neuron specific enolase immunostaining documents overall cell damage of cortical neurons in culture following exposure to NMDA either alone or in the presence of MK-801. **A.** Cells exposed to NMDA (1000 μM) appear swollen and granular. **B.** A complete salvage from NMDA toxicity by MK-801 (30 μM) (*arrowheads*) point to cell bodies of injured neurons; (*arrows*), indicate cell bodies of recovered neurons. Line scale, 50 μM. This work was conducted in the Pharmacology Department, Pharmos Ltd., Rehovot, Israel.

application of AP5. Furthermore, NMDA antagonists are also effective anticonvulsants in the kindling model of epilepsy.

NMDA and kainate receptor densities are markedly decreased in brains of Alzheimer's disease patients. The evidence for impairment of glutamatergic function in Alzheimer's disease derives from radioligand binding studies. These have indicated a drastic decrease in glutamate and TCP (see Table 1) binding (as compared to receptor densities determined in normal brains) mainly in the cortex and hippocampus. A reduction in glutamate and aspartate uptake sites and perforant path degeneration (a path known to be enriched in glutamatergic receptors) in brains of patient showing Alzheimer's disease symptoms, provides strong evidence for impairment in presynaptic glutamatergic mechanisms in Alzheimer's disease.

Specific NMDA antagonists were shown to prevent induced neuronal injury by gp120, an HIV-1 envelope glycoprotein. By contrast, non-NMDA receptor antagonists did not attenuate gp120 neurotoxicity. Whether NMDA receptors are associated with AIDS dementia is still unclear.

NMDA antagonists tested thus far exhibited several side-effects and could not be used in the clinic. Considerable effort will have to be made in the search for effective NMDA-receptor antagonists that will not cause untoward and irreversible side- effects. This is obviously of tremendous importance for therapeutic use clinically.

SUMMARY

Ligand-gated ion channels of vertebrates constitute a major family of receptor proteins that transduce rapid neurotransmission at chemical synapses of the central nervous system and at the neuromuscular junction. In response to a selective chemical neurotransmitter, the receptor channel opens and gates ion movement and generates electrical signals along the membranes of the postsynaptic neuron. Excitatory neurotransmission is induced by acetylcholine or glutamate and is mediated via cation-selective ion channels. On the other hand, inhibition of neuronal firing is induced by the neurotransmitters glycine or γ-aminobutyric acid, and is mediated by anion-selective channels. Receptor activation by neurotransmitters or ligands, and receptor desensitization to their agonists, is modulated by a variety of factors. The use of compounds (ligands) that either agonize or antagonize neurotransmitter function in numerous studies employing different technologies has contributed a plethora of data on ion channel-forming proteins in the fields of pharmacology, biochemistry, and molecular biology. Molecular cloning indicates that neurotransmitters activate unique types of multi-subunit complexes which are the products of distinct genes. Neurotransmitter channel receptors are most likely composed of five homologous subunits arranged around a central aqueous pore forming a pentagonal complex through the membrane. The different receptor subunits cloned thus far disclose structural similarity. Each subunit is composed of an amino-terminal hydrophobic signal sequence, a long extracellular amino-terminal domain, four predicted membrane-spanning segments (M1-M4), a large intracellular loop separating M3 and M4 segments, that is a mosaic of consensus sites for protein phosphorylation, and a relatively short carboxyl terminus. The carboxy-terminal of acetycholine, GABA and glycine receptor subunits is predicted to lie on the extracellular side of the lipid membrane and intracellularly in subunits of the glutamate receptors. Neurotransmitter recognition sites are located most probably on the N-terminal extracellular domain of distinct receptor subunits. Channel function appears to be controlled by the second transmembrane segment (M2) of each receptor subunit, and flow of ions, either cations or anions across the channel is determined by the type of electric charges flanking the M2 segment. Neurotransmitter-gated ion channels show great molecular and pharmacological diversity that has surpassed all expectation. Molecular cloning indicates that functional channel receptors can be formed by multisubunit associations in many different combinations, or by single subunit assemblies. The actual subunit composition of either heteromeric or homomeric receptors of most native neurotransmitter-gated ion channels expressed by a given neuron is as yet uncertain.

Ligand-gated ion channels are distributed widely in the vertebrate brain where they play crucial roles in a wide variety of neuronal functions. They are implicated in synaptic plasticity, in processes of learning and memory, and in neuritogenesis and synaptogenesis during brain ontogeny. Dysfunction and/or abnormal activation of ligand-gated ion channels is thought to play a role in the etiology of various neurological diseases and neurodegenerative disorders.

ACKNOWLEDGMENT

I thank Dr. Ronald S. Petralia for his excellent job of editing this manuscript.

REFERENCES

Curtis, D.R., Phillis, J.W., & Watkins, J.C. (1959). Chemical excitement of spinal neurones. Nature 183, 611.

Choi, D.W. (1988). Glutamate neurotoxicity and diseases of the nervous system. Neuron 1, 623–634.

Changeux, J.-P. (1990). The TiPS lecture. The nicotinic acetylcholine receptor: An allosteric prototype of ligand-gated ion channels. TiPS 11, 485–492.

Dingledine, R., Myers, S.J., & Nicholas, R.A. (1990). Molecular biology of mammalian amino acid receptors. FASEB 4, 2636–2645.

Eccles, J.C., Fatt, P., & Koketsu, K. (1954). Cholinergic and inhibitory synapses in a pathway from motor-axon collaterals to motorneurones. J. Physiol. 126, 524–562.

Hayashi, T. (1954). Keio J. Med. 3, 183–192.

Kemp, J.A., Foster, A.C., & Wong, H.E.F. (1987). Non-competitive antagonists of excitatory amino acid receptors. TINS 10, 294–299.

Meldrum, B.S. (1985). Possible therapeutic applications of antagonists of excitatory amino acid neurotransmitters. Clin. Sci. 68, 113–122.

Mayer, M.L., & Westbrook, G.L. (1987). The physiology of excitatory amino acid in the vertebrate central nervous system. Prog. Neurobiol. 28, 197–276.

Nadler, J.V., Perry, B.W., & Cotman, C.W. (1987). Intraventricular kainic acid preferentially destroys hippocampal pyramidal cells. Nature 271, 676–677.

Swope, S.L., Moss, S.J., Blackstone, C.D., & Huganir, R.L. (1992). Phosphorylation of ligand-gated ion channels: A possible mode of synaptic plasticity. FASEB J. 6, 2514–2523.

Watkins, J.C., & Evans, R.H. (1981). Excitatory amino acid transmitters, Ann. Rev. Pharmacol. Toxicol. 21, 165–204.

Watkins, J.C., & Olverman, H.J. (1987). Agonist and antagonist for excitatory amino acid receptors. Trends Neurosci. 10, 265–272.

Wood, M.W., VanDongen, M.A.H., & Vandongen, A.M.J. (1995). Structural conservation of ion conduction pathways in K channels and glutamate receptors. Proc. Natl. Acad. Sci. USA 92, 4882–4886.

RECOMMENDED READING

General Reviews

Unwin, N. (1989). The structure of ion channels in membranes of excitable cells. Neuron 3, 665–676.

Betz, H. (1990). Homology and analogy in transmembrane channel design; lessons from synaptic membrane proteins. Biochemistry 29, 3591–3599.

Betz, H. (1990). Ligand-gated ion channels in the brain: The amino acid receptor superfamily. Neuron 5, 383–392.

Huganir, R.L., & Greengard, P. (1990). Regulation of neurotransmitter receptor desensitization by protein phosphorylation. Neuron 5, 555–567.

Raymond, L.A., Blackstone, C.D., & Huganir, R.L. (1993). Phosphorylation of amino acid neurotransmitter receptors in synaptic plasticity. TINS 16, 147–159.

Westbrook, G.L. (1994). Glutamate receptors update. Curr. Opin. Neurobiol. 4, 337–346.

Dani, J.A., & Mayer, M.L. (1995). Structure and function of glutamate and nicotinic acetylcholine receptors. Curr. Opin. Neurobiol. 5, 310–317.

Bettler, B., & Mulle, C.N. (1995). Review: Neurotransmitter receptors II AMPA and kainate receptors. Neuropharmacol. 34, 123–139.

Glutamate Receptors

Cotman, C.W., & Iversen, L.L. (1987). Excitatory amino acids in the brain-focus on NMDA receptors. TINS, 10, 263–272.

Wroblewski, J.T., & Danysz, W. (1989). Modulation of glutamate receptors: Molecular mechanisms and functional implications. Ann. Rev. Pharmacol. Toxicol. 29, 441–474.

Watkins, J.C., Krogsgaard-Larsen, P., & Honore, T. (1990). Structure-activity relationships in the development of excitatory amino acid receptor agonist and competitive antagonists. TiPS 11, 25–33.

Young, A.B., & Fagg, G.E. (1990). Excitatory amino acid receptors in the brain: Membrane binding and receptor autoradiographic approaches. TiPS 11, 126–133.

Collingridge, G.L., & Singer, W. (1990). Excitatory amino acid receptors and synaptic plasticity. TIPS, 11, 290–296.

Bernard, E.A., & Henley, J.M. (1990). The non-NMDA receptors: Types, protein structure and molecular biology. TiPS 11, 500–507.

Mayer, M.L., & Miller, R.J. (1990). Excitatory amino acid receptors, second messengers and regulation of intracellular Ca^{2+} in mammalian neurons. TiPS 11, 254–260.

Keinanen, K., Wisden, W., Sommer, B., Werner, P., Herb, A., Verdoorn, T.A., Sakmann, B., & Seeburg, P.H. (1990). A family of AMPA-selective glutamate receptors. Science 249, 556–560.

Nakanishi, N., Shneider, N.A., & Axel, R. (1990). A family of glutamate receptor genes: Evidence for the formation of heteromultimeric receptors with distinct channel properties. Neuron 5, 569–581.
Nakanishi, S. (1992). Molecular diversity of glutamate receptors and implications for brain function. Science 258, 597–603.
Sato, K., Kiyama, H., & Tohyama, M. (1993). The differential expression patterns of messenger RNAs encoding non-N-methyl-D-aspartate glutamate receptor subunits (GluR1-4) in the rat brain. Neurosci. 52, 515–539.
Wisden, W., & Seeburg, P.H. (1993). Mammalian ionotropic glutamate receptors. Curr. Opin. Neurobiology 3, 291–298.
Hollmann, M., Maron, C., & Heinemann, S. (1994). N-glycosylation site tagging suggests a three transmembrane domain topology for the glutamate receptor GluR1. Neuron 13, 1331–1343.
Bennet, J.A., & Dingledine, R. (1995). Topology profile for a glutamate receptor: Three transmembrane domains and a channel-lining reentrant membrane loop. Neuron 14, 3373–3384.
Jonas, P., & Burnashev, N. (1995). Molecular mechanisms controlling calcium entry through AMPAA-type glutamate receptor channels. Neuron 15, 987–990.
Roche, K.W., O'Brien, R.J., Mammen, A.L., Bernhardt, J., & Huganir, R.L. (1996). Characterization on multiple phosphorylation sites on the AMPA receptor GluR1 subunit. Neuron 16, 1179–1188.

Glutamate Receptors in Neuronal Disorders

Dingledine, R., McBain C.J., & McNamara, J.O. (1990). Excitatory amino acid receptors in epilepsy. TiPS 11, 334–338.
Meldrum B., & Garthwaite, J. (1990). Excitatory amino acid neurotoxicity and neurodegenerative diseases. TiPS 11, 379–387.

Nicotinic Acetylcholine Receptors

Karlin, A. (1991). Exploration of the nicotinic acetylcholine receptors. The Harvey Lecture Series 85, 71–107.

nAChRs from Fish Electric Organ and Skeletal Muscle

Guy, H.R., & Hucho, F. (1987). The ion channel of the acetylcholine receptor. TINS 10, 318–321.
Cohen, J.B., Sharp, S.D., & Liu, W.S. (1991). Structure of the agonist-binding site of the nicotinic acetylcholine. J. Biol. Chem. 266, 23354–23364.
Lena, C., & Changeux, J.-P. (1993). Allosteric modulations of the nicotinic acetylcholine receptor. TINS 16, 181–186.
Unwin, N. (1993). The nicotinic acetylcholine receptor at 9Å resolution. J. Mol. Biol. 229, 1101–1124.

Neuronal Nicotinic Receptors

Luetje, C.W., Patrick, J., & Seguela, P. (1990). Nicotinic receptors in the mammalian brain. FASEB J. 4, 2753–2759.
Deneris, E.S., Connolly, J., Rogers, S.W., & Duvoisin, R. (1991). Pharmacological and functional diversity of neuronal nicotinic acetylcholine receptors. TiPS 12, 34–39.

Glycine Receptors

Langosch, D., Thomas, L., & Betz, H. (1988). Conserved quaternary structure of ligand-gated ion channels: The postsynatic glycine receptor is a pentamer. Proc. Natl. Acad. Sci. USA 85, 7394–7398.

Langosch, D., Becker, C.M., & Betz, H. (1990). The inhibitory glycine receptor: A ligand-gated chloride channel of the central nervous system. Eur. J. Biochem. 194, 1–8.

Betz, H. (1991). Glycine receptors: Heterogeneous and widespread in the mammalian brain. TINS 14, 458–461.

Disorders of the Glycine Receptor

Gundlach, A.L. (1990). Disorders of the inhibitory glycine receptors: Inherited myoclonus in Poll Herford calves. FASEB J. 4, 2761–2766.

Becker, C.M. (1990). Disorders of the inhibitory glycine receptor: The *spastic* mouse. FASEB J. 4, 2767–2773.

$GABA_A$ Receptors

Bernard, E.A., Darlison, M.G., & Seeburg, P. (1987). Molecular biology of the $GABA_A$ receptor: The receptor/channel superfamily. TINS 10, 502–509.

Pritchett, D.B., Sontheimer, H., Shivers, B.D., Yemer, S., Kettenmann, H., Schofield, P.R., & Seeburg, P.H. (1989). Importance of a novel $GABA_A$ receptor subunit for benzodiazepine pharmacology. Nature 338, 582–585.

Olsen, R.W., & Tobin, A.J. (1990). Molecular biology of $GABA_A$ receptors. FASEB J. 4, 1469–1480.

Araki, T., Kiyama, H., & Tohyama, M. (1992). $GABA_A$ receptors subunit messenger RNAs show differential expression during cortical development in the rat brain. Neurosci. 51, 583–591.

Doble, A., & Martin, I.L. (1992). Multiple benzodiazepine receptors: No reason for anxiety. TiPS 13, 76–81.

Sieghart, W. (1992). $GABA_A$ receptors; ligand-gated Cl^- ion channels modulated by multiple drug-binding sites. TiPS 13, 446–450.

Kuhse, J., Betz, H., & Kirsch, J. (1995). The inhibitory glycine receptor: Architecture, synaptic localization and molecular pathology of a postsynaptic ion-channel complex. Curr. Opin. Neurobiol. 5, 318–323.